Food Addiction, Eating Addiction and Other Forms of Addictive-Like Eating Behavior

Food Addiction, Eating Addiction and Other Forms of Addictive-Like Eating Behavior

Editor

Paul Brunault

MDPI • Basel • Beijing • Wuhan • Barcelona • Belgrade • Manchester • Tokyo • Cluj • Tianjin

Editor
Paul Brunault
University Hospital of Tours
France

Editorial Office
MDPI
St. Alban-Anlage 66
4052 Basel, Switzerland

This is a reprint of articles from the Special Issue published online in the open access journal *Nutrients* (ISSN 2072-6643) (available at: https://www.mdpi.com/journal/nutrients/special_issues/Eating_Disorders_Patients).

For citation purposes, cite each article independently as indicated on the article page online and as indicated below:

LastName, A.A.; LastName, B.B.; LastName, C.C. Article Title. *Journal Name* **Year**, *Volume Number*, Page Range.

ISBN 978-3-0365-7466-0 (Hbk)
ISBN 978-3-0365-7467-7 (PDF)

Contents

About the Editor

Paul Brunault

Paul Brunault, MD, PhD, is a psychiatrist specializing in addiction medicine and an assistant professor at the University of Tours and University Hospital of Tours (Department of Psychiatry and Addiction), France. He is member of INSERM U1253 iBrain—Équipe Psychiatrie Neuro-Fonctionnelle and an associated member of the QualiPsy Team (University of Tours, Department of Psychology, EE 1901). His main research interests focus on the relationship between ADHD and addictive disorders, eating disorders in patients with obesity (food addiction/addictive-like eating behavior/eating addiction, binge eating disorder), and behavioural addictions.

nutrients

MDPI

Editorial

Inter-Individual Differences in Food Addiction and Other Forms of Addictive-Like Eating Behavior

Paul Brunault [1,2,3,*] **and Nicolas Ballon** [1,2]

1 UMR 1253, iBrain, Université de Tours, Inserm, 37000 Tours, France; nicolas.ballon@univ-tours.fr
2 CHRU de Tours, Service d'Addictologie Universitaire, Équipe de Liaison et de Soins en Addictologie, 37000 Tours, France
3 Qualipsy EE 1901, Université de Tours, 37000 Tours, France
* Correspondence: paul.brunault@univ-tours.fr; Tel.: +33-247-478043; Fax: +33-247-478402

Citation: Brunault, P.; Ballon, N. Inter-Individual Differences in Food Addiction and Other Forms of Addictive-Like Eating Behavior. *Nutrients* **2021**, *13*, 325. https://doi.org/10.3390/nu13020325

Received: 19 January 2021
Accepted: 21 January 2021
Published: 23 January 2021

Publisher's Note: MDPI stays neutral with regard to jurisdictional claims in published maps and institutional affiliations.

The "addictive-like eating behavior" phenotype encompasses different terms or concepts, including "food addiction" (FA), "eating addiction" or "compulsive eating behavior" [1–3]. Although these terms may theoretically refer to different conceptualizations of addictive-like eating, all agree on the similarities this phenotype may share with other addictive disorders in terms of diagnostic criteria (with some core symptoms being food craving, loss of control over eating, and maintenance of the behavior despite negative consequences), epidemiology, risk factors, and treatment [1–3]. The main hypothesis underlying the "addictive-like eating behavior" phenotype is that it may help identify, among persons with obesity, eating disorders or persons with other eating symptoms, a specific and distinct subpopulation of vulnerable individuals for whom specific therapeutic management strategies may be proposed [4]. Although the FA model has the potential to open new avenues of conceptualization and management in obesity and eating disorders by providing new options to the existing treatments [3,5], some authors questions the validity and the specificity of the FA/addictive-like eating behavior phenotype [6,7]. To explain the possible inconsistencies in this evidence, and to gain insight into the possible validity and clinical utility of the addictive-like eating behavior phenotype, we argue here that we have to take into account the heterogeneous nature of FA. Beyond the identification of this phenotype, we hypothesize here that one key issue may be, as already demonstrated for other addictive disorders, that different psychobiological factors or different pathways may account for this increased vulnerability, with the identification of different clusters of vulnerable patients [8–11]. By examining the inter-individual differences that could account for this phenotype, and by disentangling the contribution of these different factors for a given individual, we may then propose tailor-based interventions based on the specific psychobiological factors involved. Before directly testing this hypothesis in interventional studies, one preliminary step is to identify what are the psychobiological factors associated with this "addictive-like eating behavior" phenotype in different clinical and non-clinical populations, and to determine how these factors may cluster together to help identify these different clusters of patients.

The aim of the special issue is to provide a modest—but hopefully constructive—contribution to this research area, by presenting research studies and literature reviews that will improve our understanding of the "addictive-like eating behavior" phenotype and of its associated factors, thus paving the way for the identification of these clusters of patients. This special issue includes eleven studies (four reviews and seven original studies). These seven original studies investigate the factors associated with this phenotype in different contexts: patients seeking treatment for an eating disorder, including anorexia nervosa, bulimia nervosa or binge eating disorder [12,13], patients seeking treatment for obesity (i.e., bariatric surgery candidates) [14], and non-clinical populations (i.e., persons not directly seeking treatment for these conditions) such as adolescents [15], female restrained eaters [16], and persons with weight-related disorders recruited in the general population [17]. This special issue also present four literature reviews that focus on: neuroimaging of sex/gender differences in obesity, in which FA is prevalent [18], involvement of the melanocortin system in binge eating, food reward and motivation [19], association between addictive-like eating behavior and attention-deficit/hyperactivity disorder (ADHD) [20], and a discussion about the FA concept and its practical implications through four complementary disciplines: addiction medicine, nutrition, health psychology, and behavioral neuroscience [21].

Obesity is the most studied population in the FA research field because of their high co-occurrence [22]. Benzerouk et al. focused on the association between binge eating, emotion dysregulation, impulsivity, depression, and anxiety in bariatric surgery candidates [14]. They confirmed that persons at risk for binge eating disorder were more prone to report limited access to emotion regulation strategies as well as higher impulsivity. More importantly, they demonstrated in multivariable models that emotional eating, external eating, and binge eating were independently associated with specific dimensions of emotion regulation and impulsivity. Although treatment-seeking persons with obesity are more frequently women, obesity is also prevalent in men. When studying the inter-individual differences, gender/sex may be an important factor to consider. In an elegant literature review, Kroll et al. shed light on the neuroimaging of sex/gender differences in persons with obesity, with a focus on structure, function, and neurotransmission [18]. They highlighted inter-individual differences based on gender: changes in somatosensory regions appeared to be associated with obesity in men, while changes in reward regions were more strongly associated with obesity in women than in men. They also found sex/gender differences in the neural response to taste among persons with obesity. These data are consistent with the idea that different neural mechanisms may be observed in obesity, and that the consideration of gender/sex may help in designing more tailored interventions based on the specific mechanisms involved. Future studies could investigate these interesting inter-individual differences in persons with obesity and binge eating disorder/FA vs. persons with obesity but without disordered eating.

To demonstrate the reliability and the relevance of the FA phenotype, one complementary area for research focusses on the biological correlates of FA in persons with obesity or overweight. In the FA field, fewer studies have been conducted using biobehavioral or genetic measures than subjective measures. To advance knowledge in this field, Aviram-Friedeman compared the brain asymmetry at rest and cue-reactivity to images of rewarding food in a Stroop Task between persons with weight-related problems and FA (group 1), persons with weight-related problems but no FA (group 2), and persons without weight-related problems and no FA (control group). The group with overweight/obesity and FA displayed a specific profile that was not observed in the two other groups: a lower resting left alpha brain asymmetry, an attenuated Stroop bias following exposure to high-calorie food relative to nonfood image, and a lower late positive potential component in frontal and occipital regions. Although association does not mean causation, these results point out the possibility of neural correlates specific to FA, as well as the need to consider how environmental stimuli may trigger or exacerbate FA. As demonstrated for addictive disorders, addictive-like eating may result from a gene-environment interaction, and Micioni Di Bonaventura et al. provide here a focused review of how alterations in the melanocortin system may be involved in obesity, binge eating behavior, and food reward/motivation [19]. In this literature review based on preclinical and clinical studies, these authors underline the pivotal role of the melanocortin system in controlling feeding behavior, appetite, energy balance, motivation for highly palatable food, and stress regulation. They discuss here how the loss of function of the melanocortin receptors (especially through the genetic variations of the MC3R and the MC4R, that are mainly located in the mesolimbic dopamine system) may lead to a breakdown of normal regulatory processes, thus increasing the risk of overeating and binge eating.

In addition to patients with obesity, patients with eating disorders are also persons for whom the FA phenotype may be useful. Fewer studies have been conducted in clinical persons seeking treatment for an eating disorder than in persons with obesity, especially in samples of persons with the full spectrum of eating disorders [22]. In a large sample of 195 adult women referred to an eating disorder treatment center for anorexia nervosa, bulimia nervosa, or binge eating disorder, Fauconnier et al. assessed the prevalence of FA and its associated factors [12]. In line with Granero [23], they confirmed the high prevalence of the FA phenotype in all types of eating disorders, with prevalence rates being respectively 61.5% for anorexia nervosa restrictive subtype, 87.9% for anorexia nervosa binge-eating/purging subtype, 93.3% for binge eating disorder, and 97.6% for bulimia nervosa. The main study result was the demonstration that the FA phenotype was independently associated with three variables: the presence of recurrent episodes of binge eating, eating disorder severity, and lower interoceptive awareness (this latter being compatible with one aspect of the three-systems neural model of addiction proposed by Noël et al., with the involvement of the insula in interoceptive awareness [10]). In another study presented here, Bou Khalil et al. tested among persons seeking treatment for an eating disorder whether FA was associated with a more severe eating disorder symptom severity and with more frequent childhood maltreatment, with the hypothesis that FA may mediate the relationship between childhood trauma and eating disorder severity [13]. They found that existence of FA was associated with a more severe eating disorder and with all types of traumas. Their data were compatible with a mediational role of FA in the relationship between childhood trauma and eating disorder severity, with largest effects emerging for physical neglect and emotional abuse. These data strengthen the hypothesis that the identification of a FA phenotype among eating disorder patients may help in identifying a distinct subpopulation of vulnerable individuals for whom specific therapeutic management strategies could be proposed, with low interoceptive awareness and childhood trauma being two potential targets. More studies in this specific population will be helpful to test (and eventually refine) this hypothesis, especially using interventional studies.

One potential target for the treatment of persons with FA is emotional eating and emotion dysregulation. Emotion dysregulation is indeed strongly associated with addictive disorders, and it is now integrated as a core component of addictive disorders treatment. In a large sample of 1142 persons recruited in the general population, Bourdier et al. used correlation and mediation analyses to compare the association between anxiety, depression, FA, emotional eating and the difficulty to rely on hunger and satiety cues between persons with versus without obesity [17]. First, they found associations between depression, anxiety, FA symptoms and the difficulty in relying on hunger and satiety cues across all weight classes. They also examined the association between emotional eating and these factors in each weight class, and they further demonstrated that emotional eating was associated with these factors but only in persons with obesity. Emotional eating may be an important factor to consider in persons with obesity: data were compatible with a meditational role of negative emotional eating in the relationship between anxiety symptoms and the difficulty in relying on internal cues to regulate food intake, as well as between depression symptoms and the difficulty in relying on these internal cues. Emotion dysregulation is also a core component of some psychiatric disorders, especially ADHD. ADHD symptoms and diagnosis are associated with addictive disorders and eating disorders, but the exact mechanisms underlying this association are currently unclear. In a literature review that included 41 papers, El Archi et al. report that ADHD and disordered eating are significantly associated, especially for binge eating and addictive-like eating behavior/FA. This review supports the idea that negative affectivity and emotion dysregulation may be mediators in the relationship between ADHD and disordered eating, providing another potential treatment option for persons with addictive-like eating behavior. As emotion dysregulation is prevalent in persons with psychiatric disorders, targeting emotion dysregulation may be especially relevant in these persons with addictive-like eating behavior and a co-occurring psychiatric disorder.

The addictive like-eating behavior phenotype may also be useful outside the obesity and eating disorders fields. Adolescence is an at-risk population for addictive disorders, given the higher impulsivity and less efficient emotional regulation processes compared to adults. In a controlled study, Rodrigue et al. examine the cognitive factors associated with FA symptoms in adolescence [15]. They compare adolescents with versus without FA symptoms in terms of "objective" sustained attention and executive functions (CANTAB neuropsychological battery) and in terms of "subjective" assessment (self-reported questionnaire assessing executive functions). Interestingly, adolescents with FA symptoms had higher subjective executive difficulties (as assessed by the self-reported questionnaires) than adolescents without FA, but there was no difference in terms of "objective" executive functions (as assessed by the neuropsychological tasks). When studying potential risk factors for FA in adolescents, the authors suggest the assessment of subjective executive difficulties (rather than only the objective ones) in addition to impulsivity, depression and anxiety. In addition to adolescents, persons with restrained eating may also be at risk for FA. As hypothesized by Herman & Polivy's theory and Fairburn et al.'s cognitive-behavioral model of bulimia nervosa, loss of control over food intake may arise from dietary/cognitive restraint (i.e., an intentional effort to achieve weight loss through reduced caloric intake) [24]. In a two-fold randomized-controlled study, Weinbach et al. assessed among female restrained eaters the impact of two different response inhibition trainings on food consumption, food related anxiety, and implicit attitudes toward food [16]. They compared two types of responses: a food-response training, in which stop cues were always associated with non-food images, and a balanced food-response/inhibition training, in which participants inhibited motor actions to food and non-food stimuli equally. Contrary to the food-response group, participants in the food-response/inhibition training group reduced their snack consumption and experienced an increase in positive attitudes toward palatable foods. Cognitive training may be helpful in restrained eaters, and future studies could investigate its effects on eating behavior and weight in clinical population with FA and restrained eating.

A comprehensive understanding of the factors associated with the addictive-like eating behavior phenotype is a preliminary step towards tailor-based treatments. In a review published in this special issue, Constant et al. argues in favor of a multidisciplinary dialogue between specialists in addiction medicine, clinical nutrition, health psychology, and behavioral neuroscience to account for and manage the complexity of FA [21]. After the presentation of each specialist's point of view as well as a literature review in each field, they proposed a multidisciplinary framework and practical implications derived from this framework in order to improve FA prevention, diagnosis, and treatment in at-risk populations. They proposed that only a multidisciplinary perspective can render the complexity of FA and how it relates to environmental, social, and individual factors, with FA being considered in a continuum ranging from normal to disordered eating.

Altogether, the studies gathered in this special issue support the idea that the "addictive-like eating behavior" phenotype, especially FA, may help in identifying a specific subpopulation of vulnerable individuals (i.e., persons with this phenotype differ from those without in different psychobiological factors). Addictive-like eating behavior was indeed associated across different populations with different psychobiological factors, including emotion dysregulation, higher prevalence for childhood traumas, lower emotional awareness, anxiety and depressive symptoms, difficulties to rely on internal cues, but also EEG abnormalities or alterations in the melanocortin system. Future challenges will be to address the questions of the validity and of the clinical utility of addictive-like eating behavior (i.e., how these different and specific psychobiological factors may cluster together to identify different subsets of vulnerable patients; how results from this field of research should be integrated into the existing models of obesity and/or eating disorders to provide treatment advances). To achieve this aim, we argue here that we need to take into account the inter-individual differences in addictive-like eating behavior. Studies conducted outside the FA field have already provided valuable insights into the inter-individual differences in addictive disorders [8–11], and these theoretical models may be useful to address the complexity of the FA puzzle and help in designing better tailor-based treatment for these persons.

Funding: This research received no external funding.

Conflicts of Interest: The authors declare no conflict of interest related to this article.

References

1. Gearhardt, A.N.; Corbin, W.R.; Brownell, K.D. Development of the Yale Food Addiction Scale Version 2.0. *Psychol. Addict. Behav.* **2016**, *30*, 113–121. [CrossRef] [PubMed]
2. Hebebrand, J.; Albayrak, Ö.; Adan, R.; Antel, J.; Dieguez, C.; de Jong, J.; Leng, G.; Menzies, J.; Mercer, J.G.; Murphy, M.; et al. "Eating Addiction", Rather than "Food Addiction", Better Captures Addictive-like Eating Behavior. *Neurosci. Biobehav. Rev.* **2014**, *47*, 295–306. [CrossRef] [PubMed]
3. Cottone, P.; Moore, C.F.; Sabino, V.; Koob, G.F. *Compulsive Eating Behavior and Food Addiction: Emerging Pathological Constructs*; Academic Press: Cambridge, MA, USA, 2019.
4. Meule, A.; Gearhardt, A.N. Ten Years of the Yale Food Addiction Scale: A Review of Version 2.0. *Curr Addict. Rep.* **2019**, *6*, 218–228. [CrossRef]
5. Fernandez-Aranda, F.; Karwautz, A.; Treasure, J. Food Addiction: A Transdiagnostic Construct of Increasing Interest. *Eur. Eat. Disord. Rev.* **2018**, *26*, 536–540. [CrossRef]
6. Fletcher, P.C.; Kenny, P.J. Food Addiction: A Valid Concept? *Neuropsychopharmacology* **2018**, *43*, 2506–2513. [CrossRef]
7. Ziauddeen, H.; Fletcher, P.C. Is Food Addiction a Valid and Useful Concept? *Obes. Rev.* **2013**, *14*, 19–28. [CrossRef]
8. Belin, D.; Belin-Rauscent, A.; Everitt, B.J.; Dalley, J.W. In Search of Predictive Endophenotypes in Addiction: Insights from Preclinical Research. *Genes Brain Behav.* **2016**, *15*, 74–88.
9. Blaszczynski, A.; Nower, L. A Pathways Model of Problem and Pathological Gambling. *Addiction* **2002**, *97*, 487–499. [CrossRef]
10. Noël, X.; Brevers, D.; Bechara, A. A Neurocognitive Approach to Understanding the Neurobiology of Addiction. *Curr. Opin. Neurobiol.* **2013**, *23*, 632–638. [CrossRef]
11. Everitt, B.J.; Robbins, T.W. Drug Addiction: Updating Actions to Habits to Compulsions Ten Years On. *Annu. Rev. Psychol.* **2016**, *67*, 23–50. [CrossRef]
12. Fauconnier, M.; Rousselet, M.; Brunault, P.; Thiabaud, E.; Lambert, S.; Rocher, B.; Challet-Bouju, G.; Grall-Bronnec, M. Food Addiction among Female Patients Seeking Treatment for an Eating Disorder: Prevalence and Associated Factors. *Nutrients* **2020**, *12*, 1897. [CrossRef] [PubMed]
13. Bou Khalil, R.; Sleilaty, G.; Richa, S.; Seneque, M.; Iceta, S.; Rodgers, R.; Alacreu-Crespo, A.; Maimoun, L.; Lefebvre, P.; Renard, E.; et al. The Impact of Retrospective Childhood Maltreatment on Eating Disorders as Mediated by Food Addiction: A Cross-Sectional Study. *Nutrients* **2020**, *12*, 2969. [CrossRef]
14. Benzerouk, F.; Djerada, Z.; Bertin, E.; Barrière, S.; Gierski, F.; Kaladjian, A. Contributions of Emotional Overload, Emotion Dysregulation, and Impulsivity to Eating Patterns in Obese Patients with Binge Eating Disorder and Seeking Bariatric Surgery. *Nutrients* **2020**, *12*, 3099. [CrossRef] [PubMed]
15. Rodrigue, C.; Iceta, S.; Bégin, C. Food Addiction and Cognitive Functioning: What Happens in Adolescents? *Nutrients* **2020**, *12*, 3633. [CrossRef] [PubMed]
16. Weinbach, N.; Keha, E.; Leib, H.; Kalanthroff, E. The Influence of Response Inhibition Training on Food Consumption and Implicit Attitudes toward Food among Female Restrained Eaters. *Nutrients* **2020**, *12*, 3609. [CrossRef] [PubMed]
17. Bourdier, L.; Fatseas, M.; Maria, A.-S.; Carre, A.; Berthoz, S. The Psycho-Affective Roots of Obesity: Results from a French Study in the General Population. *Nutrients* **2020**, *12*, 2962. [CrossRef] [PubMed]
18. Kroll, D.S.; Feldman, D.E.; Biesecker, C.L.; McPherson, K.L.; Manza, P.; Joseph, P.V.; Volkow, N.D.; Wang, G.-J. Neuroimaging of Sex/Gender Differences in Obesity: A Review of Structure, Function, and Neurotransmission. *Nutrients* **2020**, *12*, 1942. [CrossRef]
19. Micioni Di Bonaventura, E.; Botticelli, L.; Tomassoni, D.; Tayebati, S.K.; Micioni Di Bonaventura, M.V.; Cifani, C. The Melanocortin System behind the Dysfunctional Eating Behaviors. *Nutrients* **2020**, *12*, 3502. [CrossRef]
20. El Archi, S.; Cortese, S.; Ballon, N.; Réveillère, C.; De Luca, A.; Barrault, S.; Brunault, P. Negative Affectivity and Emotion Dysregulation as Mediators between ADHD and Disordered Eating: A Systematic Review. *Nutrients* **2020**, *12*, 3292. [CrossRef]
21. Constant, A.; Moirand, R.; Thibault, R.; Val-Laillet, D. Meeting of Minds around Food Addiction: Insights from Addiction Medicine, Nutrition, Psychology, and Neurosciences. *Nutrients* **2020**, *12*, 3564. [CrossRef]
22. Pursey, K.M.; Stanwell, P.; Gearhardt, A.N.; Collins, C.E.; Burrows, T.L. The Prevalence of Food Addiction as Assessed by the Yale Food Addiction Scale: A Systematic Review. *Nutrients* **2014**, *6*, 4552–4590. [CrossRef] [PubMed]

23. Granero, R.; Hilker, I.; Agüera, Z.; Jiménez-Murcia, S.; Sauchelli, S.; Islam, M.A.; Fagundo, A.B.; Sánchez, I.; Riesco, N.; Dieguez, C.; et al. Food Addiction in a Spanish Sample of Eating Disorders: DSM-5 Diagnostic Subtype Differentiation and Validation Data. *Eur. Eat. Disord. Rev.* **2014**, *22*, 389–396. [CrossRef] [PubMed]
24. Fairburn, C.G.; Cooper, P.J.; Cooper, Z. The Clinical Features and Maintenance of Bulimia Nervosa. In *Physiology, Psychology and Treatment of the Eating Disorders*; Brownell, K.D., Foreyt, J.P., Eds.; Basic Books: New York, NY, USA, 1986.

nutrients

MDPI

Article

Food Addiction among Female Patients Seeking Treatment for an Eating Disorder: Prevalence and Associated Factors

Marie Fauconnier [1], Morgane Rousselet [1,2], Paul Brunault [3,4,5], Elsa Thiabaud [1], Sylvain Lambert [1], Bruno Rocher [1], Gaëlle Challet-Bouju [1,2] and Marie Grall-Bronnec [1,2,*]

1 Addictology and Psychiatry Department, Hôpital Saint Jacques, University Hospital of Nantes, 85 rue Saint Jacques, CEDEX 1, 44 093 Nantes, France; marie.fauconnier@chu-nantes.fr (M.F.); morgane.rousselet@chu-nantes.fr (M.R.); elsa.thiabaud@chu-nantes.fr (E.T.); sylvain.lambert@chu-nantes.fr (S.L.); bruno.rocher@chu-nantes.fr (B.R.); gaelle.bouju@chu-nantes.fr (G.C.-B.)

2 Inserm, SPHERE U1246 « methodS in Patients-centered outcomes and HEalth ResEarch », Nantes University, Tours University, 22 boulevard Benoni Goullin, 44 200 Nantes, France

3 Department of Psychiatry & Addiction, University Hospital of Tours, 2 boulevard Tonnellé, 37 000 Tours, France; paul.brunault@univ-tours.fr

4 Inserm UMR 1253, iBrain, Tours University, 10 boulevard Tonnellé, 37 000 Tours, France

5 Qualipsy EE 1901, Tours University, 3 rue des Tanneurs, CEDEX 1, 37041 Tours, France

* Correspondence: marie.bronnec@chu-nantes.fr; Tel.: +33-(0)2-40-84-61-16; Fax: +33-(0)2-40-84-61-18

Received: 13 May 2020; Accepted: 24 June 2020; Published: 26 June 2020

Abstract: The concept of "food addiction" (FA) has aroused much focus because of evidence for similarities between overeating and substance use disorders (SUDs). However, few studies have explored this concept among the broad spectrum of eating disorders (ED), especially in anorexia nervosa (AN). This study aimed to assess FA prevalence in ED female patients and to determine its associated factors. We recruited a total of 195 adult women with EDs from an ED treatment center. The prevalence of FA diagnosis (Yale Food Addiction Scale) in the whole ED sample was 83.6%; AN restrictive type (AN-R), 61.5%; AN binge-eating/purging type (AN-BP), 87.9%; bulimia nervosa (BN), 97.6%; and binge-eating disorder (BED), 93.3%. The most frequently met criteria of FA were "clinically significant impairment or distress in relation to food", "craving" and "persistent desire or repeated unsuccessful attempts to cut down". An FA diagnosis was independently associated with three variables: presence of recurrent episodes of binge eating, ED severity, and lower interoceptive awareness. In showing an overlap between ED and FA, this study allows for considering EDs, and AN-R in particular, from an "addictive point of view", and thus for designing therapeutic management that draws from those proposed for addictive disorders.

Keywords: food addiction; eating disorder; anorexia nervosa; bulimia nervosa; binge eating disorder; YFAS; addictive disorder; eating addiction; addictive-like eating behavior

1. Introduction

Similarities between overeating and substance use disorder (SUD) were envisaged decades ago. In 1956, Theron Randolph mentioned for the first time the term food addiction (FA), with the hypothesis that certain food, as psychoactive substances, produces a "common pattern of symptoms descriptively similar to those of addictive processes" [1]. Subsequently, many studies have found similarities between certain forms of overeating and SUDs, especially studies conducted with animal models, notably rat models, in which the overconsumption of sweet food led to specific behavioral modifications (bingeing, withdrawal and cross-sensitization) [2,3] and neurochemical signs were also observed in models of substance dependence [2–4]. In humans, neuroimaging studies, notably those conducted with obese patients with FA, have also suggested the involvement of brain dopamine (DA) pathways and reward circuitry, and similarities with substance dependence have been observed as well [3,5,6].

The increasing prevalence of obesity, reflecting multiple factors that include the overall easy access to highly palatable energy-dense foods, linked with the food industry's efforts to boost sales, has contributed to making the concept of FA more popular. In 2009, Gearhardt et al. [7] therefore proposed an operationalization of a measure of FA by extrapolating the diagnostic criteria for substance dependence (Diagnostic and Statistical Manual of Mental Disorders, fourth edition Text Revised: DSM-IV-TR) [8] to hyperpalatable foods (i.e., foods high in fat and/or sugar). These criteria included (1) tolerance, (2) withdrawal, (3) consumption of larger amounts or over a longer period than was intended, (4) loss of control, (5) a great deal of time spent, (6) important activities are given up or reduced, and (7) persistent use despite damage. As in SUD, the presence of three (or more) of the criteria, as well as a clinically significant impairment or distress, have been suggested as necessary to characterize FA. This has led to the validation of a new evaluation tool, the Yale Food Addiction Scale (YFAS), which is a self-administered questionnaire assessing eating behavior in the past 12 months, with 25 questions exploring the 7 DSM-IV-TR extrapolated criteria. This tool has shown good internal consistency (Kuder–Richardson $\alpha = 0.86$), good convergence with measures of similar constructs (i.e., binge eating, emotional eating), good construct validity relative to dissimilar constructs (i.e., alcohol use, impulsivity), and good incremental validity toward binge-eating behavior and has been translated in several languages [9]. After publication of the DSM-5, a new version was developed in 2016, the YFAS 2.0, allowing a more dimensional approach with the exploration of the 11 DSM-5 criteria of SUD as applied to food through 35 questions [10].

The prevalence of FA, determined by the YFAS, varies greatly across samples, ranging from 0 to 25% in nonclinical samples [11,12], from 14 to 57.8% in prebariatric surgery samples [11,13], and from 70 to 90% in samples of patients suffering from eating disorders (ED), especially bulimia nervosa (BN) and binge eating disorder (BED) [11]. In most studies, YFAS symptoms were positively associated with BMI scores, and elevated YFAS scores have been observed in patients suffering from obesity [11,14]. Moreover, FA is associated with clinical characteristics that are commonly found with other addictive disorders: depression, anxiety [11], comorbid addictive disorders [15], posttraumatic stress disorders [14] and ADHD [16,17]. People with FA show more insecure attachment styles [18] and higher impulsivity [17,19,20], a classical trait of addictive disorders. In ED samples, FA has been associated with a more severe eating pathology and psychopathology, such as higher negative urgency, higher reward dependence and higher harm avoidance [20,21].

To date, the YFAS has mainly been used in overweight or obese patients (with or without BED) or in patients suffering from BN. The common factor among all these patients is overeating. To our knowledge, studies examining the links between anorexia nervosa (AN), a disorder characterized by restriction in energy intake, and FA are rare. Only two studies assessed FA prevalence in a sample of ED patients that included AN patients [20,21]. Nevertheless, AN is considered a counterpart of BN, and though FA and EDs of all types differ, they also show several similarities, as shown in Table 1. The concept of FA remains widely debated, and some authors argue that eating rather than food is addictive, underlining the behavioral dimension of this addiction. In light of these issues, we conducted a study with a sample of patients suffering from ED, characterized by eating behavioral symptoms ranging from those of AN (restricting: AN-R or binge-eating/purging: AN-BP types) to those of BN and BED.

Table 1. Discrepancies and similarities between food addiction (FA) and all types of eating disorders (Eds): comparison of different criteria between each of the types of disorders.

	AN-R	AN-BP	BN	BED	FA
Binge eating episodes		X	Xx	Xx	
Excessive food consumption		X	Xx	Xx	X
Sense of lack of control/loss of control of eating		X	Xx	Xx	X
Intense fear of gaining weight/self-evaluation unduly influenced by body shape and weight	Xx	Xx	Xx	Associated feature	
Restriction in food intake	Xx	Xx	Associated feature		
Restriction of energy intake/recurrent behaviors that interfere with weight gain/inappropriate compensatory behaviors	Xx (dieting, fasting, excessive exercise)	Xx (dieting, fasting, excessive exercise/purging behaviors)	Xx (excessive exercise, purging behaviors, or fasting)		
Obsessions related to food	Associated feature	Associated feature	Associated feature	Associated feature	X
Distress in relation to food	Associated feature	Associated feature	Associated feature	Associated feature	Xx
Social and/or professional consequences	Associated feature	Associated feature	Associated feature	Associated feature	X

AN-R: anorexia nervosa restricting type; AN-BP: anorexia nervosa binge eating/purging type; BN: bulimia nervosa; BED: binge-eating disorder; FA: food addiction; X: diagnostic feature; Xx: necessary diagnosis feature; Associated feature: symptom classically associated according to the Diagnostic and Statistical Manual (DSM) and/or the literature.

We aimed to estimate (i) the prevalence of FA among ED patients in general and according to the type of ED. We also aimed (ii) to assess the most commonly fulfilled criteria of the YFAS among ED in general and according to the type of ED and (iii) to determine the clinical and psychopathological correlates of FA in ED patients in general with an explorative approach, including the assessment of characteristics usually associated with addictive disorders and particularly with FA. It was first hypothesized that the prevalence of FA among ED patients would be important because of the similarities between those two disorders. Second, no clear hypothesis was made concerning the most fulfilled YFAS criteria that would be found in ED, except that the two physical criteria (tolerance and withdrawal) would not be very prevalent. Third, regarding the literature cited above, we hypothesize that the presence of FA would be associated with ED severity and the binge-eating episodes, and we also expected that FA would be associated with more comorbid addictive disorders, ADHD in childhood and trauma history, higher impulsivity, greater reward dependence, harm avoidance, and more insecure attachment styles.

2. Methods

2.1. Procedure and Ethics

Our Addictology and Psychiatry Department, based in the University Hospital of Nantes, France, is especially specialized in ED management (i.e., AN, BN, and BED) and is recognized as a National Reference Center in France. To receive treatment in our ED unit, patients must be referred to us by a medical professional. We provide physical, psychological and social care in accordance with the guidelines for ED management [22–24]. The care objectives of our unit are as follows: (i) to restore patients to a healthy weight, (ii) to alter core dysfunctional symptoms and attitudes related to ED (excessive concerns about body shape and weight, dietary restriction, purge and binge symptoms, etc.), and (iii) to manage all other negative features associated with ED (anxiety and depressive symptoms, low self-esteem, etc.) Treatment is primarily conducted in an outpatient format, with inpatient treatment provided only if necessary. Treatment is adapted to patient heterogeneity and often differs from one patient to another, in accordance with ED treatment guidelines.

Since September 2012, an in-depth clinical assessment has been systematically carried out for all new ED patients referred to our unit for treatment. The aforementioned assessment, which is part of the EVALuation of behavioral ADDictions (EVALADD) cohort (NCT01248767), occurs prior to the first medical consultation (at inclusion) and is then readministered at predefined intervals (at 6 months, at 12 months, and then every year). This assessment aims to highlight the risk factors involved in ED initiation and persistence. The main criteria for inclusion in the cohort were as follows: age of 15 years or older and a diagnosis of ED as defined by the DSM. Patients with cognitive impairment or difficulties reading or writing French were not included. All patients participated in a face-to-face semistructured interview and completed self-report questionnaires (see Section 2.3). Qualified and experienced staff members performed these assessments. Inclusion in the EVALADD cohort is still in progress.

The EVALADD cohort study is conducted in accordance with Good Clinical Practice Guidelines and the Declaration of Helsinki, with approval from the local ethics committee (Groupe Nantais d'Ethique dans le Domaine de la Santé, GNEDS, Nantes—Number 6 September 2012). All participants provide written informed consent, including consent from parents or guardians for participants under age 18. No compensation is given for participation.

For this specific study, we only used data collected at inclusion.

2.2. Participants

The participants were patients from the EVALADD cohort. For the present study, the specific inclusion criteria were as follows: (i) having a current diagnosis of AN (AN-R or AN-BP), BN or BED according to the DSM at inclusion; (ii) being included in the EVALADD cohort; and (iii) being a woman.

A total of 195 patients were included in this study. Sixty-five patients (33.3%) were diagnosed with AN-R, 33 (16.9%) with AN-BP, 82 (42.1%) with BN and 15 (7.7%) with BED. The flow chart of patient selection is presented in Figure 1.

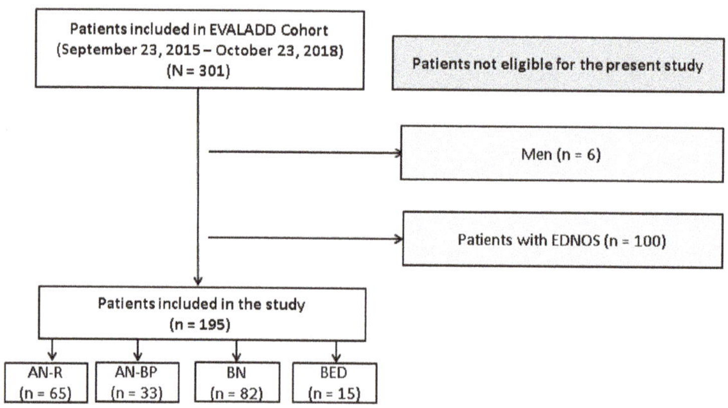

Figure 1. Flow chart of patient selection. AN-R: anorexia nervosa restricting type; AN-BP: anorexia nervosa binge eating/purging type; BN: bulimia nervosa; BED: binge-eating disorder; EDNOS: eating disorders not otherwise specified; *n*: number of patients included or excluded at each step of the inclusion process.

2.3. Measures

2.3.1. Sociodemographic Characteristics

Sociodemographic data included age and gender.

2.3.2. Eating Disorder Characteristics

- *Type of ED*

AN and BN diagnoses were made according to the ED sections of the fifth version of the Mini International Neuropsychiatric Interview (MINI). It is a structured diagnostic interview that enables rapid and systematic investigations of the main axis 1 psychiatric disorders, according to DSM-IV criteria [25,26]. From 2017, we used an adapted version of the MINI to take into account the revised diagnostic criteria of the DSM-5 [27]. Questions were added to diagnose BED according to DSM-IV (and then DSM-5) criteria. Age at ED onset and disease duration were also collected.

- *Severity of ED*

The Morgan–Russell Outcome Assessment Schedule (MROAS) is a structured interview that covers various clinical symptoms of ED and their repercussions on patient functioning in the past six months [28]. The questionnaire consists of five subscales exploring food intake and nutritional status, menstrual function, mental state, psychosexual adjustment, and socioeconomic status. Each subscale was scored from 1 to 12, with a higher score indicating a better outcome in the corresponding field. The average of these five scores was used as the MROAS total score, with potential results ranging from 1 to 12.

- *Characteristics of ED*

The Eating Disorder Inventory-2 (EDI-2) is a 91-item self-assessment questionnaire that evaluates the symptomatology and behavior associated with ED [29]. It examines 11 dimensions: "drive for thinness", "bulimia", "body dissatisfaction", "ineffectiveness", "perfectionism", "interpersonal distrust", "interoceptive awareness", "maturity fears", "asceticism", "impulse regulation" and "social insecurity". Answers are rated on a 6-point Likert-type scale ranging from "never" to "always". Each of these dimensions can be independently analyzed, and a score was calculated for each item. The internal consistency values for the EDI-2 dimensions are between 0.44 and 0.93.

- *Food addiction*

The Yale Food Addiction Scale (YFAS) was designed to identify those exhibiting addictive-like eating behavior toward certain types of foods high in fat and/or sugar [7]. The YFAS is composed of questions based upon substance dependence criteria in the DSM. The DSM-IV criteria were used for the initial version of the YFAS and showed good convergence with measures of similar constructs (i.e., binge eating, emotional eating), good construct validity relative to dissimilar constructs (i.e., alcohol use, impulsivity), and good incremental validity toward binge-eating behavior. When the fifth edition of the DSM was published, a new version of the YFAS was developed, YFAS 2.0, to take into account the changes made to the substance-related and addictive disorders section and to extrapolate them to food [10]. The YFAS 2.0 version also showed good internal consistency, as well as convergent, discriminant and incremental validity. For the present study, the French version of the initial version of YFAS [30] was used until 31 October 2017, then replaced by the French version of YFAS 2.0 once it was validated [31]. According to Meule and Gearhardt (2019), prevalence rates and correlates of YFAS 2.0 diagnoses are largely similar to those observed with the original YFAS [32].

2.3.3. Other Clinical Characteristics

- *Psychiatric comorbidities*

The fifth version of the MINI (described above) was used. For the purposes of this study, mood disorders (major depressive episode, dysthymia, (hypo)manic episodes), anxiety disorders (panic disorder, agoraphobia, social phobia, obsessive-compulsive disorder, posttraumatic stress disorder, generalized anxiety disorder), psychotic syndrome and SUD (alcohol, psychoactive substances)), current or past, were considered. We also assessed the presence of behavioral addictions with the Minnesota Impulsive Disorders Interview (MIDI) [33,34]. The MIDI is a structured interview that enables rapid and systematic investigations of pathological gambling, hypersexuality, and compulsive buying. In the framework of the EVALADD cohort, we adapted the MIDI to screen other behavioral addictions (videogame, internet, exercise and work addictions). When a behavioral addiction was screened by the MIDI, its diagnosis was confirmed using a specific diagnostic interview. Finally, the Wender Utah Rating Scale-Child (WURS-C) was used to retrospectively screen for childhood attention-deficit/hyperactivity disorder (ADHD) [35,36]. A threshold of 46/100 was defined to identify probable childhood ADHD.

- *Impulsivity*

The French version [37] of the Impulsivity Behavior Scale (UPPS) [38] was used to measure impulsivity. During the data collection, we transitioned to the UPPS-P French short version of the scale [39], which included a fifth new dimension, "positive urgency". To standardize the results, we reconstructed the four available scores of the new UPPS-P ("negative urgency," (lack of) "premeditation," (lack of) "perseverance" and "sensation seeking") based on the initial UPPS for the first patients.

- *Temperament*

The 125-item version of the Temperament and Character Inventory (TCI-125) is a validated self-report questionnaire. It is used to briefly evaluate 4 temperament dimensions (novelty seeking, harm avoidance, reward dependence and persistence) and 3 character dimensions (self-directedness, cooperativeness and self-transcendence) [40]. For the present study, only temperament dimensions were considered.

- *Attachment*

The Relationship Scales Questionnaire (RS-Q) is a 30-item self-assessment questionnaire that was developed in 1991 [41] and validated in French in 2010 [42]. It is based on the theoretical principles of Bowlby and, more specifically, on the concept of an internal working model to examine four different types of attachments: "secure", "fearful", "preoccupied" and "dismissing". For all items, answers were given on a 5-point Likert-type scale ranging from "not at all like me" to "just like me". In the French translation study, Cronbach's alpha coefficient was moderate ($\alpha > 0.60$), and the intraclass coefficients were good (>0.75).

- *History of Traumatic Events*

We used a revised version of the French Life Events questionnaire (EVE) [43], which was previously used in another study from the EVALADD cohort [44]. The revised EVE questionnaire explores 6 areas (family, professional life, social life, marital and emotional life, health, and other traumatic events). For this study, we focused on the history of physical abuse and sexual abuse.

2.4. Outcome Measure

The primary outcome measure was the diagnosis of FA (yes or no) according to the YFAS. When the YFAS version 1.0 was used, FA was diagnosed when 3 or more criteria out of 7 were present during the last 12 months and when clinically significant impairment or distress was endorsed. When the YFAS

2.0 version was used, FA was diagnosed when 2 or more criteria out of 11 were present during the last 12 months and when clinically significant impairment or distress was endorsed.

2.5. Statistical Analysis

A descriptive statistical analysis was conducted for the entire sample. Continuous variables are described by means and standard deviations, while categorical variables are presented as numbers and percentages. The prevalence of FA according to YFAS 1.0 and 2.0 was computed for the whole sample and for each ED diagnosis, as well as the frequency of each YFAS criterion for each ED diagnosis. We divided the sample into two groups ("FA" and "No FA") according to FA diagnosis. Bivariate analyses were conducted to explore the associations between FA and other collected data (sociodemographic, ED and other clinical characteristics). We focused on patients with recurrent episodes of binge eating (at least once a week). We used chi^2 tests or Fisher's tests, if necessary, to analyze the categorical variables. For the continuous variables, we used Student's tests for variables with a normal distribution and Wilcoxon nonparametric tests for variables with a non-Gaussian distribution. For both types of variables (categorical and continuous), differences were statistically significant when the p-value was less than or equal to 0.05.

A multiple logistic regression was performed using an iterative selection procedure to identify the variables that were significantly associated with FA, as assessed by the likelihood ratio test. Variables were entered as candidates for the model if they were associated with the presence of FA in the bivariate analysis with a $p < 0.20$ [45]. Then, nonsignificant variables were removed one at a time starting with the least significant variable (backward procedure), to retain only the variables that provided significant information to the model ($p < 0.05$) [46]. The corresponding odds ratios (OR) and associated 95% confidence intervals were estimated. Discrimination of the final logistic model, which describes the model's ability to differentiate between the presence and absence of FA, was assessed using the area under the receiver operating characteristic (ROC) curve, and the goodness-of-fit of the model was assessed using the Hosmer–Lemeshow test. The statistical analysis was carried out with TIBCO Statistica® 13.3.0 (Statsoft, Inc. 2300 East 14th Street. Tulsa, OK 74104, USA.). The conditions of validity were verified for all tests and the final model.

3. Results

3.1. Description of the Sample

The mean age was 23.1 (+/−7.4) years. The characteristics of the sample are shown in Tables 2 and 3. p-values ≤ 0.05 are in bold in Tables 2 and 3. A description of the sample according to the type of ED is available in Supplementary Material (Table S1).

Table 2. Description of the sample and comparison of the patients with "Food Addiction" or "No Food Addiction" (n = 195)—Sociodemographic and Eating Disorder characteristics.

	Entire Sample (N = 195)	**"Food Addiction" (n = 163)**	**"No Food Addiction" (n = 32)**	***p*-Value**	**Statistical Test**
	n (%) or m (sd)				
Sociodemographic characteristics					
Age (years)	23.1 (7.4)	23.3 (7.8)	22.1 (5.2)	0.903	Wilcoxon
Eating disorder characteristics					
Type of ED				**<0.001**	Chi^2
AN-R	65 (33.3%)	40 (24.5%)	25 (78.1%)		
AN-BP	33 (16.9%)	29 (17.8%)	4 (12.5%)		
BN	82 (42.1%)	80 (49.1%)	2 (6.3%)		
BED	15 (7.7%)	14 (8.6%)	1 (3.1%)		
Recurrent episodes of binge eating (yes)	114 (58.5%)	110 (67.5%)	4 (12.5%)	**<0.001**	Chi^2
Age of disease onset (years)	15.9 (5.0)	15.6 (5.1)	17.4 (4.0)	**0.006**	Wilcoxon
Disease duration (years)	7.2 (7.6)	7.6 (7.9)	4.8 (5.3)	**0.050**	Wilcoxon
Severity of ED (MROAS total score)	6.4 (2.0)	6.3 (2.0)	6.9 (1.9)	0.111	Student's t
Dimensions associated with ED (EDI-2)					
Ineffectiveness	13.4 (7.0)	14.0 (7.0)	10.1 (6.0)	**0.004**	Student's t
Interoceptive awareness	13.6 (6.8)	14.8 (7.4)	6.5 (4.7)	**<0.001**	Wilcoxon
Asceticism	8.5 (4.5)	8.9 (4.5)	6.6 (3.5)	**0.007**	Student's t
Drive for thinness	15.3 (5.1)	16.0 (4.4)	11.6 (6.5)	**<0.001**	Wilcoxon
Bulimia	8.2 (6.7)	9.4 (6.4)	2.0 (3.4)	**<0.001**	Wilcoxon
Body dissatisfaction	18.0 (7.2)	18.9 (7.0)	13.3 (6.4)	**<0.001**	Student's t
Perfectionism	7.3 (4.5)	7.7 (4.4)	5.0 (3.7)	**<0.001**	Student's t
Interpersonal distrust	7.7 (4.5)	7.9 (4.7)	6.7 (3.6)	0.165	Student's t
Maturity fears	7.9 (5.9)	8.1 (5.9)	6.8 (6.0)	0.263	Student's t
Impulse regulation	8.4 (6.6)	9.2 (6.7)	4.4 (3.8)	**<0.001**	Wilcoxon
Social insecurity	9.9 (4.8)	10.3 (4.8)	7.6 (4.2)	**0.003**	Student's t

%: percentage; m: mean; sd: standard deviation; AN-BP: anorexia nervosa binge-eating/purging type; AN-R: anorexia nervosa restricting type; BED: binge eating disorder; BN: bulimia nervosa; ED: eating disorder; EDI: Eating Disorders Inventory; MROAS: Morgan–Russell Outcome Assessment Schedule; sd: standard deviation; YFAS: Yale Food Addiction Scale.

Table 3. Description of the sample and comparison of the patients with "Food Addiction" or "No Food Addiction" (n = 195)—Other clinical characteristics.

	Entire Sample (N = 195)	"Food Addiction" (n = 163)	"No Food Addiction" (n = 32)	p-Value	Statistical Test
		n (%) or m (sd)			
Comorbidities (current or past)					
Mood disorders (MINI)	156 (80.0%)	135 (83.8%)	21 (65.6%)	**0.026**	Chi^2
Anxiety disorders (MINI)	141 (72.3%)	124 (76.1%)	17 (53.1%)	**0.008**	Chi^2
Psychotic syndrome (MINI)	12 (6.2%)	10 (6.0%)	2 (5.9%)	0.981	Chi^2
Addictive disorders (MINI and MIDI)	90 (46.2%)	79 (48.5%)	11 (34.4%)	0.144	Chi^2
ADHD in childhood (WURS-C)	66 (33.8%)	63 (38.7%)	3 (9.4%)	**0.001**	Chi^2
Impulsivity					
UPPS-Urgency	10.5 (3.0)	10.8 (2.9)	9.3 (3.1)	**0.009**	Student's t
UPPS-Premeditation (lack)	7.5 (2.5)	7.6 (2.6)	7.2 (2.0)	0.438	Student's t
UPPS-Perseverance (lack)	7.4 (2.9)	7.6 (2.9)	6.8 (2.6)	0.140	Student's t
UPPS-Sensation seeking	9.8 (3.2)	9.8 (3.1)	10.0 (3.5)	0.803	Student's t
Temperament Comorbidities (current or past)					
TCI-Novelty seeking	40.8 (19.3)	41.9 (19.0)	35.3 (19.7)	0.077	Student's t
TCI-Harm avoidance	74.2 (20.6)	75.1 (20.1)	69.4 (22.7)	0.150	Student's t
TCI-Reward dependence	60.4 (17.5)	59.4 (17.9)	65.1 (15.1)	0.094	Student's t
TCI-Persistence	72.6 (28.7)	72.0 (28.8)	75.6 (28.2)	0.518	Student's t
Attachment					
RSQ-Secure	2.7 (0.6)	2.7 (0.6)	2.7 (0.5)	0.896	Student's t
RSQ-Fearful	2.9 (0.6)	2.9 (0.6)	2.7 (0.6)	0.091	Student's t
RSQ-Preoccupied	2.6 (0.7)	2.6 (0.6)	2.4 (0.7)	0.106	Student's t
RSQ-Dismissing	3.3 (0.8)	3.3 (0.8)	3.1 (0.9)	0.172	Student's t
Life events					
History of physical abuse	20 (10.3%)	20 (12.3%)	0	-	-
History of sexual abuse	27 (13.8%)	24 (14.7%)	3 (9.4)	0.423	Chi^2

%: percentage; m: mean; sd: standard deviation; ADHD: attention-deficit/hyperactivity disorder; MIDI: Minnesota Impulsive Disorders Interview; MINI: Mini International Neuropsychiatric Interview; RSQ: Relationship Scales Questionnaire; sd: standard deviation; TCI: Temperament and Character Inventory; UPPS: Impulsive behavior scale; WURS-C: Wender Utah Rating Scale-Child.

3.2. Prevalence of Food Addiction

Of the 195 patients included in the study, 163 displayed "FA" (83.6%) and 32 exhibited "No FA" at inclusion. There was no significant difference (p = 0.067) between the prevalence of FA in the total sample according to the version of YFAS used (1.0 or 2.0). The prevalence of FA according to ED diagnosis was 61.5% for AN-R, 87.9% for AN-BP, 97.6% for BN and 93.3% for BED diagnoses. FA prevalence was significantly different across ED diagnoses (p < 0.001). There was no significant difference between the prevalence of FA according to YFAS version for AN-R (p = 0.129) and AN-BP (p = 0.948). For BN and BED diagnoses, the number of patients was insufficient to conduct a comparison analysis. Regarding the presence of recurrent episodes of binge eating (at least once a week), patients (n = 114, 58.5%) with this characteristic (82 with BN diagnosis, 15 with BED diagnosis and 17 with AN-BP diagnosis) were significantly more likely to have FA (p < 0.001).

3.3. Frequency of YFAS Criteria

Regarding the whole sample, the most prevalent criteria were (i) clinically significant impairment or distress in relation to food (90.8%); (ii) craving (79.2%); and (iii) persistent desire or repeated unsuccessful attempts to cut down (78.5%). Table 4 shows the percentage of each YFAS criterion met in the total sample, as well as for each type of ED. "Clinically significant impairment or distress in relation to food" was one of the most prevalent diagnostic criteria regardless of the type of ED, but each type of ED was associated with specific criteria: "Use in physically hazardous situations" with ED associated with under- or overweight, "craving" with ED characterized by recurrent episodes of binge eating.

Table 4. Percentage of each YFAS criterion met for ED patients and according to the type of ED.

	Total Sample (N = 195)		**AN-R (n = 65)**		**AN-BP (n = 33)**		**BN (n = 82)**		**BED (n = 15)**	
YFAS Criteria	Valid N	Number of patients (%) with positive YFAS criteria	Valid N	Number of patients (%) with positive YFAS criteria	Valid N	Number of patients (%) with positive YFAS criteria	Valid N	Number of patients (%) with positive YFAS criteria	Valid N	Number of patients (%) with positive YFAS criteria
1- Loss of control	195	127 (65.1%)	65	16 (24.6%)	33	21 (63.6%)	82	76 (92.7%)	15	14 (93.3%)
2- Persistent desire or repeated unsuccessful attempts to cut down	195	153 (78.5%)	65	42 (64.6%)	33	22 (66.7%)	82	75 (91.5%)	15	14 (93.3%)
3- Much time spent	195	110 (56.4%)	65	18 (27.7%)	33	17 (51.5%)	82	61 (74.4%)	15	14 (93.3%)
4- Craving *	77	61 (79.2%)	23	14 (60.9%)	16	12 (75.0%)	34	33 (97.1%)	4	2 (50.0%)
5- Continued used despite social or interpersonal problem *	77	54 (70.1%)	23	10 (43.5%)	16	10 (62.5%)	34	30 (88.2%)	4	4 (100%)
6- Impaired daily functioning *	77	46 (59.7%)	23	8 (34.8%)	16	8 (50.0%)	34	28 (82.4%)	4	2 (50.0%)
7- Important activities given up	195	141 (72.3%)	65	36 (55.4%)	33	23 (69.7%)	82	69 (84.2%)	15	13 (86.7%)
8- Use in physically hazardous situations *	77	56 (72.7%)	23	17 (73.9%)	16	12 (75.0%)	34	23 (67.7%)	4	4 (100%)
9- Use despite knowledge of adverse consequences	195	100 (51.3%)	65	25 (38.5%)	33	16 (48.5%)	82	49 (59.8%)	15	10 (66.7%)
10- Tolerance	195	95 (48.7%)	65	18 (27.7%)	33	17 (51.5%)	82	52 (63.4%)	15	8 (53.3%)
11- Withdrawal symptoms	195	119 (61.0%)	65	19 (29.2%)	33	23 (69.7%)	82	66 (80.5%)	15	11 (73.3%)
12- Clinically significant impairment or distress	195	177 (90.8%)	65	52 (80.0%)	33	31 (93.9%)	82	80 (97.6%)	15	14 (93.3%)

* Criteria that are present only in the second version of the YFAS (modeled on DSM-5 criteria), assessed in only 77 ED patients, 23 AN-R patients, 16 AN-BP patients, 33 BN patients, and 4 BED patients. %: percentage; AN-BP: anorexia nervosa binge-eating/purging type; AN-R: anorexia nervosa restricting type; BED: binge-eating disorder; BN: bulimia nervosa; ED: eating disorder; YFAS: Yale Food Addiction Scale.

3.4. Bivariate Comparison of the "FA" Group and the "No FA" Group

No differences were observed in the sociodemographic data and severity of ED, but several significant differences were observed for ED and other clinical characteristics. Tables 2 and 3 show the results of the comparisons of the two groups.

3.5. Factors Associated with Food Addiction

Sample sizes for BN and BED in the "no FA" group were too small. We therefore have chosen to include the variable "recurrent episodes of binge eating" (at least once a week) instead of "type of ED" in the multiple regression model. Among the variables selected based on the bivariate analysis to be included as candidates in the multiple regression model, high correlations were found, leading to the exclusion of four variables ("EDI-feeling of ineffectiveness", "EDI-impulse regulation", "EDI-interpersonal distrust" and "EDI-social insecurity") from the multivariate analysis.

Following multiple logistic regression, only three variables remained independently associated with FA: presence of recurrent episodes of binge eating (OR = 28.2), lower MROAS total score

(OR = 0.67) corresponding to a higher severity of ED, and higher "EDI-interoceptive awareness" score (OR = 1.22) corresponding to a higher lack of interoceptive awareness (Table 5). The Hosmer–Lemeshow goodness-of-fit test was non-significant (p = 0.60; Chi-squared = 6,460 and df = 8), showing that the final model was well calibrated. The area under the ROC curve was 0.91, indicating that the model discriminated well between patients who had FA (N = 165) and those who did not have FA (N = 32).

Table 5. Multiple logistic regression analysis (final model)—factors associated with "food addiction" (N = 195).

Variables	OR	$CI_{95\%}$ (OR)	p-Value
MROAS total score	0.67	[0.50; 0.89]	<0.01
EDI-2 interoceptive awareness	1.22	[1.10; 1.34]	<0.001
Recurrent episodes of binge eating (yes)	28.20	[7.00; 113.7]	<0.0001

EDI: Eating Disorder Inventory; $CI_{95\%}$: 95% confidence interval; MROAS: Morgan–Russel Outcome Assessment Schedule; OR: odds ratio.

4. Discussion

4.1. Main Results

The aim of this study was to estimate the prevalence of FA in a sample of ED patients. We also aimed to assess the most commonly met criteria of the YFAS and to determine the factors associated with the presence of FA.

First, our study confirmed the hypothesis of a strong link between an FA diagnosis according to YFAS and an ED diagnosis, with a prevalence of 83.6% in a cohort of 195 patients suffering from ED. These results are fully in line with previous work, with a prevalence ranging from 70 to 90% (1). The data seem to demonstrate an overlap between FA and ED, with a gradient according to the type of ED: prevalence of FA appeared to be important in BN (97.6%) and BED (93.3%), as other studies had previously demonstrated [21,47–49], as well as in AN-R (61.5%) and AN-BP (87.9%). The umbrella term ED encompasses a broad spectrum of disorders, with AN at one end and BED at the other, and includes BN and other specified feeding and eating disorders (OSFEDs). The high FA prevalence in BED, BN patients and, to a lesser extent, AN-BP patients is not surprising given that EDs defined by the presence of binge eating share behavioral, clinical and neurobiological characteristics with other types of addictive disorders [50–61]. However, conceptualizing AN-R as overlapping with FA is somewhat more debatable. For example, Barbarich-Marsteller et al. (2011) stated that AN is not an addiction [62]. Indeed, people with AN-R seem not to be addicted to food but quite the opposite, i.e., addicted to food deprivation, and they show real determination instead of losing control. In their recent paper, Mallorquí-Bagué et al. (2020) concluded that patients with AN exhibited a successful down-regulation of food craving, despite the presence of food addiction symptomatology [63]. In the present study, we found a prevalence of FA in patients with AN-R that was far more substantial than in nonclinical samples (0 to 25%) [11]. Our results are in line with those of Granero et al. [21] and Wolz et al. [20] The YFAS was built to screen addictive symptoms, but it is important to note that the "object" of addiction is not clearly specified. Despite the fact that the YFAS putatively explores eating behaviors toward specific hyperpalatable foods high in fat and/or sugar (i.e., pizza, sweets, soda, chips, etc.), it has been debated whether any of these foods comprise different "substances" [4]. Then, the substance of abuse is not defined, which raises an important question in the explanation of addictive-like eating: are the addictive properties intrinsic to some foods or associated with eating behavior? As has been shown for rodents and humans, certain types of food, such as high sugar and high fat palatable foods, have rewarding properties [4]. From an evolutionary point of view, these foods promote survival by increasing the motivation to eat nutrients with a substantial energy value. In our societies, which are characterized by easy access to highly palatable food, these specific properties could overwhelm cognitive inhibition and homeostatic mechanisms and lead to overweight [4]. However, as explained by Hebebrand et al. [4], characterizing a food or nutriment as an addictive substance implies that it has intrinsic addictive properties with the capacity to make vulnerable individuals addicted to it. In their recent article, Fletcher and Kenny wrote that no clear consensus has yet emerged on the validity of the concept of food addiction, and they presented arguments and counterarguments [64]. Regarding human research, Ahmed et al. concluded in their review that sugar and sweet rewards could not only substitute for addictive drugs such as cocaine, but also could potentially be more rewarding and attractive [65]. However, apart from caffeine, human research has found no clear evidence that any specific food, ingredient, micronutrient or combination is addictive and thus that some individuals would crave some foods akin to ingesting a specific substance. Hebebrand et al. therefore proposed the term eating addiction rather than food addiction to better capture eating addiction-related disorders [4], going beyond the substance-based view assumed in the YFAS. This term eating addiction might partially help explain why the prevalence of FA is so high in patients with AN-R because of their relationship with food. The notable prevalence of FA in AN-R but also in AN-BP patients might then be linked with natural consequences of chronic food deprivation, as shown in the Minnesota Semistarvation Experiment, which resulted in preoccupation with food and conversations centered around food, recipes and food production among healthy volunteers submitted to severe and prolonged dietary restriction [66]. However, these results are also in line with the clinical experience of AN (AN-R as AN-BP) patients showing restrictive eating behaviors to combat impulses of hunger and a loss of control over eating. The classic shift from restriction to binge eating is one argument, among others, that supports this notion. In that way, a study using functional magnetic resonance imaging (fMRI) in women recovered from AN-R showed an increased neural response to pleasant food stimuli in the ventral striatum, a brain region implicated in the motivational salience of stimuli [67]. According to the authors, these results support the idea that AN-R patients may restrict their eating in order to control exposure to food stimuli because of a hypersensitive neural response to them. However, it is difficult to determine whether this neural dysfunction is a stable trait characteristic preceding the development of AN-R, supporting the theory of addiction-like eating tendencies in AN-R patients, or a scar effect. Longitudinal studies are needed to answer that question.

Second, the analysis of each criterion revealed that the most prevalent ones in our sample were (i) "clinically significant impairment or distress in relation to food" (90.8%); *(ii) "craving"* (79.2%); and (iii) "persistent desire or repeated unsuccessful attempts to cut down" (78.5%). Previous studies have found similar results, with the criteria "clinically significant impairment or distress in relation to food" and "persistent desire or repeated unsuccessful attempts to cut down" being the most important criteria in ED patients [21,47,48]. Nevertheless, *"craving"* has not been evaluated in previous studies because this criterion was not present in the first version of the YFAS. The frequency of the first criterion is not surprising given that a significant impairment or distress in relation to food is a core feature in ED. The importance of craving is in line with the evolution in the addiction-related diagnostic criteria according to the DSM: whereas the presence of tolerance or withdrawal symptoms was necessary to confirm a diagnosis of Alcohol Dependence in the DSM-III [68], it was no longer the case with the publication of the DSM-IV [8]. In the DSM-5 [27], craving appeared as a new diagnostic criterion and has been progressively viewed as a relevant and core symptom in addiction. In our study, we observed that "classic" symptoms such as tolerance and withdrawal were not among the three most frequent criteria fulfilled in our sample, irrespective of the type of ED. This finding is in line with the conceptual evolution of the definition of addictive disorders but again calls into question the relevance of the substance-based model of FA. However, craving was the second most fulfilled criterion in AN-BP and BN patients, suggesting once again an overlap with addictive disorders, and the fourth most fulfilled criterion in AN-R patients, which might indicate a natural response to chronic food restriction but might also be linked to a natural affinity for eating as mentioned previously. Regarding the frequency of the criterion "persistent desire or repeated unsuccessful attempts to cut down", it is in line with previous studies conducted with both clinical and general population samples, in which this criterion was the most frequently endorsed FA symptom [48]. It reflects a behavioral control failure typically observed in EDs as well as in addictive disorders. Furthermore, it is noteworthy that this criterion was the third most frequently fulfilled in AN-R patients (64.6%), possibly due to a misunderstanding related to their subjective feeling of eating too much. Overall, the relevance of modeling FA criteria based on SUD criteria to better conceptualize overeating has been debated [69], and should be considered with caution when energy intake is restricted, as in AN-R, but also in AN-BP and, to a lesser extent, in BN. It is thus difficult to conclude firmly that these criteria can be considered symptoms of FA.

Third, the presence of FA in our sample appeared to be independently correlated with three variables: illness severity, the presence of binge-eating episodes and a more pronounced lack of interoceptive awareness assessed by the EDI-2. As noted in previous studies, the presence of addiction-related symptoms is associated with a more severe eating pathology and psychopathology among ED patients [11,20,21,47]. The association between FA and the presence of binge eating episodes is also in line with previous research. In the study conducted by Granero et al. [21], higher dimensional scores in the YFAS were associated with the number of binge episodes per week (and not with the number of purging behaviors per week). More generally, several authors have highlighted that FA represents an extreme state of overeating (with a correlation between the number of YFAS symptoms and BMI in most of the studies [11]) and a more severe variant of BED [52,70,71], since binge eating has been consistently correlated with YFAS scores [7,17,71,72]. Given this, a high-risk population might be identified, and made-to-measure treatment approaches might be proposed based on the existence of FA. Indeed, a potential therapeutic implication would be to tailor SUD interventions to individuals exhibiting binge-eating episodes. This could involve motivational interviewing, psychoeducational programs, cognitive behavioral therapy to cope with cravings and cognitive remediation focusing on executive function and inhibitory control, classically proposed for SUDs. Moreover, the present findings support the development of drug therapy targeting the reward circuitry such as mu opiate receptor antagonists, for these patients [73]. Regarding the greater lack of interoceptive awareness found in the patients with FA, this could constitute a bias suggesting that FA may have been overestimated, especially in patients with AN-R. The lack of interoceptive awareness reflects one's lack of confidence in recognizing and accurately identifying emotions and sensations of hunger or satiety and was labeled fundamental to AN by Bruch and Selvini-Palazzoli [74–76]. Thus, some items could have been coded as positive by patients because of difficulties in recognizing sensations of hunger or satiety. We could also consider that a lack of interoceptive awareness might truly predict FA. ED patients with FA might exhibit a different profile than ED patients without FA. Therefore, a more specific treatment program, notably based on body-oriented psychotherapy aimed at improving interoceptive skills, could be proposed according to the presence of FA. In our sample, FA was not associated with the expected factors that are typically correlated with FA. This might be due to the ED sample heterogeneity and to the existence of associations between these factors and certain types of EDs, as found in the literature [77–83], displaying comorbidity and personality traits shared between FA and ED.

According to the DSM-5 [27], there is an overlap between Substance-related and Addictive Disorders and Feeding and Eating Disorders, given that "control" plays a major role in these two categories of disorders. Whereas "impaired control" (which may reflect impairments in brain inhibitory mechanisms) appears to be a key feature in SUDs, a "sense of lack of control over eating during the (binge eating) episode" is presented in the DSM as more central in both BN and BED [27]. Thus, in BN and BED, as noted by Hebebrand et al. [4], the focus is made by the DSM on subjective feelings of the loss of control. The importance of FA in ED patients questions the pertinence of this distinction between objective impaired control in the field of addictive disorders on the one hand and a subjective sense of lack of control in the field of Feeding and Eating Disorders on the other hand. Some studies have suggested that disturbances in the inhibitory control pathway, occurring in particular rewarding conditions, may favor ED, in particular BED and BN [6,84,85]. Moreover, in AN patients, the literature has also suggested the pivotal role of the reward system in the context of exposure to particular stimuli, such as underweight stimuli for patients presenting acute AN [86], that support theories of starvation dependence, and food stimuli for patients presenting recovered AN as previously cited [67], that supports a particular affinity for eating, which persists even after starvation. That being said, in addictive disorders, subjective feelings about the "object" of addiction need to be taken into account as much as the objective impaired control. In that sense, a study demonstrated that the subscale of the Food Cravings Questionnaire-Trait that assesses the anticipation of positive reinforcement that may result from eating had negatively predicted FA symptoms, contrary to the other subscales [87]. According to the authors, people with FA symptoms may want craved foods but were also aware that the food will not make them feel better. Similarly, they experienced feelings of guilt after giving in to cravings. According to the authors, these results illustrate the ambivalence associated with food craving experiences, which seem to be especially important in individuals with addictive-like eating behaviors. Then, patients with FA might experience craving for food associated with a substantial sentiment of ambivalence and guilt. It is noteworthy that these clinical aspects are particularly observed in AN.

In this way, an integrative treatment approach inspired on the one hand by classic ED treatment based on nutrition rehabilitation, body-oriented psychotherapy, and cognitive therapy aimed at reducing cognitive distortions about eating, body shape and weight and on the other hand by traditional SUD treatment as cited above should be developed for ED patients with FA, taking into account the presence of addictive tendencies.

4.2. Strengths and Weaknesses

The results must be viewed in the context of some limitations. First, compared with the AN and BN groups, the BED group was small (n = 15), which could have minimized the power of the study. Second, the cognitive distortions that usually affect ED patients, notably AN patients, could have skewed the way they answered the questionnaires. Some items, such as "I continued to eat certain foods even though I was no longer hungry", "I spend a lot of time feeling sluggish or fatigued from overeating", "I felt so bad about overeating that I didn't do other important things", *or* "I didn't do well at work or school because I was eating too much", could have been coded as positive by the patients because of difficulties in recognizing sensations of hunger or satiety and because of particular beliefs about eating. However, as previously stated, the YFAS does not measure objective overeating but a particular relationship with food and eating; thus AN patients could satisfy the criteria for FA even if it is quite difficult to determine whether this tendency stems from starvation or a natural affinity for eating. Other limitations include the cross-sectional design of the study and the definition of the FA concept in itself, which is still a debated topic (i.e., does the YFAS truly measure what it is designed to measure?).

These limits are compensated by the strengths of the study. First, we want to emphasize the sample size. Two hundred and one patients were recruited, and such a large number allowed for a good representativeness of patients seeking treatment for ED. Moreover, ED diagnoses were established by structured clinical interviews and were based on DSM criteria. All patients were assessed at the beginning of the care in our specialized department. Finally, to the best of our knowledge, only a few studies have evaluated FA in AN [20,21], and we provided original results.

4.3. Perspectives

In showing an overlap between ED and FA, this study allows consideration of ED, including AN-R, from an addictive perspective, thus paving the way for therapeutic management that draws from those proposed for addictive disorders. Given that patients with ED and FA exhibit a different profile than patients with ED and no FA, tailor-made treatment might be proposed based on the existence of FA. Because the object of addiction is not clearly defined in the YFAS, the relevance of modeling its criteria on diagnostic criteria for SUD can be questioned, and further studies are needed to evaluate the intrinsic nature of some food addictive properties. The term eating addiction rather than FA might be most appropriate because it includes the behavioral component of the disorder. It would be of interest in further studies to specifically assess eating addiction in ED samples, with a specific scale such as the Addiction-like Eating Behaviour Scale [88]. This could be considered a clinical entity that needs to be better characterized.

Supplementary Materials: The following are available online at http://www.mdpi.com/2072-6643/12/6/1897/s1, Table S1: Description of the sample according to ED (n = 195).

Author Contributions: Conceptualization, M.G.-B. and M.F.; Methodology, M.G.-B., M.F. and M.R.; Validation, G.C.-B. and M.G.-B.; Formal analysis, M.R.; Investigation, E.T., M.R., S.L., B.R. and M.F.; Data curation, E.T. and M.R., Writing—Original draft preparation, M.F.; Writing—Review and Editing, M.F., M.R., P.B., G.C.-B. and M.G.-B.; Supervision, M.G.-B.; Project Administration, M.G.-B. All authors have read and agreed to the published version of the manuscript.

Funding: This research did not receive any specific grant from funding agencies in the public, commercial, or not-for-profit sectors.

Acknowledgments: We would like to warmly thank other staff members who performed the assessment of the EVALADD cohort.

Conflicts of Interest: The authors declare no conflict of interest.

References

1. Randolph, T.G. The descriptive features of food addiction; addictive eating and drinking. *Q. J. Stud. Alcohol.* **1956**, *17*, 198–224. [CrossRef]
2. Avena, N.M.; Bocarsly, M.E.; Hoebel, B.G. Animal models of sugar and fat bingeing: Relationship to food addiction and increased body weight. *Methods Mol. Biol.* **2012**, *829*, 351–365. [CrossRef] [PubMed]
3. Gordon, E.L.; Ariel-Donges, A.H.; Bauman, V.; Merlo, L.J. What Is the Evidence for "Food Addiction?" A Systematic Review. *Nutrients* **2018**, *10*, 477. [CrossRef] [PubMed]
4. Hebebrand, J.; Albayrak, Ö.; Adan, R.; Antel, J.; Dieguez, C.; de Jong, J.; Leng, G.; Menzies, J.; Mercer, J.G.; Murphy, M.; et al. "Eating addiction", rather than "food addiction", better captures addictive-like eating behavior. *Neurosci. Biobehav. Rev.* **2014**, *47*, 295–306. [CrossRef] [PubMed]
5. Wang, G.-J.; Volkow, N.D.; Thanos, P.K.; Fowler, J.S. Imaging of brain dopamine pathways: Implications for understanding obesity. *J. Addict. Med.* **2009**, *3*, 8–18. [CrossRef]
6. Gearhardt, A.N.; Yokum, S.; Orr, P.T.; Stice, E.; Corbin, W.R.; Brownell, K.D. Neural correlates of food addiction. *Arch. Gen. Psychiatry* **2011**, *68*, 808–816. [CrossRef]
7. Gearhardt, A.N.; Corbin, W.R.; Brownell, K.D. Preliminary validation of the Yale Food Addiction Scale. *Appetite* **2009**, *52*, 430–436. [CrossRef]
8. American Psychiatric Association. *Diagnostic and Statistical Manual of Mental Health Disorders: DSM-IV-TR*, 4th ed.; American Psychiatric Publishing: Washington, DC, USA, 2000.
9. Meule, A.; Gearhardt, A.N. Five years of the Yale Food Addiction Scale: Taking stock and moving forward. *Curr. Addict. Rep.* **2014**, *1*, 193–205. [CrossRef]
10. Gearhardt, A.N.; Corbin, W.R.; Brownell, K.D. Development of the Yale Food Addiction Scale Version 2.0. *Psychol. Addict. Behav.* **2016**, *30*, 113–121. [CrossRef]
11. Penzenstadler, L.; Soares, C.; Karila, L.; Khazaal, Y. Systematic Review of Food Addiction as Measured With the Yale Food Addiction Scale: Implications for the Food Addiction Construct. *Curr. Neuropharmacol.* **2018**. [CrossRef]
12. Pursey, K.M.; Stanwell, P.; Gearhardt, A.N.; Collins, C.E.; Burrows, T.L. The prevalence of food addiction as assessed by the Yale Food Addiction Scale: A systematic review. *Nutrients* **2014**, *6*, 4552–4590. [CrossRef] [PubMed]
13. Ivezaj, V.; Wiedemann, A.A.; Grilo, C.M. Food addiction and bariatric surgery: A systematic review of the literature. *Obes. Rev.* **2017**, *18*, 1386–1397. [CrossRef] [PubMed]
14. Brewerton, T.D. Food addiction as a proxy for eating disorder and obesity severity, trauma history, PTSD symptoms, and comorbidity. *Eat. Weight Disord.* **2017**, *22*, 241–247. [CrossRef] [PubMed]
15. Mies, G.W.; Treur, J.L.; Larsen, J.K.; Halberstadt, J.; Pasman, J.A.; Vink, J.M. The prevalence of food addiction in a large sample of adolescents and its association with addictive substances. *Appetite* **2017**, *118*, 97–105. [CrossRef] [PubMed]

16. Brunault, P.; Frammery, J.; Montaudon, P.; De Luca, A.; Hankard, R.; Ducluzeau, P.H.; Cortese, S.; Ballon, N. Adulthood and childhood ADHD in patients consulting for obesity is associated with food addiction and binge eating, but not sleep apnea syndrome. *Appetite* **2019**, *136*, 25–32. [CrossRef]
17. Davis, C.; Curtis, C.; Levitan, R.D.; Carter, J.C.; Kaplan, A.S.; Kennedy, J.L. Evidence that "food addiction" is a valid phenotype of obesity. *Appetite* **2011**, *57*, 711–717. [CrossRef]
18. Pipová, H.; Kaščáková, N.; Fürstová, J.; Tavel, P. Development of the Modified Yale Food Addiction Scale Version 2.0 summary version in a representative sample of Czech population. *J. Eat. Disord.* **2020**, *8*, 16. [CrossRef]
19. Brunault, P.; Ducluzeau, P.-H.; Courtois, R.; Bourbao-Tournois, C.; Delbachian, I.; Réveillère, C.; Ballon, N. Food Addiction is Associated with Higher Neuroticism, Lower Conscientiousness, Higher Impulsivity, but Lower Extraversion in Obese Patient Candidates for Bariatric Surgery. *Subst. Use Misuse* **2018**, *53*, 1919–1923. [CrossRef]
20. Wolz, I.; Hilker, I.; Granero, R.; Jiménez-Murcia, S.; Gearhardt, A.N.; Dieguez, C.; Casanueva, F.F.; Crujeiras, A.B.; Menchón, J.M.; Fernández-Aranda, F. "Food Addiction" in Patients with Eating Disorders is Associated with Negative Urgency and Difficulties to Focus on Long-Term Goals. *Front. Psychol.* **2016**, *7*. [CrossRef]
21. Granero, R.; Hilker, I.; Agüera, Z.; Jiménez-Murcia, S.; Sauchelli, S.; Islam, M.A.; Fagundo, A.B.; Sánchez, I.; Riesco, N.; Dieguez, C.; et al. Food addiction in a Spanish sample of eating disorders: DSM-5 diagnostic subtype differentiation and validation data. *Eur. Eat. Disord. Rev.* **2014**, *22*, 389–396. [CrossRef]
22. Haute Autorité de Santé (HAS). Anorexie Mentale: Prise en Charge. June 2010. Available online: https://www.has-sante.fr/jcms/c_985715/fr/anorexie-mentale-prise-en-charge (accessed on 25 April 2020).
23. Haute Autorité de Santé (HAS). Boulimie et Hyperphagie Boulimique: Repérage et Éléments Généraux de Prise en Charge. September 2010. Available online: https://www.has-sante.fr/jcms/c_2581436/fr/boulimie-et-hyperphagie-boulimique-reperage-et-elements-generaux-de-prise-en-charge (accessed on 25 April 2020).
24. National Collaborating Centre for Mental Health (UK). Eating Disorders: Core Interventions in the Treatment and Management of Anorexia Nervosa, Bulimia Nervosa and Related Eating Disorders. 2004. Available online: http://www.ncbi.nlm.nih.gov/books/NBK49304 (accessed on 25 April 2020).
25. Sheehan, D.V.; Lecrubier, Y.; Sheehan, K.H.; Amorim, P.; Janavs, J.; Weiller, E.; Hergueta, T.; Baker, R.; Dunbar, G.C. The Mini-International Neuropsychiatric Interview (M.I.N.I.): The development and validation of a structured diagnostic psychiatric interview for DSM-IV and ICD-10. *J. Clin. Psychiatry* **1998**, *59* (Suppl. 20), 22–33, quiz 34–57.
26. Lecrubier, Y.; Sheehan, D.; Weiller, E.; Amorim, P.; Bonora, I.; Harnett Sheehan, K.; Janavs, J.; Dunbar, G. The Mini International Neuropsychiatric Interview (MINI). A short diagnostic structured interview: Reliability and validity according to the CIDI. *Eur. Psychiatry* **1997**, *12*, 224–231. [CrossRef]
27. American Psychiatric Association. *Diagnostic and Statistical Manual of Mental Health Disorders: DSM-5*, 5th ed.; American Psychiatric Publishing: Washington, DC, USA, 2013.
28. Morgan, H.G.; Hayward, A.E. Clinical assessment of anorexia nervosa. The Morgan-Russell outcome assessment schedule. *Br. J. Psychiatry* **1988**, *152*, 367–371. [CrossRef]
29. Garner, D.M. *Eating Disorder Inventory 2—Professional Manual*; Psychological Assessment Ressources: Odessa, FL, USA, 1991.
30. Brunault, P.; Ballon, N.; Gaillard, P.; Réveillère, C.; Courtois, R. Validation of the French version of the yale food addiction scale: An examination of its factor structure, reliability, and construct validity in a nonclinical sample. *Can. J. Psychiatry* **2014**, *59*, 276–284. [CrossRef] [PubMed]
31. Brunault, P.; Courtois, R.; Gearhardt, A.N.; Gaillard, P.; Journiac, K.; Cathelain, S.; Réveillère, C.; Ballon, N. Validation of the French Version of the DSM-5 Yale Food Addiction Scale in a Nonclinical Sample. *Can. J. Psychiatry* **2017**, *62*, 199–210. [CrossRef] [PubMed]
32. Meule, A.; Gearhardt, A.N. Ten Years of the Yale Food Addiction Scale: A Review of Version 2.0. *Curr. Addict. Rep.* **2019**, *6*, 218–228. [CrossRef]
33. Christenson, G.A.; Faber, R.J.; de Zwaan, M.; Raymond, N.C.; Specker, S.M.; Ekern, M.D.; Mackenzie, T.B.; Crosby, R.D.; Crow, S.J.; Eckert, E.D. Compulsive buying: Descriptive characteristics and psychiatric comorbidity. *J. Clin. Psychiatry* **1994**, *55*, 5–11. [PubMed]
34. Lee, C.-K.; Lee, B.; Bernhard, B.J.; Lee, T.K. A Comparative Study of Involvement and Motivation among Casino Gamblers. *Psychiatry Investig.* **2009**, *6*, 141–149. [CrossRef]

35. Caci, H.M.; Bouchez, J.; Baylé, F.J. An aid for diagnosing attention-deficit/hyperactivity disorder at adulthood: Psychometric properties of the French versions of two Wender Utah Rating Scales (WURS-25 and WURS-K). *Compr. Psychiatry* **2010**, *51*, 325–331. [CrossRef]
36. Ward, M.F.; Wender, P.H.; Reimherr, F.W. The Wender Utah Rating Scale: An aid in the retrospective diagnosis of childhood attention deficit hyperactivity disorder. *Am. J. Psychiatry* **1993**, *150*, 885–890. [CrossRef] [PubMed]
37. Van der Linden, M.; d'Acremont, M.; Zermatten, A.; Jermann, F.; Larøi, F.; Willems, S.; Juillerat Van der Linden, A.-C.; Bechara, A. A French Adaptation of the UPPS Impulsive Behavior Scale: Confirmatory Factor Analysis in a Sample of Undergraduate Students. *Eur. J. Psychol. Assess.* **2006**, *22*, 38–42. [CrossRef]
38. Whiteside, S.P.; Lynam, D.R. The Five Factor Model and impulsivity: Using a structural model of personality to understand impulsivity. *Personal. Individ. Differ.* **2001**, *30*, 669–689. [CrossRef]
39. Billieux, J.; Rochat, L.; Ceschi, G.; Carré, A.; Offerlin-Meyer, I.; Defeldre, A.-C.; Khazaal, Y.; Besche-Richard, C.; Van der Linden, M. Validation of a short French version of the UPPS-P Impulsive Behavior Scale. *Compr. Psychiatry* **2012**, *53*, 609–615. [CrossRef] [PubMed]
40. Cloninger, C.R.; Svrakic, D.M.; Przybeck, T.R. A psychobiological model of temperament and character. *Arch. Gen. Psychiatry* **1993**, *50*, 975–990. [CrossRef]
41. Bartholomew, K.; Horowitz, L.M. Attachment styles among young adults: A test of a four-category model. *J. Personal. Soc. Psychol.* **1991**, *61*, 226–244. [CrossRef]
42. Guédeney, N.; Fermanian, J.; Bifulco, A. La version française du Relationship Scales Questionnaire de Bartholomew (RSQ, Questionnaire des échelles de relation): Étude de validation du construit. *L'Encéphale* **2010**, *36*, 69–76. [CrossRef]
43. Ferreri, M.; Vacher, J.; Tawils, S.; Alby, J.M. Life events and depression. A new methodological approach [Facteurs évènementiels et dépression: Le questionnaire EVE, une nouvelle approche de méthodologie prédictive]. *Psychol. Méd.* **1987**, *19*, 2441–2448.
44. Guillou-Landreat, M.; Guilleux, A.; Sauvaget, A.; Brisson, L.; Leboucher, J.; Remaud, M.; Challet-Bouju, G.; Grall-Bronnec, M. Factors associated with suicidal risk among a French cohort of problem gamblers seeking treatment. *Psychiatry Res.* **2016**, *240*, 11–18. [CrossRef]
45. Mickey, R.M.; Greenland, S. The impact of confounder selection criteria on effect estimation. *Am. J. Epidemiol.* **1989**, *129*, 125–137. [CrossRef]
46. Shtatland, E.; Cain, E.; Barton, M. *The Perils of Stepwise Logistic Regression and How to Escape Them Using Information Criteria and the Output Delivery System*; Harvard Pilgrim Health Care, Harvard Medical School: Boston, MA, USA, 2001; Volume 26.
47. Meule, A.; von Rezori, V.; Blechert, J. Food addiction and bulimia nervosa. *Eur. Eat. Disord. Rev.* **2014**, *22*, 331–337. [CrossRef]
48. Hilker, I.; Sánchez, I.; Steward, T.; Jiménez-Murcia, S.; Granero, R.; Gearhardt, A.N.; Rodríguez-Muñoz, R.C.; Dieguez, C.; Crujeiras, A.B.; Tolosa-Sola, I.; et al. Food Addiction in Bulimia Nervosa: Clinical Correlates and Association with Response to a Brief Psychoeducational Intervention. *Eur. Eat. Disord. Rev.* **2016**, *24*, 482–488. [CrossRef] [PubMed]
49. Gearhardt, A.N.; Boswell, R.G.; White, M.A. The association of "food addiction" with disordered eating and body mass index. *Eat. Behav.* **2014**, *15*, 427–433. [CrossRef] [PubMed]
50. Volkow, N.D.; O'Brien, C.P. Issues for DSM-V: Should obesity be included as a brain disorder? *Am. J. Psychiatry* **2007**, *164*, 708–710. [CrossRef] [PubMed]
51. Kinzl, J.F.; Biebl, W. Are eating disorders addictions? *Neuropsychiatrie* **2010**, *24*, 200–208.
52. Gearhardt, A.N.; White, M.A.; Potenza, M.N. Binge Eating Disorder and Food Addiction. *Curr. Drug Abuse Rev.* **2011**, *4*, 201–207. [CrossRef] [PubMed]
53. Speranza, M.; Revah-Levy, A.; Giquel, L.; Loas, G.; Venisse, J.-L.; Jeammet, P.; Corcos, M. An investigation of Goodman's addictive disorder criteria in eating disorders. *Eur. Eat. Disord. Rev.* **2012**, *20*, 182–189. [CrossRef]
54. Davis, C. A narrative review of binge eating and addictive behaviors: Shared associations with seasonality and personality factors. *Front. Psychiatry* **2013**, *4*, 183. [CrossRef]
55. Curtis, C.; Davis, C. A qualitative study of binge eating and obesity from an addiction perspective. *Eat. Disord.* **2014**, *22*, 19–32. [CrossRef]

56. Cota, D.; Tschöp, M.H.; Horvath, T.L.; Levine, A.S. Cannabinoids, opioids and eating behavior: The molecular face of hedonism? *Brain Res. Rev.* **2006**, *51*, 85–107. [CrossRef]
57. Corwin, R.L.; Avena, N.M.; Boggiano, M.M. Feeding and reward: Perspectives from three rat models of binge eating. *Physiol. Behav.* **2011**, *104*, 87–97. [CrossRef]
58. Avena, N.M.; Bocarsly, M.E. Dysregulation of brain reward systems in eating disorders: Neurochemical information from animal models of binge eating, bulimia nervosa, and anorexia nervosa. *Neuropharmacology* **2012**, *63*, 87–96. [CrossRef]
59. Umberg, E.N.; Shader, R.I.; Hsu, L.K.G.; Greenblatt, D.J. From disordered eating to addiction: The "food drug" in bulimia nervosa. *J. Clin. Psychopharmacol.* **2012**, *32*, 376–389. [CrossRef] [PubMed]
60. Kaye, W.H.; Wierenga, C.E.; Bailer, U.F.; Simmons, A.N.; Wagner, A.; Bischoff-Grethe, A. Does a shared neurobiology for foods and drugs of abuse contribute to extremes of food ingestion in anorexia and bulimia nervosa? *Biol. Psychiatry* **2013**, *73*, 836–842. [CrossRef] [PubMed]
61. Volkow, N.D.; Wang, G.-J.; Tomasi, D.; Baler, R.D. The addictive dimensionality of obesity. *Biol. Psychiatry* **2013**, *73*, 811–818. [CrossRef] [PubMed]
62. Barbarich-Marsteller, N.C.; Foltin, R.W.; Walsh, B.T. Does anorexia nervosa resemble an addiction? *Curr. Drug Abuse Rev.* **2011**, *4*, 197–200. [CrossRef]
63. Mallorquí-Bagué, N.; Lozano-Madrid, M.; Testa, G.; Vintró-Alcaraz, C.; Sánchez, I.; Riesco, N.; César Perales, J.; Francisco Navas, J.; Martínez-Zalacaín, I.; Megías, A.; et al. Clinical and Neurophysiological Correlates of Emotion and Food Craving Regulation in Patients with Anorexia Nervosa. *J. Clin. Med.* **2020**, *9*, 960. [CrossRef]
64. Fletcher, P.C.; Kenny, P.J. Food addiction: A valid concept? *Neuropsychopharmacology* **2018**, *43*, 2506–2513. [CrossRef]
65. Ahmed, S.H.; Guillem, K.; Vandaele, Y. Sugar addiction: Pushing the drug-sugar analogy to the limit. *Curr. Opin. Clin. Nutr. Metab. Care* **2013**, *16*, 434–439. [CrossRef]
66. Tobey, J.A. The Biology of Human Starvation. *Am. J. Public Health Nations Health* **1951**, *41*, 236–237. [CrossRef]
67. Cowdrey, F.A.; Park, R.J.; Harmer, C.J.; McCabe, C. Increased neural processing of rewarding and aversive food stimuli in recovered anorexia nervosa. *Biol. Psychiatry* **2011**, *70*, 736–743. [CrossRef]
68. American Psychiatric Association. *Diagnostic and Statistical Manual of Mental Health Disorders: DSM-III*, 3th ed.; American Psychiatric Publishing: Washington, DC, USA, 1980.
69. Ziauddeen, H.; Farooqi, I.S.; Fletcher, P.C. Obesity and the brain: How convincing is the addiction model? *Nat. Rev. Neurosci.* **2012**, *13*, 279–286. [CrossRef] [PubMed]
70. Davis, C. From passive overeating to "food addiction": A spectrum of compulsion and severity. *ISRN Obes.* **2013**, *2013*, 435027. [CrossRef] [PubMed]
71. Gearhardt, A.N.; White, M.A.; Masheb, R.M.; Morgan, P.T.; Crosby, R.D.; Grilo, C.M. An Examination of the Food Addiction Construct in Obese Patients with Binge Eating Disorder. *Int. J. Eat. Disord.* **2012**, *45*, 657–663. [CrossRef] [PubMed]
72. Burmeister, J.M.; Hinman, N.; Koball, A.; Hoffmann, D.A.; Carels, R.A. Food addiction in adults seeking weight loss treatment. Implications for psychosocial health and weight loss. *Appetite* **2013**, *60*, 103–110. [CrossRef] [PubMed]
73. Beitscher-Campbell, H.; Blum, K.; Febo, M.; Madigan, M.A.; Giordano, J.; Badgaiyan, R.D.; Braverman, E.R.; Dushaj, K.; Li, M.; Gold, M.S. Pilot clinical observations between food and drug seeking derived from fifty cases attending an eating disorder clinic. *J. Behav. Addict.* **2016**, *5*, 533–541. [CrossRef]
74. Bruch, H. Perceptual and conceptual disturbances in anorexia nervosa. *Psychosom. Med.* **1962**, *24*, 187–194. [CrossRef]
75. Bruch, H. *The Golden Cage*; Basic Books: New York, NY, USA, 1978.
76. Selvini-Palazzoli, M. *Self-Staruation—From Individual to Family Therapy in the Treatment of Anorexia Neruosa*; Jason Aronson: New York, NY, USA, 1978.
77. Kaisari, P.; Dourish, C.T.; Higgs, S. Attention Deficit Hyperactivity Disorder (ADHD) and disordered eating behaviour: A systematic review and a framework for future research. *Clin. Psychol. Rev.* **2017**, *53*, 109–121. [CrossRef]
78. Caslini, M.; Bartoli, F.; Crocamo, C.; Dakanalis, A.; Clerici, M.; Carrà, G. Disentangling the Association between Child Abuse and Eating Disorders: A Systematic Review and Meta-Analysis. *Psychosom. Med.* **2016**, *78*, 79–90. [CrossRef]

79. Steward, T.; Mestre-Bach, G.; Vintró-Alcaraz, C.; Agüera, Z.; Jiménez-Murcia, S.; Granero, R.; Fernández-Aranda, F. Delay Discounting of Reward and Impulsivity in Eating Disorders: From Anorexia Nervosa to Binge Eating Disorder. *Eur. Eat. Disord. Rev.* **2017**, *25*, 601–606. [CrossRef]
80. Frank, G.K.W.; DeGuzman, M.C.; Shott, M.E.; Laudenslager, M.L.; Rossi, B.; Pryor, T. Association of Brain Reward Learning Response with Harm Avoidance, Weight Gain, and Hypothalamic Effective Connectivity in Adolescent Anorexia Nervosa. *JAMA Psychiatry* **2018**, *75*, 1071–1080. [CrossRef]
81. Tasca, G.A. Attachment and eating disorders: A research update. *Curr. Opin. Psychol.* **2019**, *25*, 59–64. [CrossRef] [PubMed]
82. Gat-Lazer, S.; Geva, R.; Gur, E.; Stein, D. Reward Dependence and Harm Avoidance among Patients with Binge-Purge Type Eating Disorders. *Eur. Eat. Disord. Rev.* **2017**, *25*, 205–213. [CrossRef] [PubMed]
83. Bahji, A.; Mazhar, M.N.; Hudson, C.C.; Nadkarni, P.; MacNeil, B.A.; Hawken, E. Prevalence of substance use disorder comorbidity among individuals with eating disorders: A systematic review and meta-analysis. *Psychiatry Res.* **2019**, *273*, 58–66. [CrossRef] [PubMed]
84. Berner, L.A.; Marsh, R. Frontostriatal circuits and the development of bulimia nervosa. *Front. Behav. Neurosci.* **2014**, *8*, 395. [CrossRef]
85. Oliva, R.; Morys, F.; Horstmann, A.; Castiello, U.; Begliomini, C. The impulsive brain: Neural underpinnings of binge eating behavior in normal-weight adults. *Appetite* **2019**, *136*, 33–49. [CrossRef] [PubMed]
86. Fladung, A.-K.; Grön, G.; Grammer, K.; Herrnberger, B.; Schilly, E.; Grasteit, S.; Wolf, R.C.; Walter, H.; von Wietersheim, J. A neural signature of anorexia nervosa in the ventral striatal reward system. *Am. J. Psychiatry* **2010**, *167*, 206–212. [CrossRef]
87. Meule, A.; Kübler, A. Food cravings in food addiction: The distinct role of positive reinforcement. *Eat. Behav.* **2012**, *13*, 252–255. [CrossRef]
88. Ruddock, H.K.; Christiansen, P.; Halford, J.C.G.; Hardman, C.A. The development and validation of the Addiction-like Eating Behaviour Scale. *Int. J. Obes. (Lond.)* **2017**, *41*, 1710–1717. [CrossRef]

nutrients

MDPI

Article

The Impact of Retrospective Childhood Maltreatment on Eating Disorders as Mediated by Food Addiction: A Cross-Sectional Study

Rami Bou Khalil [1,2,*], Ghassan Sleilaty [3], Sami Richa [1], Maude Seneque [2,4], Sylvain Iceta [5,6,7], Rachel Rodgers [4,8], Adrian Alacreu-Crespo [2], Laurent Maimoun [9,10], Patrick Lefebvre [11], Eric Renard [11,12], Philippe Courtet [2,4] and Sebastien Guillaume [2,4]

1 Department of Psychiatry, Saint Joseph University-Hôtel Dieu de France Hospital, Beirut BP 166830, Lebanon; sami.richa@usj.edu.lb
2 PSNREC, University of Montpellier, INSERM, CHU de Montpellier, 34295 Montpellier, France; maude.seneque@gmail.com (M.S.); adrian.alacreu@uv.es (A.A.-C.); philippecourtet@gmail.com (P.C.); s-guillaume@chu-montpellier.fr (S.G.)
3 Clincal Research Center and Department of Cardiac and Thoracic Surgery, Saint Joseph University-Hôtel Dieu de France Hospital, Beirut BP 166830, Lebanon; ghassan.sleilaty@usj.edu.lb
4 Department of Psychiatric Emergency and Acute Care, Lapeyronie Hospital, CHRU, 34295 Montpellier, France; rodgers.rachel@gmail.com
5 Quebec Heart and Lung Institute (IUCPQ), Québec, QC G1V 4G5, Canada; sylvain.iceta.1@ulaval.ca
6 School of Nutrition, Laval University, Québec, QC G1V 0A6, Canada
7 Referral Center for Eating Disorder, Hospices Civils de Lyon, F-69677 Bron, France
8 APPEAR, Department of Applied Psychology, Northeastern University, Boston, MA 02115, USA
9 PhyMedExp, University of Montpellier, INSERM, CNRS, 34295 Montpellier, France; l-maimoun@chu-montpellier.fr
10 Département de Médecine Nucléaire, Hôpital Lapeyronie, Centre Hospitalier Régional Universitaire (CHRU) Montpellier, 34295 Montpellier, France
11 Department of Endocrinology, Diabetes, and Nutrition, CHRU, 34295 Montpellier, France; p-lefebvre@chu-montpellier.fr (P.L.); e-renard@chu-montpellier.fr (E.R.)
12 UMR CNRS 5203, INSERM U1191, Institute of Functional Genomics, University of Montpellier, 34295 Montpellier, France
* Correspondence: rami.boukhalil@usj.edu.lb; Tel.: +96-1142-1000

Received: 3 August 2020; Accepted: 25 September 2020; Published: 28 September 2020

Abstract: Background: The current study aimed to test whether food addiction (FA) might mediate the relationship between the presence of a history of childhood maltreatment and eating disorder (ED) symptom severity. Methods: Participants were 231 patients with ED presenting between May 2017 and January 2020 to a daycare treatment facility for assessment and management with mainly the Eating Disorder Inventory-2 (EDI-2), the Child Trauma Questionnaire (CTQ), and the Yale Food Addiction Scale (YFAS 2.0). Results: Participants had a median age of 24 (interquartile range (IQR) 20–33) years and manifested anorexia nervosa (61.47%), bulimia nervosa (16.88%), binge-eating disorders (9.09%), and other types of ED (12.55%). They were grouped into those likely presenting FA ($N = 154$) and those without FA ($N = 77$). The group with FA reported higher scores on all five CTQ subscales, as well as the total score of the EDI-2 ($p < 0.001$). Using mediation analysis; significant indirect pathways between all CTQ subscales and the EDI-2 total score emerged via FA, with the largest indirect effect emerging for physical neglect (standardized effect = 0.208; 95% confidence interval (CI) 0.127–0.29) followed by emotional abuse (standardized effect = 0.183; 95% CI 0.109–0.262). Conclusion: These results are compatible with a model in which certain types of childhood maltreatment, especially physical neglect, may induce, maintain, and/or exacerbate ED symptoms via FA which may guide future treatments.

Keywords: eating disorders; food addiction; childhood trauma; maltreatment; physical neglect

1. Introduction

According to the World Health Organization, "childhood maltreatment is the abuse and neglect that occurs to children under 18 years of age. It includes all types of physical and/or emotional ill-treatment, sexual abuse, neglect, negligence, and commercial or other exploitation, which results in actual or potential harm to the child's health, survival, development, or dignity in the context of a relationship of responsibility, trust, or power" [1]. On the other hand, eating disorders (EDs) are multifactorial mental disorders affecting young individuals and are associated with a mortality rate higher than that of the general population of the same age [2]. The relationship between a history of childhood maltreatment and the later development of an ED is well established, as supported by two major meta-analyses [3,4]. Abused and/or neglected children who have experienced any type of maltreatment (i.e., emotional, sexual, and physical) are at least threefold more likely to develop a future ED [3,4]. Furthermore, a dose–effect relationship between the number of subtypes of childhood trauma experienced and the severity of ED clinical features has been evidenced, suggesting a consistent and partly independent association between these traumatic events and more severe clinical and functional characteristics of ED [5]. While childhood maltreatment may be reported by a high proportion of patients with ED, only a minority of those previously exposed to one or more traumatic events (9–24%) may subsequently present a comorbid post-traumatic stress disorder (PTSD) [6]. Accordingly, beyond the simple comorbidity with PTSD, it is not yet understood how different types of childhood maltreatment impact the clinical presentation of ED, with emotion dysregulation being consistently considered as an important factor mediating this effect [6,7].

In addition to emotion dysregulation, PTSD, and depression as known mediators of ED development in patients who have been exposed to childhood maltreatment, food addiction (FA) may constitute a yet unexplored contributing mediator [6–9]. FA is characterized by poorly controlled intake of preferred foods, which are postulated to act via similar mechanisms as both illicit and licit drugs of abuse in the brain [10]. An increasing amount of evidence of biological and behavioral changes in response to preferred foods (such as brain reward changes, impaired control, genetic susceptibility, substance sensitization and cross-sensitization, and impulsivity) has been sufficiently convincing to conceptualize FA as an addiction disorder [11]. FA has been increasingly considered as an important psychological dimension that leads, in patients with a history of complex trauma, to ED and more specifically binge-eating disorders (BED) and bulimia nervosa (BN) [10]. Despite being a clinical manifestation of addiction to food, as much as 61.5% of patients with anorexia nervosa (AN) of the restrictive type were found to suffer from FA, which translates how much the addictive behavior related to food can be a common pathological dimension to all EDs, as well as a possible accompanying manifestation of other forms of behavioral addiction to fasting, physical exercising, etc. [12].

Although data on the relationships between childhood maltreatment and FA are lacking to date, evidence from clinical studies examining closely related dimensions suggests that such an association might exist. Although not measuring childhood maltreatment per se, in a cohort study of 49,408 female nurse participants, the prevalence of FA increased with the number of lifetime PTSD symptoms, and women with the greatest number of PTSD symptoms reported more than twice the prevalence of FA compared to women with neither PTSD symptoms nor trauma histories. Interestingly, however, in this study, the relationship between FA and PTSD did not differ by trauma type [13]. Furthermore, the co-occurrence of FA symptoms with emotional dysregulation symptoms has led to the suggestion that these might share common characteristics and, potentially, risk factors [14]. In further support of this, when compared to individuals without addictive behaviors, both women with FA and women with substance use presented higher levels of depressive and PTSD symptoms, as well as greater emotion dysregulation [15]. It has been proposed that childhood maltreatment might also be associated with decreased emotional regulation, as well as a greater propensity to and severity of addictive-like behaviors due to structural brain changes (mainly diminished hippocampus volume) [16,17]. Taken together, converging evidence, therefore, seems to exist for an indirect relationship such that FA might constitute an intervening factor in the cross-link between childhood maltreatment and ED symptom severity.

To our knowledge, no study has yet assessed this proposed indirect pathway in patients with ED. We, therefore, hypothesized that retrospective childhood maltreatment would be indirectly related to ED symptom severity via FA among a transdiagnostic sample of ED patients. In this cross-sectional study design, we aim to establish a conceptual model to further the understanding of how different types of childhood maltreatment might impact ED in order to guide future assessment strategies and treatment development models.

2. Materials and Methods

2.1. Participants

All consecutive outpatients with all types of ED according to the DSM-5 (Diagnostic and Statistical Manual of Mental Disorders) criteria who were assessed in an eating disorders unit in Montpellier, France, between May 2017 and January 2020 were eligible for the study. Patients with ED are referred to this unit for multidisciplinary assessment, diagnostic confirmation, and management. The data utilized here are drawn from a large study approved by the Ethics Committee of CPP Sud-Est VI of Clermont Ferrand University (CPP: AU 1313; ID-RCB: 2017-A00269-44; N° Clinical Trial: NCT03160443). Signed informed consent was obtained from all participants (and from parents of underage participants). All research procedures were conducted according to the Declaration of Helsinki. Inclusion criteria were as follows: age superior to 15 years (15–70 years), speaking French, and having an ED diagnosis according to DSM-5 criteria. Exclusion criteria were as follows: refusing to consent ($n = 3$), having a mental disability such as intellectual deficiency ($n = 4$), and having a physical comorbidity that prevented study participation (severe hypokalemia that necessitated a transfer to an intensive care unit $n = 4$; very low nutritional state $n = 9$; other physical disorders having an important secondary impact on cognitive functions $n = 5$).

2.2. Measures

The multidisciplinary clinical assessment was carried out during a full day at the outpatient unit by experienced mental health professionals. The ED diagnosis was established on the basis of a nonstructured clinical assessment by psychiatrists, psychologists, and nutritionists, as well as a structured evaluation with the Mini-International Neuropsychiatric Interview (MINI, Version 5.0.0). All investigators were trained beforehand to use the MINI. Body weight and height were collected in a standardized way during the clinical examination. Among other psychometric and biometric assessments, participants completed the questionnaires below.

The Eating Disorder Inventory (EDI-2) is a self-report diagnostic tool designed for use in a clinical setting to assess the clinical dimensions of EDs. It contains 11 subscales (drive for thinness, bulimia, body dissatisfaction, ineffectiveness, perfectionism, interpersonal distrust, interoceptive awareness, maturity fears, asceticism, impulse regulation, social insecurity) that evaluate the symptoms of the ED, as well as its relationship with personality traits and emotions [18,19]. The total EDI-2 score used in this study consists of the sum of all 11 subscales scores [20]. Cronbach's alpha coefficient for internal consistency was 0.84 for EDI-2 total score, ranging from 0.64 (for the ascetism subscale) to 0.92 (for the bulimic tendency subscale).

The Yale Food Addiction Scale 2.0 (YFAS 2.0) is a 35-item self-report Likert-type scale that assesses food and eating regulation during the past 12 months. Items are scored on an eight-point scale with frequency response options ranging from "never" to "every day." The items assess clinical impairment/distress according to the DSM-5 criteria for substance use disorder. For the diagnosis of food addiction, the clinically significant impairment/distress criterion has to be met along with two or more diagnostic criteria [21,22]. Cronbach's alpha coefficient for internal consistency was 0.96 for YFAS 2.0.

The Child Trauma Questionnaire (CTQ) is a 28-item self-report instrument for the retrospective assessment of trauma exposure during childhood. The CTQ consists of five subscales representing different types of trauma (physical abuse, emotional abuse, sexual abuse, physical neglect, and emotional neglect) with multiple items according to a five-point Likert scale ranging from 1 (never true) to 5 (very often true). A higher score on a subscale indicates more severe childhood trauma [23]. Cronbach's alpha coefficients for internal consistency were 0.89 for the CTQ emotional abuse subscale, 0.94 for the CTQ physical abuse subscale, 0.96 for the CTQ sexual abuse subscale, 0.93 for CTQ emotional neglect, and 0.74 for physical neglect.

2.3. Statistical Analyses

In a first analysis, participants were divided into two groups: those with food addiction (FA(+)) and those without food addiction (FA(−)). Quantitative variables significantly departing from normality assumptions (as assessed by Kolmogorov–Smirnov test and quantile–quantile (Q–Q) plots) were expressed as medians with interquartile ranges (IQR: Q1–Q3). A bivariate comparison between the groups' characteristics was conducted using the Mann–Whitney U test and Pearson's chi-square (or Fisher correction as appropriate) test. Cronbach's alpha was computed for EDI-2 total score, YFAS 2.0, and CTQ subscales.

In a second analysis, the distributions of CTQ subscales, EDI-2 subscales, and YFAS 2.0 were assessed with normality tests (Kolmogorov–Smirnov test and Q–Q plots). Pearson's correlations were used to estimate index zero-order relationships among childhood maltreatment, FA and ED symptom severity, and their 95% confidence intervals (95% CI) calculated using Yates transform. Mediation analyses examining the hypothesis that food addiction underlies the relationship between childhood maltreatment and ED were tested using the PROCESS Model 4. Bias-corrected bootstrapped confidence intervals (CI) according to 10,000 bootstrap samples were built for the indirect effect (i.e., effect of child trauma (CT) on ED symptoms through FA). FA was considered to exert a mediation effect between childhood maltreatment and ED clinical symptoms when 95% CIs for indirect effects did not overlap with zero [24].

3. Results

Overall, of the 247 participants assessed, 231 provided YFAS 2.0 data and, accordingly, were included in the study. The majority of participants were women (n = 213; 92.2%), with a median age of 24 (IQR 20–33) years, and the most frequent diagnosis was anorexia nervosa (AN) (n = 142; 61.47%) followed by BN (n = 39; 16.88%), BED (n = 21; 9.09%), and other types of ED which were collapsed into a category, including the following DSM-5 diagnoses: (1) avoidant/restrictive food intake disorder; (2) pica; (3) merycism; (4) other specified feeding or eating disorder; (5) unspecified feeding or eating disorder (n = 29; 12.55%) (Table 1).

Participants were separated into two groups: 154 (66.66%) with a food addiction (FA(+)) and 77 (33.33%) with no food addiction (FA(−)). The comparison between the FA(+) and FA(−) groups revealed no differences in terms of age, gender, past history of depression, and current diagnosis of AN, BED, and other types of ED. However, FA(+) presented BN (27.5% in FA(+) vs. 8.1% in FA(−); p = 0.012) more frequently. Furthermore, actual body mass index (BMI) was higher in the group of patients with FA (20.97 (IQR 16.9–22.1) in FA(+) vs. 17.8 (IQR 16.1–19.9) in FA(−); p = 0.005). Moreover, actual and/or past history of PTSD was higher in the FA(+) group (15.58% in FA(+) vs. 2.59% in FA(−); p = 0.006). In addition, a current diagnosis of depression was more frequent in patients with FA (36.87% in FA(+) vs. 14.7% in FA(−); p = 0.001) (Table 2). In the comparison between both groups, patients in FA(+) presented a higher score on all five subscales of the CTQ. All EDI-2 subscales, as well as EDI-2 total scores, were significantly higher in FA(+) patients except for social insecurity, which was higher in FA(−) patients (p < 0.001), and interpersonal distrust, which did not significantly differ between groups (Table 3).

Table 1. Sociodemographic and clinical parameters, as well as Child Trauma Questionnaire (CTQ) and Eating Disorder Inventory (EDI-2) scores in the entire study population.

Variable	Category	Statistic	All Participants
		N	231
Age (years)		Me (IQR)	24 (19–33)
Gender	Male	*N* (%)	18 (7.8%)
	Female	*N* (%)	213 (92.2%)
ED	Diagnosis of AN	*N* (%)	142 (61.5%)
	Diagnosis of BN	*N* (%)	39 (16.9%)
	Diagnosis of BED	*N* (%)	21 (9.1%)
	Other diagnosis	*N* (%)	29 (12.5%)
Current and/or past history of PTSD		*N* (%)	33 (14.3%)
BMI (kg/m^2)	Current	Me (IQR)	18.7 (16.8–21.5)
CTQ	Emotional abuse	Me (IQR)	9 (6–13)
	Physical abuse	Me (IQR)	5 (5–7)
	Sexual abuse	Me (IQR)	5 (5–7)
	Emotional neglect	Me (IQR)	12 (8–16)
	Physical neglect	Me (IQR)	7 (5–9)
EDI-2	Drive for thinness	Me (IQR)	22 (17–28)
	Bulimia	Me (IQR)	15 (5–24)
	Body dissatisfaction	Me (IQR)	21 (17–24)
	Ineffectiveness	Me (IQR)	23 (20–27)
	Perfectionism	Me (IQR)	18 (12–22)
	Interpersonal distrust	Me (IQR)	18 (15–20)
	Interoceptive awareness	Me (IQR)	28 (20–34)
	Maturity fears	Me (IQR)	19 (16–22)
	Asceticism	Me (IQR)	19 (13–24)
	Impulse regulation	Me (IQR)	20 (13–28)
	Social insecurity	Me (IQR)	18 (16–21)
	Total score	Me (IQR)	220 (188–254)

ED, eating disorder; PTSD, post-traumatic stress disorder; AN, anorexia nervosa; BN, bulimia nervosa; BED, binge-eating disorder; BMI, body mass index; Me, median; IQR, interquartile range.

Table 2. Comparison between food addiction (FA(−) and FA(+)) groups with regard to sociodemographic parameters, diagnosis, and BMI.

Variable	Category	Statistic	FA(−) Group	FA(+) Group	Test	*p*-Value
		N	**77**	**154**		
Age (years)		Me (IQR)	28 (19–34)	27.84 (20–32)	U	0.263
Gender	Male	*N* (%)	7 (38.9%)	11 (61.1%)		
	Female	*N* (%)	70 (32.9%)	143 (67.1%)	Chi^2	0.603
ED	Diagnosis of AN	*N* (%)	50 (35.2%)	92 (64. 8%)	Chi^2	0.339
	Diagnosis of BN	*N* (%)	6 (15.4%)	33 (84.6%)	Chi^2	0.012
	Diagnosis of BED	*N* (%)	5 (23.8%)	16 (76.2%)	Chi^2	0.352
	Other diagnosis	*N* (%)	7 (50%)	7 (50%)	Chi^2	0.172
Current and/or past history of PTSD		*N* (%)	2 (7.7%)	24 (92.3%)	Y	0.006
BMI (kg/m^2)	Current	Me (IQR)	17.8 (16.1–19.9)	20.97 (16.9–22.1)	U	0.005

Data are presented as frequency and percentage (*N* (%)) or as median and interquartile range (Me (IQR)). The statistical comparisons in Table 1 were carried out with the Mann–Whitney U test (U), the chi-square test (Chi^2), or the Yates test (Y). FA: food addiction.

Table 3. Comparison between FA(−) and FA(+) with regard to CTQ and EDI-2 scores.

		FA(−) Group	FA(+) Group	Test	*p*-Value
	N	77	154		
Scale	Subscale				
CTQ	Emotional abuse	7 (5–10)	10 (7–14.25)	U	<0.001
	Physical abuse	5 (5–5)	5 (5–8)	U	0.005
	Sexual abuse	5 (5–5)	5 (5–8)	U	0.014
	Emotional neglect	10 (7–13.75)	13 (9–17)	U	0.005
	Physical neglect	6 (5–8)	7 (6–10)	U	0.006
EDI-2	Drive for thinness	18 (2–22)	25 (10–29)	U	<0.001
	Bulimia	4 (0–14)	19 (2–25)	U	<0.001
	Body dissatisfaction	18 (7–21)	22 (13–25)	U	<0.001
	Ineffectiveness	21 (13–24)	25 (17–29)	U	<0.001
	Perfectionism	15 (4–21)	20 (7–23)	U	0.001
	Interpersonal distrust	17 (11–20)	18 (13–20)	U	0.357
	Interoceptive awareness	21 (6–27)	32 (17–36)	U	<0.001
	Maturity fears	17 (8–20)	19 (13–23)	U	<0.001
	Asceticism	13 (3–18)	21 (8–26)	U	<0.001
	Impulse regulation	13 (2–19)	25 (10–30)	U	<0.001
	Social insecurity	20 (13–23)	18 (13–20)	U	0.001

Data are presented as the median and interquartile range (Q1–Q3).

Correlation between CTQ subscales and YFAS 2.0 total score showed a small to moderate effect size with a positive statistically significant correlation with all CTQ subscales, the largest being for emotional abuse ($r = 0.314$; $p < 0.001$) and physical neglect ($r = 0.307$; $p < 0.001$). The YFAS 2.0 scores and EDI-2 total score were positively and significantly correlated with a large effect size ($r = 0.608$; $p < 0.001$). The EDI-2 total score evidenced significant correlations with moderate effect sizes with all CTQ subscales, the highest being for emotional abuse ($r = 0.349$; $p < 0.001$) (Table 4).

Table 4. Correlations among Yale Food Addiction Scale (YFAS) total score, EDI-2 total scores, and all CTQ subscales. CI, confidence interval.

	CTQ *r* (95% CI); *p*-Value				
	Emotional Abuse	Physical Abuse	Sexual Abuse	Emotional Neglect	Physical Neglect
YFAS *r* (95% CI) *p*-Value	0.314 (0.19–0.428) <0.001	0.246 (0.12–0.365) <0.001	0.16 (0.028–0.286) 0.018	0.208 (0.079–0.331) 0.002	0.307 (0.183–0.421) <0.001
0.608 (0.519–0.684) <0.001					
EDI-2 *r* (95% CI) *p*-Value	0.349 (0.228–0.459) <0.001	0.199 (0.071–0.322) 0.003	0.25 (0.121–0.37) <0.001	0.227 (0.098–0.348) <0.001	0.161 (0.031–0.287) 0.016

Findings from the mediation analyses are summarized in Table 5, and significant effects are represented in Figure 1. A direct effect between the CTQ subscales and the EDI-2 total score (unmediated by YFAS 2.0) was found for emotional and sexual abuse only ($p = 0.002$ and 0.003, respectively). A consistent indirect mediation effect was present between all CTQ subscales and the EDI-2 total score via YFAS 2.0. The strongest indirect mediation effect was found in relation to the CTQ physical neglect subscale (standardized effect = 0.208; 95% CI 0.127–0.29), followed by emotional abuse (standardized effect = 0.183; 95% CI 0.109–0.262). Accordingly, the mediation or indirect effect of FA related to the impact of childhood maltreatment on clinical symptoms of ED seems to be more specific to physical neglect since, in addition to exerting the highest indirect effect among all CTQ subscales, it did not exert any direct effect on EDI-2 total score (Figure 2).

Table 5. Analysis of total, direct, and indirect (via YFAS 2.0 mediation) effect of different CTQ subscales on EDI-2 total score.

		CTQ Subscales Type of Effect (CI); *p*-Value				
		Emotional Abuse	**Physical Abuse**	**Sexual Abuse**	**Emotional Neglect**	**Physical Neglect**
EDI-2 total score	Total effect (Direct and indirect)	3.401 (2.2–4.6) <0.001	2.433 (0.83–4.03) 0.003	2.934 (1.42–4.44) <0.001	2.24 (1–3.49) <0.001	2.944 (0.52–5.35) 0.017
	Direct effect	1.64 (0.6–2.68) 0.002	0.543 (−0.77–1.85) 0.417	1.809 (0.6–3.01) 0.003	0.947 (−0.07–1.96) 0.069	−0.873 (−2.89–1.14) 0.396
	Indirect effect	1.761 (1–2.66)	1.89 (1.04–2.88)	1.125 (0.11–2.28)	1.303 (0.54–2.16)	3.817 (2.26–5.53)
	Standardized indirect effect	0.183 (0.1–0.26)	0.153 (0.08–0.22)	0.097 (0.01–0.18)	0.136 (0.05–0.21)	0.208 (0.12–0.29)

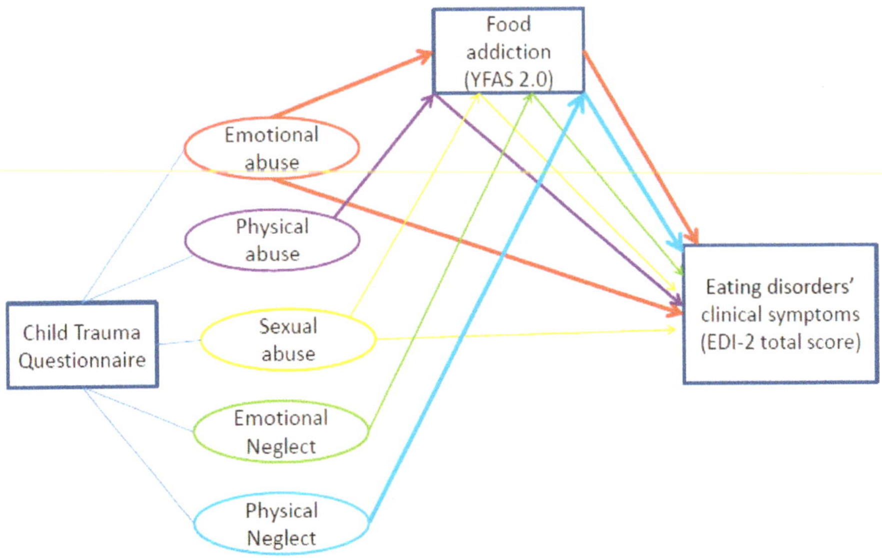

Figure 1. Direct and indirect pathways between childhood maltreatment types and the EDI-2 total score in the mediation analysis. The largest indirect effect emerged for physical neglect (standardized effect = 0.208; 95% CI [0.127-0.29]) followed by emotional abuse (standardized effect=0.183; 95% CI [0.109-0.262]. Arrows width is proportional to the effect size. YFAS 2.0: Yale Food Addiction Scale 2.0; EDI-2: Eating Disorder Inventory-2.

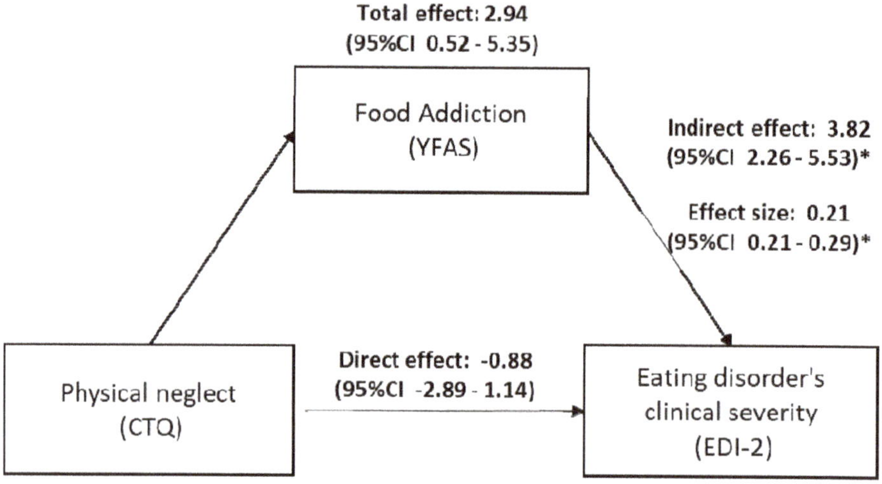

Figure 2. Triangular scheme depicting the results of mediation analysis, with food addiction (measured by YFAS 2.0 score) mediating the effect of physical neglect (CTQ physical neglect component) on eating disorder severity (EDI-2 score). Total, direct, and indirect effects correspond to the beta coefficients obtained from the mediation analysis. The effect size corresponds to the standardized indirect effect of YFAS on EDI-2. 95% CI denotes the 95% confidence interval. YFAS: Yale Food Addiction Scale 2.0, CTQ: Child Trauma Questionnaire, EDI-2: Eating Disorder Inventory-2; (*) confidence limits derived by bootstrapping 10,000 samples.

4. Discussion

This study investigated the relationship among self-reported history of childhood maltreatment, FA, and ED symptom severity in a large sample of consecutively recruited patients with ED. Patients with FA reported more frequent histories of childhood maltreatment and presented with more severe ED symptoms as assessed by the EDI-2. In addition, the strong correlation between YFAS and EDI-2 scores suggested that FA and ED symptom severity may be tightly related, while childhood maltreatment was less strongly related to FA. Emotional abuse seems to be the most important type of childhood maltreatment affecting ED symptom severity. In addition, our findings revealed evidence of an indirect effect between all types of childhood maltreatment and ED symptom severity via FA. The strongest of these indirect or mediated effects was related to physical neglect followed by emotional abuse. Moreover, a direct effect relationship between childhood maltreatment and ED symptom severity emerged for the emotional and sexual abuse dimensions. Accordingly, our findings highlight the specific importance of FA in mediating the impact of physical neglect of the clinical severity of patients with any type of ED. Although cross-sectional, these findings are consistent with a model in which retrospective childhood maltreatment, especially physical neglect, might precipitate or constitute a risk factor for FA which may later predispose for, maintain, and/or exacerbate ED symptoms. The role of FA as a mediating factor of in the relationship between retrospective childhood maltreatment and ED warrants further exploration in longitudinal observational studies.

Previous work examined the relationship between trauma exposure and FA. The large cross-sectional cohort study by Mason et al. described above revealed that the likelihood of reporting FA increased with the number of lifetime PTSD symptoms, with the prevalence of FA in women with the greatest number of PTSD symptoms more than twice that of women with neither PTSD symptoms nor trauma histories [13]. Moreover, cross-sectional evidence of relationships between childhood maltreatment and FA and between FA and binge eating was found individuals with higher weight [25]. Further support for this relationship was provided by a comparative study in which individuals with FA reported greater severity of PTSD symptoms as compared to controls [17]. Finally, Stojek et al. revealed that FA severity mediated the association between childhood maltreatment and insulin resistance in women with type 2 diabetes [26]. These cumulative findings, together with ours, suggest a relationship between traumatic exposure and history and FA, as well as other dimensions of disinhibition related to food or substances. Previous work suggested that, at the neurobiological level, both the emotional and the motivational circuits in the brain seem to be affected after exposure to childhood maltreatment [27]. These effects may be associated with disruptions to the experience of inner cues related to the regulation of food and eating, for example, through the effects of stress hormones such as glucocorticoids on the cerebral cortex and limbic system, which may affect the patient's impulse control [27]. These disruptions and the development of maladaptive behavioral patterns related to food may then increase risk for several types of ED especially those that include bingeing behaviors [28,29].

Consistent with this, in our sample, BN was more prevalent in the group of patients reporting FA as compared to the group without FA. This is further in line with findings of relatively recent studies in which FA was found in as many as 96% of patients with BN with a tendency for FA severity to decrease over the course of effective management of BN [30,31]. In a recent study aiming to characterizing FA as a phenotypical construct in patients with different types of EDs and obesity via a factor analysis, results suggested that patients with FA and BN presented with more severe ED psychopathology [32]. Moreover, other work suggested that, among women with a high BMI, the presence of a relationships between early life adversities and FA may be underpinned by specificities in brain regions implicated in reward and emotional regulation [33–35]. Similarly, patients with PTSD were described to be at higher risk for FA and EDs due to the potential mediating role of emotional dysregulation [36]. In addition to affecting the activation pattern and connectivity in brain reward circuits, acute and chronic exposure to stress is considered to affect the hypothalamic–pituitary–adrenal axis, leading to multiple pathological cascades that may induce the development of food craving and addiction, as well as symptoms of depression [37–39].

The varying findings across the dimensions of childhood maltreatment in our mediation analyses suggest that the relationship between childhood maltreatment and ED symptoms may follow a different pathway depending on the subtype of maltreatment. Indeed, this is consistent with evidence from studies conducted in the past few decades indicating that specific types of childhood maltreatment may be differentially associated with particular types of disordered eating [5,8,40–43]. Thus, for example, in a large cohort of American young adults, individuals with a history of physical abuse only displayed a higher tendency toward fasting and skipping meals [39]. In contrast, emotional abuse seems to be most consistently related to EDs symptoms, with evidence supporting a mediated pathway via emotional dysregulation [42,43]. In addition, emotional abuse was found to predict higher eating, shape and weight concerns, and poorer functioning in patients with EDs independently of the presence of other comorbidities [5]. Consistent with this, in our study, of the five dimensions of childhood maltreatment assessed, emotional abuse presented the strongest relationship with ED symptom severity. Furthermore, our findings from the mediation analysis revealed that emotional abuse presented the highest total effect of childhood maltreatment on ED symptoms with direct and indirect effects being globally in the same range. The direct effect of emotional abuse in our mediation model predicting ED symptom severity may also reflect the presence of other contributing factors, such as emotional dysregulation, which was not been included here. Further research examining the role of emotional dysregulation in these relationships is warranted.

Our most interesting finding is related to the fact that the strongest indirect effect of retrospective childhood maltreatment on current ED symptom severity via FA emerged for physical neglect. Moreover, physical neglect's effect on ED symptom severity was only mediated via FA. Physical neglect refers to the failure to provide a child with basic necessities of life such as food and clothing [44]. Brain maturation via myelination, synaptic plasticity, and the release of neurotransmitters depends largely on the prenatal and postnatal nutritional status of children and adolescents [45]. Indeed, it was shown that parental neglect is an intervening factor in the association between food-approaching appetitive traits and higher weight in children [46]. Accordingly, we can speculate that physical neglect may lead to brain maturation difficulties that may increase risk for FA and, subsequently, an ED.

Patients with EDs and a history of childhood maltreatment may benefit from care that specifically targets this history [47]. Furthermore, the symptom pathway leading to an ED in individuals with a history of childhood maltreatment has been described as specific to this group. Thus, overvaluation of weight and body shape may lead to feelings of loss of control followed by depressive symptoms and, subsequently, overeating [48]. Tailoring of usual treatment protocols to account for these pathways may help to improve clinical outcomes. Given the evidence found in this study for the mediating role of FA, it would be interesting to evaluate whether, in addition to the usual treatment, therapeutic strategies specifically targeting FA might improve overall prognosis. The presence of a history of physical neglect should raise the clinicians' index of suspicion for the presence of FA. In this regard, in case FA is confirmed as a comorbid clinical entity accompanying the ED (after using screening tools such as YFAS), the treatment of FA clinical dimensions and its overall impact on ED symptoms should be assessed in future studies. Accordingly, known suggested treatment protocols such as combining pharmacotherapies (opiate antagonists) and psychotherapies (such as cognitive behavioral therapy and psychodynamic group treatments) may be successful in targeting FA clinical dimensions and, subsequently, ED symptom severity [49,50].

The current study includes several limitations. First, all assessments (other than ED diagnosis) were self-reported which might constitute a source of bias. Indeed, participants with more severe clinical dimensions of ED may more easily recall incidents of childhood maltreatment. Second, the study is cross-sectional and retrospective in its assessment of childhood maltreatment, which limits the extent to which the directionality of relationships can be inferred from the findings. Furthermore, the lack of a nonclinical control group of individuals without EDs limited the extent to which confounding factors could be controlled. Finally, EDs were considered as a spectrum of disorders manifesting in different psychopathological dimensions as reflected by the EDI-2 score. However, in the current study only a composite score of ED symptom severity was used, and future work aiming to clarify the relationships among childhood maltreatment, FA, and different dimensions of disordered eating would be valuable.

5. Conclusions

In conclusion, although cross-sectional, our findings support the existence of a mediated relationship between retrospective childhood maltreatment and ED, via FA, especially in the presence of a history of physical neglect. Patients with severe ED symptoms and a history of childhood maltreatment should be systematically assessed for the presence of FA. Moreover, when childhood maltreatment is documented in patients with ED, tailoring treatment plans to specifically address FA should be considered.

Author Contributions: Conceptualization, R.B.K. and S.G.; methodology, R.B.K., G.S., A.A.-C., and S.G.; software, R.B.K., G.S., M.S., and A.A.-C.; validation, S.R., S.I., R.R., P.C., and S.G.; formal analysis, R.B.K., G.S., and S.G.; investigation, M.S., L.M., P.L., and E.R.; resources, S.G. and M.S.; data curation, M.S. and G.S.; writing—original draft preparation, R.B.K., G.S., and S.G.; writing—review and editing, R.B.K., S.R., S.I., R.R., P.C., and S.G.; visualization, R.B.K. and G.S.; supervision, S.G.; project administration, M.S. and S.G.; funding acquisition, S.G. All authors have read and agreed to the published version of the manuscript.

Funding: This study received financial support from CHRU Montpellier grant number [UF 9804] And The APC was funded by CHRU Montpellier. The CHRU Montpellier had no role in the design, analysis, interpretation, or publication of this study.

Acknowledgments: This study received financial support from CHRU Montpellier. CHRU Montpellier had no role in the design, analysis, interpretation, or publication of this study.

Conflicts of Interest: The authors declare no conflict of interest.

References

1. Child Maltreatment. Available online: https://www.who.int/news-room/fact-sheets/detail/child-maltreatment#:~:text=Child%20maltreatment%20is%20the%20abuse,under%2018%20years%20of%20age (accessed on 16 September 2020).
2. Treasure, J.; Claudino, A.M.; Zucker, N. Eating disorders. *Lancet* **2010**, *375*, 583–593. [CrossRef]
3. Caslini, M.; Bartoli, F.; Crocamo, C.; Dakanalis, A.; Clerici, M.; Carrà, G. Disentangling the association between child abuse and eating disorders: A systematic review and meta-analysis. *Psychosom. Med.* **2016**, *78*, 79–90. [CrossRef] [PubMed]
4. Molendijk, M.L.; Hoek, H.W.; Brewerton, T.D.; Elzinga, B.M. Childhood maltreatment and eating disorder pathology: A systematic review and dose-response meta-analysis. *Psychol. Med.* **2017**, *47*, 1–15. [CrossRef] [PubMed]
5. Guillaume, S.; Jaussent, I.; Maimoun, L.; Ryst, A.; Seneque, M.; Villain, L.; Hamroun, D.; Lefebvre, P.; Renard, E.; Courtet, P. Associations between adverse childhood experiences and clinical characteristics of eating disorders. *Sci. Rep.* **2016**, *6*, 3576. [CrossRef] [PubMed]
6. Rijkers, C.; Schoorl, M.; van Hoeken, D.; Hoek, H.W. Eating disorders and posttraumatic stress disorder. *Curr. Opin. Psychiatry* **2019**, *32*, 510–517. [CrossRef]
7. Racine, S.E.; Wildes, J.E. Emotion dysregulation and anorexia nervosa: An exploration of the role of childhood abuse. *Intern. J. Eat. Dis.* **2015**, *48*, 55–58. [CrossRef]
8. Rai, T.; Mainali, P.; Raza, A.; Rashid, J.; Rutkofsky, I. Exploring the link between emotional child abuse and anorexia nervosa: A psychopathological correlation. *Cureus* **2019**, *11*, e5318. [CrossRef]
9. Kong, S.; Bernstein, K. Childhood trauma as a predictor of eating psychopathology and its mediating variables in patients with eating disorders. *J. Clin. Nurs.* **2009**, *18*, 1897–1907. [CrossRef]
10. Brewerton, T.D. Food addiction as a proxy for eating disorder and obesity severity, trauma history, PTSD symptoms, and comorbidity. *Eat. Weight Disord.* **2017**, *22*, 241–247. [CrossRef]
11. Gordon, E.L.; Ariel-Donges, A.H.; Bauman, V.; Merlo, L.J. What is the evidence for "food addiction?" A systematic review. *Nutrients* **2018**, *10*, 477. [CrossRef]
12. Fauconnier, M.; Rousselet, M.; Brunault, P.; Thiabaud, E.; Lambert, S.; Rocher, B.; Challet-Bouju, G.; Grall-Bronnec, M. Food addiction among female patients seeking treatment for an eating disorder: Prevalence and associated factors. *Nutrients* **2020**, *12*, 1897. [CrossRef] [PubMed]
13. Mason, S.M.; Flint, A.J.; Roberts, A.L.; Agnew-Blais, J.; Koenen, K.C.; Rich-Edwards, J.W. Posttraumatic stress disorder symptoms and food addiction in women by timing and type of trauma exposure. *JAMA Psychiatry* **2014**, *71*, 1271–1278. [CrossRef] [PubMed]
14. Carlson, L.; Steward, T.; Agüera, Z.; Mestre-Bach, G.; Magaña, P.; Granero, R.; Jiménez-Murcia, S.; Claes, L.; Gearhardt, A.N.; Menchón, J.M.; et al. Associations of food addiction and nonsuicidal self-injury among women with an eating disorder: A common strategy for regulating emotions? *Eur. Eat. Disord. Rev.* **2018**, *26*, 629–637. [CrossRef] [PubMed]
15. Hardy, R.; Fani, N.; Jovanovic, T.; Michopoulos, V. Food addiction and substance addiction in women: Common clinical characteristics. *Appetite* **2018**, *120*, 367–373. [CrossRef]
16. Elton, A.; Smitherman, S.; Young, J.; Kilts, C.D. Effects of childhood maltreatment on the neural correlates of stress- and drug cue-induced cocaine craving. *Addict. Biol.* **2015**, *20*, 820–831. [CrossRef]
17. Van Dam, N.T.; Rando, K.; Potenza, M.N.; Tuit, K.; Sinha, R. Childhood maltreatment, altered limbic neurobiology, and substance use relapse severity via trauma-specific reductions in limbic gray matter volume. *JAMA Psychiatry* **2014**, *71*, 917–925. [CrossRef] [PubMed]
18. Garner, D.M.; Olmstead, M.P.; Polivy, J. Development and validation of a multidimensional eating disorder inventory for anorexia nervosa and bulimia. *Intern. J. Eat. Disord.* **1983**, *2*, 15–34. [CrossRef]
19. Criquillion-Doublet, S.; Divac, S.; Dardennes, R.; Guelfi, J. *Le 'Eating Disorder Inventory' (EDI)*; Guelfi, J.D., Gaillac, V., Dardennes, R., Eds.; Psychopathologie Quantitative: Paris, Franch, 1995.
20. Steward, T.; Mestre-Bach, G.; Agüera, Z.; Granero, R.; Martín-Romera, V.; Sánchez, I.; Riesco, N.; Tolosa-Sola, I.; Fernández-Formoso, J.A.; Fernández-García, J.C.; et al. Enduring changes in decision making in patients with full remission from anorexia nervosa. *Eur. Eat. Disord. Rev.* **2016**, *24*, 523–527. [CrossRef]
21. Gearhardt, A.N.; Corbin, W.R.; Brownell, K.D. Development of the yale food addiction scale Version 2.0. *Psychol. Addict. Behav.* **2016**, *30*, 113–121. [CrossRef]

22. Brunault, P.; Ballon, N.; Gaillard, P.; Réveillère, C.; Courtois, R. Validation of the French version of the yale food addiction scale: An examination of its factor structure, reliability, and construct validity in a nonclinical sample. *Can. J. Psychiatry* **2014**, *59*, 276–284. [CrossRef]
23. Bernstein, D.P.; Fink, L. *Childhood Trauma Questionnaire: A Retrospective Self-Report*; Psychological Corporation: Orlando, FL, USA, 1998.
24. Hayes, A.F. *Introduction to Mediation, Moderation, and Conditional Process Analysis*; Guildford Press: New York, NY, USA, 2013.
25. Imperatori, C.; Innamorati, M.; Lamis, D.A.; Farina, B.; Pompili, M.; Contardi, A.; Fabbricatore, M. Childhood trauma in obese and overweight women with food addiction and clinical-level of binge eating. *Child. Abuse. Negl.* **2016**, *58*, 180–190. [CrossRef] [PubMed]
26. Stojek, M.M.; Maples-Keller, J.L.; Dixon, H.D.; Umpierrez, G.E.; Gillespie, C.F.; Michopoulos, V. Associations of childhood trauma with food addiction and insulin resistance in African-American women with diabetes mellitus. *Appetite* **2019**, *141*, 104317. [CrossRef] [PubMed]
27. Sinha, R. Role of addiction and stress neurobiology on food intake and obesity. *Biol. Psychol.* **2018**, *131*, 5–13. [CrossRef] [PubMed]
28. Pursey, K.M.; Stanwell, P.; Gearhardt, A.N.; Collins, C.E.; Burrows, T.L. The prevalence of food addiction as assessed by the Yale Food Addiction Scale: A systematic review. *Nutrients* **2014**, *6*, 4552–4590. [CrossRef] [PubMed]
29. Wiss, D.A.; Brewerton, T.D. Incorporating food addiction into disordered eating: The disordered eating food addiction nutrition guide (DEFANG). *Eat. Weight Disord.* **2017**, *22*, 49–59. [CrossRef] [PubMed]
30. Meule, A.; von Rezori, V.; Blechert, J. Food addiction and bulimia nervosa. *Eur. Eat. Disord. Rev.* **2014**, *22*, 331–337. [CrossRef] [PubMed]
31. De Vries, S.K.; Meule, A. Food addiction and bulimia nervosa: New data based on the yale food addiction scale 2.0. *Eur. Eat. Disord. Rev.* **2016**, *24*, 518–522. [CrossRef]
32. Jiménez-Murcia, S.; Agüera, Z.; Paslakis, G.; Munguia, L.; Granero, R.; Sánchez-González, J.; Sánchez, I.; Riesco, N.; Gearhardt, A.N.; Dieguez, C.; et al. Food addiction in eating disorders and obesity: Analysis of clusters and implications for treatment. *Nutrients* **2019**, *11*, 2633. [CrossRef]
33. Loxton, N.J.; Tipman, R.J. Reward sensitivity and food addiction in women. *Appetite* **2017**, *115*, 28–35. [CrossRef]
34. Beyer, F.; García-García, I.; Heinrich, M.; Schroeter, M.L.; Sacher, J.; Luck, T.; Riedel-Heller, S.G.; Stumvoll, M.; Villringer, A.; Witte, A.V. Neuroanatomical correlates of food addiction symptoms and body mass index in the general population. *Hum. Brain. Mapp.* **2019**, *40*, 2747–2758. [CrossRef]
35. Osadchiy, V.; Mayer, E.A.; Bhatt, R.; Labus, J.S.; Gao, L.; Kilpatrick, L.A.; Liu, C.; Tillisch, K.; Naliboff, B.; Chang, L.; et al. History of early life adversity is associated with increased food addiction and sex-specific alterations in reward network connectivity in obesity. *Obes. Sci. Pract.* **2019**, *5*, 416–436. [CrossRef] [PubMed]
36. Mitchell, K.S.; Wolf, E.J. PTSD, food addiction, and disordered eating in a sample of primarily older veterans: The mediating role of emotion regulation. *Psychiatry Res.* **2016**, *243*, 23–29. [CrossRef] [PubMed]
37. Wenzel, K.R.; Weinstock, J.; McGrath, A.B. The clinical significance of food addiction. *J. Addict. Med.* **2020**. [CrossRef] [PubMed]
38. Kalon, E.; Hong, J.Y.; Tobin, C.; Schulte, T. Psychological and neurobiological correlates of food addiction. *Int. Rev. Neurobiol.* **2016**, *129*, 85–110. [PubMed]
39. Mills, J.G.; Thomas, S.J.; Larkin, T.A.; Deng, C. Overeating and food addiction in Major Depressive Disorder: Links to peripheral dopamine. *Appetite* **2020**, *148*, 104586. [CrossRef] [PubMed]
40. Monteleone, A.M.; Cascino, G.; Pellegrino, F.; Ruzzi, V.; Patriciello, G.; Marone, L.; De Felice, G.; Monteleone, P.; Maj, M. The association between childhood maltreatment and eating disorder psychopathology: A mixed-model investigation. *Eur. Psychiatry* **2019**, *61*, 111–118. [CrossRef]
41. Hazzard, V.M.; Bauer, K.W.; Mukherjee, B.; Miller, A.L.; Sonneville, K.R. Associations between childhood maltreatment latent classes and eating disorder symptoms in a nationally representative sample of young adults in the United States. *Child Abuse. Negl.* **2019**, *98*, 104171. [CrossRef]
42. Burns, E.E.; Fischer, S.; Jackson, J.L.; Harding, H.G. Deficits in emotion regulation mediate the relationship between childhood abuse and later eating disorder symptoms. *Child Abuse. Negl.* **2012**, *36*, 32–39. [CrossRef]
43. Grilo, C.M.; Masheb, R.M. Childhood maltreatment and personality disorders in adult patients with binge eating disorder. *Acta. Psychiatr. Scand.* **2002**, *106*, 183–188. [CrossRef]

44. Zeanah, C.H.; Humphreys, K.L. Child abuse and neglect. *J. Am. Acad. Child Adolesc. Psychiatry* **2018**, *57*, 637–644. [CrossRef]
45. Arain, M.; Haque, M.; Johal, L.; Mathur, P.; Nel, W.; Rais, A.; Sandhu, R.; Sharma, S. Maturation of the adolescent brain. *Neuropsychiatr. Dis. Treat.* **2013**, *9*, 449–461. [PubMed]
46. Rodenburg, G.; Kremers, S.P.; Oenema, A.; van de Mheen, D. Associations of children's appetitive traits with weight and dietary behaviours in the context of general parenting. *PLoS ONE* **2012**, *7*, e50642. [CrossRef] [PubMed]
47. Hicks White, A.A.; Pratt, K.J.; Cottrill, C. The relationship between trauma and weight status among adolescents in eating disorder treatment. *Appetite* **2018**, *129*, 62–69. [CrossRef]
48. Rodgers, R.F.; DuBois, R.; Thiebaut, S.; Jaussent, I.; Maimoun, L.; Seneque, M.; Lefebvre, P.; Renard, E.; Courtet, P.; Guillaume, S. Structural differences in eating disorder psychopathology after history of childhood abuse: Insights from a Bayesian network analysis. *J. Abnorm. Psychol.* **2019**, *128*, 795–805. [CrossRef] [PubMed]
49. Schwartz, D.C.; Nickow, M.S.; Arseneau, R.; Gisslow, M.T. A substance called food: Long-term psychodynamic group treatment for compulsive overeating. *Int. J. Group Psychother.* **2015**, *65*, 386–409. [CrossRef]
50. Valbrun, L.P.; Zvonarev, V. The opioid system and food intake: Use of opiate antagonists in treatment of binge eating disorder and abnormal eating behavior. *J. Clin. Med. Res.* **2020**, *12*, 41–63. [CrossRef] [PubMed]

nutrients

MDPI

Article

Contributions of Emotional Overload, Emotion Dysregulation, and Impulsivity to Eating Patterns in Obese Patients with Binge Eating Disorder and Seeking Bariatric Surgery

Farid Benzerouk [1,2,*], Zoubir Djerada [3], Eric Bertin [4], Sarah Barrière [1], Fabien Gierski [1,2,5] and Arthur Kaladjian [1,2]

1 Psychiatry Department, Reims University Hospital, EPSM Marne, 51100 Reims, France; sbarriere@chu-reims.fr (S.B.); fabien.gierski@univ-reims.fr (F.G.); kaladjiana@epsm-marne.fr (A.K.)
2 Cognition Health and Society Laboratory (EA 6291), Université de Reims Champagne-Ardenne, 51100 Reims, France
3 Department of Pharmacology E.A.3801, SFR CAP-santé, Reims University Hospital, 51100 Reims, France; zdjerada@chu-reims.fr
4 Champagne Ardenne Specialized Center in Obesity, University Hospital Center, 51100 Reims, France; ebertin@chu-reims.fr
5 INSERM U1247 GRAP, Research Group on Alcohol and Pharmacodependences, Université de Picardie Jules Verne, 80000 Amiens, France
* Correspondence: benzeroukf@yahoo.fr

Received: 30 August 2020; Accepted: 6 October 2020; Published: 12 October 2020

Abstract: Background: Binge eating disorder (BED) is very frequently observed in patients considered for weight loss surgery and seems to influence their outcome critically. Literature highlights a global emotional overload in individuals with BED, but little is known on the mechanisms involved. The present study aimed to focus on emotion regulation, impulsivity, depression, and anxiety in people with and without BED and fulfilling inclusion criteria for bariatric surgery. Doing so, we sought to individualize factors related to BED. Then, we examined the contribution of depression, anxiety, emotion regulation difficulties, and impulsivity to inappropriate eating behaviors observed in patients with BED. Methods: A sample of 121 individuals (79.3% female, mean age: 40.82 ± 9.26, mean current body mass index (BMI): 44.92 kg/m^2 ± 7.55) seeking bariatric surgery were recruited at the Champagne Ardenne Specialized Center in Obesity in Reims, France from November 2017 to October 2018. They were stratified as with or without BED according to the binge eating scale. Characteristics identified in univariate analyses as differentiating the two groups were then included in multivariable analyses. Results: Multivariable analyses showed that limited access to emotional regulation strategies was significantly associated with BED. Furthermore, inappropriate eating behaviors were independently associated with age, depression severity, anxiety, emotional dysregulation, and impulsivity in BED group. Conclusions: The present findings are indicative of an association between emotion deficit and BED in obese patients seeking bariatric surgery. Patients with BED could benefit from the addition of an emotion regulation intervention.

Keywords: obesity; bariatric surgery; binge eating disorder; emotion dysregulation; emotional eating; external eating; strategies; non-acceptance; impulsivity

1. Introduction

Binge eating disorder (BED) involves frequent overeating during a discreet period of time (at least once a week for three months), combined with a lack of control, and is associated with three or more of the following items: eating more rapidly than normal; eating until feeling uncomfortably full; eating large amounts of food when not feeling physically hungry; eating alone because of feeling embarrassed by how much one is eating; and feeling disgusted with oneself, depressed, or very guilty afterward [1]. BED also causes significant distress [1] and is associated with various inappropriate eating behaviors. It is more common in females (3.5%) than in males (2.0%) and in obese individuals (5% to 30%) [2,3], especially those who are severely obese and those seeking obesity treatment: 17% at the time of surgery [4,5]. Moreover, BED seems to influence success after weight loss surgery [6]. Accordingly, it should be of interest to assess why people suffering from this disorder engage in various inappropriate eating behaviors in order to find a way to help them give up these harmful behaviors.

Inappropriate eating behaviors are like engaging in emotional eating or in binge eating. As a matter of fact, emotional eating behavior, the tendency to overeat in response to negative emotions, appears to be common in bariatric candidates (see for review [7]). Moreover, obese people with BED who are candidates for bariatric surgery are more likely to have severe binge eating symptoms than obese non-surgical individuals [7]. Bariatric surgery candidates also have more objective and subjective binge eating episodes per month than non-surgical weight loss patients [8].

According to the Diagnostic and Statistical Manual of Mental Disorders, DSM-5 criteria, binge eating and negative emotions are interconnected [1]. In their review, Dingemans et al. (2017) pointed that (1) several authors, using experimental studies, emphasized a relationship between emotional factors and overeating in individuals with BED; (2) these individuals were characterized by a higher prevalence of psychiatric comorbidities, exhibited higher levels of depression and anxiety; (3) they reported poorer mood especially prior to binge eating and can experience more negative stressors than subjects without BED; and (4) they can also feel more negative emotions (i.e., anger and/or frustration) related to interpersonal experiences [9]. Taken together, these results highlight a global emotional overload in individuals with BED. This emotional overload might increase the occurrence of binge eating. According to Polivy and Herman's (1993) affect regulation model of binge eating, this behavior could be implemented to decrease emotional distress or negative affects [10].

Rather than focusing on emotions themselves in individuals with BED, other works have focused on emotion regulation. According to Gross [11], emotion regulation refers to "shaping which emotion one has, when one has them, and how one experiences or expresses these emotions". Emotion regulation is conceptualized as involving emotion regulation abilities (i.e., the awareness and understanding of emotions, the acceptance of emotions, the ability to control impulsive behaviors and behave in accordance with desired goals when experiencing negative emotions), and emotion regulation strategies (i.e., use situationally appropriate emotion regulation strategies flexibly to modulate emotional responses as desired in order to meet individual goals and situational demands). Strategies can include adaptive ones such as reappraisal, problem-solving, and acceptance and maladaptive ones such as avoidance, rumination, and suppression. The relative absence of any or all of these abilities and strategies would indicate the presence of difficulties in emotion regulation, or an emotion dysregulation [12]. Regarding eating disorders, a recent meta-analysis by Prefit et al. [13] identified a transdiagnostic character of emotion regulation problems. Furthermore, compared to a control group without obesity, people suffering from obesity use significantly fewer cognitive emotion regulation strategies considered as adaptive regardless of their body mass index (BMI) [14] and they report using more emotional suppression [15]. Data also revealed that only emotional dysregulation significantly predicted binge eating vulnerability in a study involving 63 obese patients seeking surgical treatment [16]. Moreover, many studies identified lack of skills and strategies required to regulate negative affect adaptively and effectively (i.e., poorer emotional awareness and clarity, nonacceptance, difficulties with reappraisal, and with problem-solving) as being associated with eating disorders. Accordingly, individuals with disordered eating may have a greater vulnerability to using maladaptive emotion regulation strategies (i.e., rumination, avoidance of emotions, and suppression) [13].

In sum, the global emotional overload and dysfunctional emotion regulation abilities and strategies are then increasingly thought to be co-occurring risk factors in the onset and maintenance of BED by promoting maladaptive behaviors such as overeating and binge eating. However, according to Dingemans et al. [9], studies investigating the use of emotion regulation strategies amongst individuals with BED have found mixed results. Moreover, beyond emotional regulation, other researchers have outlined the role of impulsivity in this disorder. According to Giel et al. (2017), BED is also considered as a distinct phenotype, within the obesity spectrum, characterized by increased impulsivity and by an increased rash–spontaneous behavior in general and specifically toward food [17]. However, the simultaneous consideration of emotion regulation and impulsivity remains to be deepened. Therefore, there is a major interest to understand the impact of both emotion regulation as well as impulsivity in patients suffering from BED and seeking bariatric surgery.

Altogether, emotional overload (i.e., depression, anxiety), emotion regulation, and impulsivity may be associated with BED and could predispose individuals to developing and/or maintaining inappropriate eating behaviors among patients suffering from BED who are candidates for bariatric surgery. Given this, the primary aim of our study was to examine the associations with BED of emotional overload (depression, anxiety), emotion regulation, and impulsivity in obese people with and without BED. We expected people suffering from obesity with BED to present more depression, more anxiety, more emotion regulation difficulties, and more impulsivity than people suffering from obesity without BED (wBED). The second aim of our study was to examine the contribution of depression, anxiety, emotion regulation difficulties, and impulsivity to eating patterns observed in patients with BED. In this population, we sought to individuate which factors were significantly related to the assessed eating behaviors. More precisely, we expected that high levels of depression, anxiety, emotion regulation difficulties, and impulsivity were significantly related to emotional eating, external eating, and bulimic symptomatology. Improving our knowledge on BED is a necessary step to then develop indications of specific therapeutic strategies before surgery and, thus, allow their access to bariatric surgery and improve their outcomes after bariatric surgery.

2. Materials and Methods

2.1. Participants and Procedure

A sample of 121 obese candidates for bariatric surgery was recruited at the Champagne Ardenne Specialized Center in Obesity in Reims, France from November 2017 to October 2018. Participants were 79.3% female (n = 96), and ranged in age from 19 to 58, with a mean age of 40.82 (SD = 9.26). Mean BMI was 44.92 kg/m^2 (SD = 7.55; range: 35.63–75.72).

The inclusion criteria were to be a candidate for bariatric surgery: obesity grade 2 (BMI 35.0–39.9 kg/m^2) and at least one obesity-related comorbidity (e.g., hypertension, type 2 diabetes mellitus, dyslipidemia, sleep apnea), or obesity grade 3 (BMI $\geq$ 40.0 kg/m^2). Obese patients had to be French-speaking females or males between the ages of 18 and 60. Participants with present or past drug or alcohol abuse or dependence were not included (as assessed by the Mini International Neuropsychiatric Interview (MINI) [18,19]), as these conditions would have been able to alter the assessments of emotion regulation and impulsivity, as well as participants having a bariatric procedure previously or a severe comorbid disorder such as neurologic impairments. The BMI was calculated by dividing the weight in kilograms by the square of the height in meters (BMI = weight (kg)/height (m^2)) [20].

All study procedures were reviewed and approved by the local Institutional Review Board (IRB 2016-12). The study was carried out according to the Helsinki Declaration [21] and every patient included into the study provided written informed consent.

2.2. Measures

2.2.1. Eating Behaviors

The Binge Eating Scale (BES; [22]) is a 16-item self-administered questionnaire used to assess the presence of binge eating behavior indicative of an eating disorder. Eight items describe behavioral manifestations (for example, eating fast or consuming large amounts of food) and eight items on associated feelings and cognitions (for example, fear of not stopping eating). Each item has a response range from 0 to 3 points (0 = no severity of the BES symptoms, 3 = serious problems on the BES symptoms). Marcus et al. (1988) created a range of scores for the BES from 0 to 46 points: a score of less than 17 points indicates minimal binge eating (BE) problems; a score between 18 and 26 points indicates moderate BE problems, and a score of more than 27 points indicates severe BE problems [23]. We considered binge eating as a categorical variable (significant binge eating if BES score ≥18). We used the validated French version [24]. The Cronbach's alpha for the current study was 0.83.

The Bulimic Investigatory Test, Edinburgh (BITE; [25]) is a self-report questionnaire used to evaluate the presence and severity of bulimic symptomatology. It is composed of 33 items divided into two different subscales: a symptom subscale (30 items) and a severity subscale (3 items). Henderson and Freeman (1987) considered a BITE score under 10 points as indicative of no problem with eating behavior, a score between 10 and 20 points as indicative of abnormal eating patterns (from 15 to 20 points warns us of the presence of a possible subthreshold bulimia nervosa), and a score higher than 20 points constitutes altered eating patterns with a possible bulimia nervosa. The Cronbach's alpha for the current study was 0.80.

The Dutch Eating Behavior Questionnaire (DEBQ) was administered using the French version to assess three components of eating behavior: emotional, external, and restrained eating [26,27]. It is a self-report measure that contains 33 items. Thirteen items assess emotional eating, 10 items assess external eating, and 10 items assess restrained eating. Each item is rated on a 5-point Likert scale. In the current study, Cronbach's alpha was 0.95 for emotional eating, 0.86 for external eating, and 0.86 for restrained eating.

2.2.2. Emotion Regulation and Impulsivity

The Difficulty in Emotion Regulation Scale (DERS; [12]) is a 36-item self-report questionnaire. It assesses six different aspects of emotional regulation including non-acceptance of emotional responses (Non-Acceptance), difficulty engaging in goal-directed behavior (Goals), impulse control difficulties (Impulse), lack of emotional awareness (Awareness), limited access to emotional regulation strategies (Strategies), and lack of emotional clarity (Clarity). This scale has demonstrated good internal consistency, construct and predictive validity, and test–retest reliability. We used the validated French version [28]. The total score demonstrated a Cronbach's alpha of 0.92 in the present study. It was 0.90, 0.78, 0.84, 0.66, 0.86, and 0.66 respectively for the six different subscales described above.

The UPPS-P Impulsive Behavior Scale [29] consists of 20 items that evaluate different facets of impulsivity labeled Negative Urgency (4 items), Positive Urgency (4 items), (lack of) Premeditation (4 items), (lack of) Perseverance (4 items), and Sensation seeking (4 items). Cronbach's alphas in our sample were 0.83, 0.74, 0.77, 0.79, 0.77, respectively.

2.3. *Other Assessments (Depression Severity and Anxiety Levels)*

Depression severity was assessed with the shortened Beck Depression Inventory (BDI). This is a widely used self-report scale consisting of 13 items [30], which has been validated in French [31]. The total score is obtained by adding the scores of the 13 items and ranges from 0 to 39, with higher scores indicating greater depression symptoms. The Cronbach's alpha for the current study was 0.83.

Anxiety severity was assessed with the Spielberger State–Trait Inventory (STAI; [32]) which is a 40-item scale, using a 4-point Likert scale for each item. This scale was used to measure both trait anxiety (how dispositionally anxious a person is across time and situations) and state anxiety (how anxious a person is feeling at a particular moment). The Cronbach's alpha for the current study was 0.93 for the trait anxiety scale and 0.73 for the state anxiety scale.

2.4. Statistical Analysis

Statistical analyses were performed with R 3.1.4 (The R Foundation for Statistical Computing, Vienna, Austria, http://www.r-project.org). Data summaries were presented as the mean and standard deviation (sd) for continuous measurements and as the frequency (percentage) for categorical variables. Categorical variables were analyzed using overall chi-squared (χ^2).

Cronbach's alpha was calculated for each scale and subscale (package psych).

Univariate analyses were done for continuous variables using the non-parametric Mann–Whitney U test. To avoid computational issues (model convergence failure due to sparse data), only covariates with at least five cases were considered in the model. Univariate analysis was performed to screen potential variables for inclusion in the final multivariable model.

Multivariable analyses were performed using logistic regression modeling (BED and wBED categories as the dependent variables), and the association between the identified variables and eating patterns in the BED group was assessed by multiple linear regression analysis while controlling the other covariates for confounding effects. Adjusted β coefficients (βadj) were estimated for all significant associations [33]. Among the variable selection procedures, backward elimination is preferred as it starts with the assumed unbiased global model [34,35]. The potential prognostic factors were established, and a multivariable model was derived by backward selection according to Akaike's Information Criterion. For sensitivity, all identified associated covariates in the different model were also determined using the appropriate high-dimensional procedure as random forest (package randomForest) and sparse partial least squares discriminant analysis (package mixomics) and Lasso (package glmnet). The goodness-of-fit and appropriateness of the logistic regression model were evaluated using the Nagelkerke R squared and Hosmer–Lemeshow values and by the correct overall percentage of prediction. Multicollinearity was checked for all analyses. Variables significant at $p = 0.05$ at final multivariable analysis were retained as independent predictive factors. The Wald test was used for hypothesis testing. The stability and robustness of the model were validated using the technique of "bootstrap" resampling.

All p-values were two-tailed, with statistical significance indicated by a value of $p < 0.05$.

3. Results

3.1. Group Comparisons of Study Variables in Obese Adults Seeking Bariatric Surgery (with Binge Eating Disorder (BED) and without BED (wBED))

The stratification procedure based on the BES score (cutoff = 18) led to the identification of a group of 27 participants with BED and a group of 94 participants without BED (wBED) and to a prevalence of BED of 22.31%. There were no significant differences between BED and wBED groups with regard to current body mass index (kg/m^2; mean ± SD) (45.83 ± 8.21 in the BED group and 44.66 ± 7.38 in the wBED group, $p = 0.513$) and maximum body mass index (kg/m^2; mean ± SD) (48.19 ± 9.65 in the BED group and 46.87 ± 7.74 in the wBED group, $p = 0.806$). The comparison of demographic and clinical characteristics of the two groups is presented in Table 1.

Table 1. Comparison of sociodemographic and clinical characteristics in obese adults seeking bariatric surgery using univariate analyses (with binge eating disorder (BED) and without BED (wBED)).

Variables	BED (*n* = 27)	wBED (*n* = 94)	Univariate Analysis
			p
Age	43.19 ± 9.80	40.14 ± 9.04	0.095
Female/Male	18/9	78/16	0.066
Beck Depression Inventory	10.96 ± 6.22	5.47 ± 4.81	<0.001
State anxiety inventory (STAI-A)	42.33 ± 13.34	32.41 ± 9.16	0.001
Trait anxiety inventory (STAI-B)	47.74 ± 5.95	41.26 ± 9.27	0.002
BITE total score	18.07 ± 6.36	8.61 ± 4.81	<0.001
BITE Symptom subscale	14.67 ± 4.25	6.88 ± 3.97	<0.001
BITE Severity subscale	3.41 ± 3.75	1.72 ± 2.02	0.006
DEBQ total score	96.81 ± 16.13	78.91 ± 15.38	<0.001
Emotional eating	39.56 ± 12.50	26.30 ± 9.64	<0.001
External eating	28.33 ± 5.52	23.83 ± 6.18	0.001
Restrained eating	28.93 ± 5.42	28.79 ± 8.27	0.822
DERS			
Total score	93.11 ± 18.92	76.27 ± 16.26	<0.001
Non-acceptance	14.67 ± 6.48	10.32 ± 4.32	<0.001
Goals	13.93 ± 4.05	11.16 ± 3.82	0.003
Impulse	14.26 ± 4.60	10.96 ± 3.65	0.001
Awareness	18.07 ± 5.05	18.14 ± 4.49	0.854
Strategies	19.59 ± 5.79	14.54 ± 4.34	<0.001
Clarity	12.59 ± 3.52	11.15 ± 3.25	0.097
UPPS-P			
Total score	31.11 ± 6.18	27.10 ± 8.31	0.005
Negative Urgency	6.22 ± 2.95	4.65 ± 2.87	0.008
Positive Urgency	6.67 ± 2.13	5.69 ± 2.62	0.077
Lack of Premeditation	6.81 ± 1.75	6.16 ± 1.92	0.082
Lack of Perseverance	6.26 ± 2.03	5.69 ± 1.83	0.165
Sensation Seeking	5.15 ± 2.75	4.90 ± 2.64	0.718

Note. Data presented as mean ± sd for quantitative variables and percentages for qualitative variables. BITE: Bulimic Investigatory Test, Edinburgh; DEBQ: Dutch Eating Behavior Questionnaire; DERS: Difficulty in Emotion Regulation Scale; UPPS-P: UPPS Impulsive Behavior Scale.

In the univariate analysis, patients with BED showed significantly higher levels of depression, anxiety, eating-disorder symptoms (i.e., Emotional eating, External eating, BITE Symptom score, and BITE Severity score), emotion dysregulation, and impulsivity than patients without BED. In the multivariable analysis, the BITE symptom subscale and limited access to emotional regulation Strategies were significantly associated with BED with adjusted odds ratios (OR) of 1.517 (1.241–1.990), $p < 0.001$, and 1.176 (1.023–1.389), $p = 0.03$, respectively (Tables 1 and 2).

Table 2. Comparison of sociodemographic and clinical characteristics in obese adults seeking bariatric surgery using multivariable analyses (with binge eating disorder (BED) and without BED (wBED)).

Variables	BED (n = 27)	wBED (n = 94)	Multivariable Analysis	
			p	Adjusted OR
Age	43.19 ± 9.80	40.14 ± 9.04	0.314	
Female/Male	18/9	78/16	0.259	
Beck Depression Inventory	10.96 ± 6.22	5.47 ± 4.81	0.520	
State anxiety inventory (STAI-A)	42.33 ±13.34	32.41 ± 9.16	0.987	
Trait anxiety inventory (STAI-B)	47.74 ± 5.95	41.26 ± 9.27	0.439	
BITE total score	18.07 ± 6.36	8.61 ± 4.81		
BITE Symptom subscale	14.67 ± 4.25	6.88 ± 3.97	<0.001	1.517 (1.241–1.9900)
BITE Severity subscale	3.41 ± 3.75	1.72 ± 2.02	0.719	
DEBQ total score	96.81 ±16.13	78.91 ± 15.38		
Emotional eating	39.56 ±12.50	26.30 ± 9.64	0.076	
External eating	28.33 ± 5.52	23.83 ± 6.18	0.880	
Restrained eating	28.93 ± 5.42	28.79 ± 8.27		
DERS				
Total score	93.11 ±18.92	76.27 ± 16.26		
Non-acceptance	14.67 ± 6.48	10.32 ± 4.32	0.206	
Goals	13.93 ± 4.05	11.16 ± 3.82	0.685	
Impulse	14.26 ± 4.60	10.96 ± 3.65	0.296	
Awareness	18.07 ± 5.05	18.14 ± 4.49		
Strategies	19.59 ± 5.79	14.54 ± 4.34	0.03	1.176 (1.023–1.389)
Clarity	12.59 ± 3.52	11.15 ± 3.25	0.662	
UPPS-P				
Total score	31.11 ± 6.18	27.10 ± 8.31		
Negative Urgency	6.22 ± 2.95	4.65 ± 2.87	0.661	
Positive Urgency	6.67 ± 2.13	5.69 ± 2.62	0.833	
Lack of Premeditation	6.81 ± 1.75	6.16 ± 1.92	0.650	
Lack of Perseverance	6.26 ± 2.03	5.69 ± 1.83	0.449	
Sensation Seeking	5.15 ± 2.75	4.90 ± 2.64		

Note. Data presented as mean ± sd for quantitative variables and percentages for qualitative variables. OR: odds ratio; BITE: Bulimic Investigatory Test, Edinburgh; DERS: Difficulty in Emotion Regulation Scale; UPPS-P: UPPS Impulsive Behavior Scale. In the final multivariable model using logistic regression modeling (BED and wBED categories as the dependent variables), potential covariates were age (years), sex (female: 2, male: 1), Beck Depression Inventory score, STAI-A score, STAI-B score, BITE subscales scores, Emotional eating score, External eating score, Non-acceptance score, Goals score, Impulse score, Strategies score, Clarity score, Negative Urgency score, Positive Urgency score, Lack of Premeditation score, and Lack of Perseverance score. Only significant adjusted ORs are presented.

3.2. *Associations between Emotional Overload, Emotion Regulation Difficulties, Impulsivity, and Eating Patterns in Patients with BED*

Univariate analyses concerning Emotional eating and External eating are shown in Tables 3 and 4. Multivariable analysis showed that Emotional eating was independently associated with age (βadj = −0.415 (−0.792; −0.0390), DERS Non-Acceptance subscale score (βadj = 0.883 (0.060–1.705)), and UPPS-P Lack of premeditation subscale score (βadj = 3.750 (1.838–5.661)) (Table 3). Multivariable analysis showed that External eating was independently associated with the anxiety trait (βadj = 0.319 (0.090–0.547)), DERS Impulse control difficulties subscale score (βadj = 0.431 (0.115–0.746)), and UPPS-P Negative Urgency subscale score (βadj = 0.996 (0.496–1.497)) (Table 4).

Table 3. Multiple linear regression analysis of the association between DEBQ subscale (Emotional eating) and sociodemographic and clinical characteristics in the BED group.

Covariates	Univariate Analysis	Multivariable Analysis
	p	βadj (95% CI), p
Age	0.043	−0.415 (−0.792; −0.039), $p = 0.042$
Female/Male	0.290	
Beck Depression Inventory	0.181	
State anxiety inventory (STAI-A)	0.472	
Trait anxiety inventory (STAI-B)	0.751	
DERS		
Non-Acceptance	0.067	0.883 (0.060–1.705), $p = 0.048$
Goals	0.648	
Impulse	0.067	
Awareness	0.571	
Strategies	0.810	
Clarity	0.529	
UPPS-P		
Negative Urgency	0.404	
Positive Urgency	0.881	
Lack of Premeditation	0.004	3.750 (1.838–5.661), $p < 0.001$
Lack of Perseverance	0.157	
Sensation Seeking	0.691	

Note. Data are presented as p-value in univariate analysis and in adjusted regression β coefficients (βadj), 95% confident interval (CI), and p-value in multivariable analysis (only significant β coefficients (βadj), 95% confident interval are presented). OR: Odds ratio; DERS: Difficulty in Emotion Regulation Scale; UPPS-P: UPPS Impulsive Behavior Scale. In the final multivariable model of Emotional eating, potential covariates were age (years), Beck Depression Inventory score, Non-acceptance score, Impulse score, Lack of Premeditation score and Lack of Perseverance score.

Table 4. Multiple linear regression analysis of the association between DEBQ subscale (External eating) and sociodemographic and clinical characteristics in the BED group.

Covariates	Univariate Analysis	Multivariable Analysis
	p	βadj (95% CI), p
Age	0.366	
Female/Male	0.666	
Beck Depression Inventory	0.308	
State anxiety inventory (STAI-A)	0.608	
Trait anxiety inventory (STAI-B)	0.184	0.319 (0.090–0.547), $p < 0.05$
DERS		
Non-Acceptance	0.035	
Goals	0.557	
Impulse	0.003	0.431 (0.115–0.746), $p < 0.05$
Awareness	0.624	
Strategies	0.178	
Clarity	0.690	
UPPS-P		
Negative Urgency	<0.001	0.996 (0.496–1.497), $p < 0.001$
Positive Urgency	0.053	
Lack of Premeditation	0.175	
Lack of Perseverance	0.532	
Sensation Seeking	0.874	

Note. Data are presented as p-value in univariate analysis and in adjusted regression β coefficients (βadj), 95% confident interval, and p-value in multivariable analysis (only significant β coefficients (βadj), 95% confident interval are presented). OR: odds ratio; DERS: Difficulty in Emotion Regulation Scale; UPPS-P: UPPS Impulsive Behavior Scale. In the final multivariable model of External eating, potential covariates were STAI-B score, Non-acceptance score, Impulse score, Strategies score, Negative Urgency score, Positive Urgency score, and Lack of Premeditation score.

Univariate analyses concerning BITE symptoms and BITE severity are shown in Tables 5 and 6. Multivariable analysis showed that BITE symptom score was independently associated with depression severity score (βadj = 0.255 (0.082–0.428)), DERS Non-Acceptance subscale score (βadj = 0.299 (0.131–0.466)), and UPPS-P Lack of premeditation subscale score (βadj = 1.469 (0.851–2.087)) (Table 5). Multivariable analysis showed that BITE Severity score was independently associated with the trait anxiety (βadj = 0.176 (0.061–0.290)), DERS Impulse control difficulties subscale score (βadj = 0.587 (0.415–0.759)), DERS Clarity subscale score (βadj = 0.304 (0.094–0.514)), and UPPS Negative Urgency subscale score (βadj = −0.832 (−1.083; −0.582)) (Table 6).

Table 5. Multiple linear regression analysis of the association between BITE subscale (symptom score) and sociodemographic and clinical characteristics in the BED group.

Covariates	Univariate Analysis	Multivariable Analysis
	p	βadj (95% CI), p
Age	0.734	
Female/Male	0.347	
Beck Depression Inventory	0.085	0.255 (0.082–0.428), $p < 0.01$
State anxiety inventory (STAI-A)	0.015	
Trait anxiety inventory (STAI-B)	0.194	
DERS		
Non-Acceptance	0.068	0.299 (0.131–0.466), $p < 0.01$
Goals	0.745	
Impulse	0.085	
Awareness	0.774	
Strategies	0.279	
Clarity	0.160	
UPPS-P		
Negative Urgency	0.297	
Positive Urgency	0.72	
Lack of Premeditation	0.003	1.469 (0.851–2.087), $p < 0.001$
Lack of Perseverance	0.797	
Sensation Seeking	0.398	

Note. Data are presented as p-value in univariate analysis and in adjusted regression β coefficients (βadj), 95% confident interval, and p-value in multivariable analysis (only significant β coefficients (βadj), 95% confident interval are presented). OR: odds ratio; BITE: Bulimic Investigatory Test, Edinburgh; DERS: Difficulty in Emotion Regulation Scale; UPPS-P: UPPS Impulsive Behavior Scale. In the final multivariable model of BITE symptom subscale, potential covariates were Beck Depression Inventory score, STAI-A score, STAI-B score, Non-acceptance score, Impulse score, Clarity score, and Lack of Premeditation score.

Table 6. Multiple linear regression analysis of the association between BITE subscale (severity score) and sociodemographic and clinical characteristics in the BED group.

Covariates	Univariate Analysis	Multivariable Analysis
	p	βadj (95% CI), p
Age	0.561	
Female/Male	0.698	
Beck Depression Inventory	0.064	
State anxiety inventory (STAI-A)	0.053	0.176 (0.061–0.290), $p < 0.01$
Trait anxiety inventory (STAI-B)	0.118	
DERS		
Non-Acceptance	0.146	
Goals	0.443	
Impulse	0.003	0.587 (0.415–0.759), $p < 0.001$
Awareness	0.052	
Strategies	0.089	
Clarity	0.034	0.304 (0.094–0.514), p <0.01

Table 6. *Cont.*

Covariates	Univariate Analysis	Multivariable Analysis
	p	βadj (95% CI), p
UPPS-P		
Negative Urgency	0.078	−0.832 (−1.083; −0.582), $p < 0.001$
Positive Urgency	0.93	
Lack of Premeditation	0.907	
Lack of Perseverance	0.806	
Sensation Seeking	0.547	

Note. Data are presented as p-value in univariate analysis and in adjusted regression β coefficients (βadj), 95% confident interval, and p-value in multivariable analysis (only significant β coefficients (βadj), 95% confident interval are presented). OR: odds ratio; BITE: Bulimic Investigatory Test, Edinburgh; DERS: Difficulty in Emotion Regulation Scale; UPPS-P: UPPS Impulsive Behavior Scale. In the final multivariable model of BITE severity subscale, potential covariates were Beck Depression Inventory score, STAI-A score, STAI-B score, Non-acceptance score, Impulse score, Awareness score, Strategies score, Clarity score, and Negative Urgency score.

4. Discussion

The primary aim of the present study was to examine the contributions of emotional overload (depression and anxiety), emotion regulation, and impulsivity in female and male obese people with and without BED and seeking bariatric surgery. Moreover, this study aimed to examine the contribution of emotional overload (depression and anxiety), emotion regulation difficulties, and impulsivity to eating patterns observed in patients with BED. Our two main findings, discussed below, were as follows: (1) limited access to emotional regulation strategies and bulimic symptoms were significant predictors of BED; (2) emotional eating, external eating, the degree of binge eating symptoms and the severity of bingeing and purging behaviors in patients with BED were associated with specific dimensions of emotion regulation and impulsivity as well as anxiety and depression scores. To the best of our knowledge, our study is the first to assess these contributions in this population.

More anecdotal, in our sample, the prevalence of BED was of 22.31%, which is consistent with the prevalence reported by the literature [36,37].

4.1. Emotional Overload, Emotion Regulation, and Impulsivity

The findings showed that emotion dysregulation (i.e., limited access to emotional regulation strategies) was a significant predictor of BED. The DERS Strategies subscale reflects limited access to the flexible use of adaptive emotion regulation skills to modulate (vs. eliminate) the intensity and/or temporal features of emotional responses [12]. This finding is consistent with results reported in a review from Dingemans et al. (2017) [9]. These authors suggest that individuals with BED are more likely to engage in maladaptive emotional strategies (e.g., suppression, rumination) and less likely to engage in adaptive ones (e.g., acceptance, reappraisal). Moreover, in a prospective study, Svaldi et al. (2019) demonstrated that, in individuals with BED, rumination was a significant predictor of binge eating and that from a clinical perspective, ruminations were correlated with the probability of a binge episode by approximately 28% [38]. In the specific population of patients seeking bariatric surgery, Cella et al. (2019) reported that patients suffering from BED exhibited more emotional dysregulation, as assessed by the Eating Disorders Inventory-3 (EDI-3), than patients without BED [16]. Moreover, Gianini et al. (2013) reported that, in treatment-seeking obese adults with BED, limited access to emotion regulation strategies was strongly associated with emotional overeating [39].

When examining eating-related behaviors, as assessed by the BITE, only bulimic symptoms were associated with BED, which is not surprising as binge eating is the essential feature of this disorder [1]. The severity subscale score was not associated with BED. It outlines that individuals with BED, in our sample, as was logically expected, do not show marked or sustained dietary restriction designed to influence body weight and shape between binge eating episodes.

In our univariate analysis, patients with BED showed significantly higher levels of depression, anxiety, and impulsivity than patients without BED. We expected that these dimensions would also be significant predictors to BED in the multivariable analysis, but this was not the case. This result underlines, in our population, the major contribution of the emotional regulation dimension to the disorder.

4.2. Emotional Overload, Emotion Regulation Difficulties, Impulsivity, and Eating Patterns in Patients with BED

The second aim of our study was to examine the contribution of emotional overload, emotion regulation difficulties, and impulsivity to eating patterns observed in patients with BED. The evaluated eating patterns were emotional eating and external eating as assessed by the DEBQ [26,27] and the degree of binge eating symptoms and the severity of bingeing and purging behaviors (as defined by their frequency) as assessed by the BITE [25].

Emotional eating was independently associated with age, DERS Non-acceptance subscale score, and UPPS-P Lack of premeditation subscale score. In our study, younger-old adults exhibited more emotional eating compared to older-old ones. This tendency to overeat in response to negative emotions appears to be more frequent in young adults than in older adults and it may be due to an increase in the use of emotion regulation skills with age [40]. The DERS Non-acceptance subscale is related to coping style [11] (p. 337). Coping is generally viewed as an individual's effort (cognitively and/or behaviorally) to adapt to or reduce distress in response to stressful events [41]. Hence, a maladaptive coping style to emotions in patients with BED could trigger an emotional eating pattern. However, in the present study, we did not have the information about what stressful events or negative thoughts patients needed to deal with and what coping styles they usually used. Another dimension contributing to emotional eating was lack of premeditation. This dimension is defined as the tendency to act without thinking and is viewed as presenting deficits in conscientiousness [42]. These deficits could lead to decision-making with little regard to past outcomes or forethought for possible future outcomes. It could also reflect a high tolerance for punishment from maladaptive behaviors (i.e., the negative consequences of these behaviors may not be sufficient to deter individuals with high scores on this dimension) [43].

External eating, corresponding to overeating in response to food-related cues such as the sight and smell of attractive food, was independently associated with the anxiety trait, DERS Impulse control difficulties subscale score, and UPPS-P Negative urgency subscale score. Interestingly, Heeren et al. (2018) found that the trait anxiety can be conceptualized as a single and coherent network system of interacting elements [44]. Noteworthily, they reported that the presence of intrusive thoughts and being unable to get disappointments out of one's mind emerged as the most central features of the trait anxiety network. It could mean that craving induced by food cues and negative affectivity may predispose to this eating style. These features could be linked to the "food addiction" hypothesis [45]. The difficulties maintaining behavioral control when distressed, assessed by the DERS Impulse control difficulties subscale, describe individuals who have very strong feelings that are hard to control [11] (p. 337). Moreover, this subscale specifically focuses on feeling "out of control" in emotionally distressing situations. In our sample, it is another dimension that predisposes patients with BED to eat in response to food-related cues. Negative urgency refers to acting rashly and impulsively when in extreme distress and involves impaired inhibitory control [46]. Patients with BED seem to use palatable food to compensate for negative affect or use food in a comforting fashion to cope with life distress. Taken together, negative affect and cravings induced by food cues increase the likelihood of external eating.

The degree of binge eating symptoms was independently associated with depression score, DERS Non-acceptance subscale score and UPPS-P Lack of premeditation subscale score. Interestingly, two dimensions (i.e., DERS Non-acceptance subscale score and UPPS-P Lack of premeditation subscale score) are the same that for emotional eating. Decision-making with little regard for past outcomes or forethought for possible future outcomes and a high tolerance for punishment from maladaptive behaviors seem then to also contribute to the likelihood of having a severe binge eating behavior. These results contribute to enriching our understanding of this eating behavior. Moreover, they raise the question of the links between emotional eating and binge eating and of the underlying psychopathology of these eating patterns. Depression severity was also associated with the binge eating behavior as expected based on literature data [22,47].

Finally, the severity of bingeing and purging behaviors was independently associated with the trait anxiety, DERS Impulse control difficulties subscale score, DERS Clarity subscale score, and UPPS-P Negative urgency subscale score. Three dimensions are common with those identified as being associated with External eating (i.e., anxiety trait, DERS Impulse subscale score, and UPPS Negative urgency subscale score). These dimensions are then risk factors for developing a highly disordered eating pattern and a presence of binge eating. However, unexpectedly, there was a reversed link with negative urgency. This may be due to the behaviors assessed by this score. In fact, it provides an index of the severity as defined by the frequency of binge eating and purging behavior. Among the purging behaviors, is the use of fasting. Therefore, we could make the assumption that patients with BED and with low levels of negative urgency may be more prone to using fasting to control their weight. It could be a reason why our two populations of patients (BED vs. wBED) did not differ in current and past BMI. The emotional clarity subscale predicted the severity of bingeing and purging behaviors. This dimension was strongly associated with emotional overeating [39] and could be, in our patients with BED, a risk factor of a high frequency of binge eating. However, this result is to be taken with caution, given the low Cronbach's alpha observed for this dimension in our population.

4.3. Limitations

This study has several limitations. The main limitation is about the power. If we use the number of cases to estimate the a priori power, we estimate that we can study between 2 to 3 variables (epv = 27/9 = 3; 27/10 = 2.7). Calculating the a posteriori power (post hoc) using the IBM SPSS sample Power software, or using simulations, allows us to determine a power at 88% to capture the effect of the three most influential covariates in our multivariable model. Clearly, we lack the power to study the numerous covariates. To check the result of our main endpoint, we have therefore proposed methods suitable for multivariable analyses on databases with a lack of power, such as penalized regression methods like Lasso [48]. These sensitivity analyses confirm our results (Supplementary Materials Tables S1 and S2), but do not exclude, that the other covariates are not significant partly due to a lack of power. Further studies with more power will be needed to estimate the association of the other covariates with our main endpoint. The same results are also confirmed by the use of Bayesian statistical analysis performed by the BRMS package [49] (Supplementary Materials Table S3) and, further, by selecting the covariates by bootstrapping using the rms package according to the methodology previously described [50]. The three most important parameters retained in the model are BITE symptom subscale (78.69%), DERS strategies (46.03%), and Emotional eating (34.92%). A second limitation is related to its cross-sectional nature; therefore, caution is needed in inferring causality. A third limitation is based on the assessment of BED, depression, anxiety, emotion regulation, impulsivity, and eating behavior styles through self-reports, which are subject to possible biases such as desirability or response bias. However, the validity of these questionnaires has been well supported in previous studies and our reliability indices were satisfactory except for DERS Lack of emotional clarity as mentioned earlier. Moreover, in future studies, these limitations could be overcome by using ecological momentary assessments considering patients' natural environment. Such tools are, for example, validated in nutritional epidemiology and in psychiatry (e.g., depression) [51,52].

5. Conclusions

The results of the present study suggest that obese patients with BED who are seeking bariatric surgery are characterized by a limited global access to the flexible use of adaptive emotion regulation skills to modulate features of emotional responses. Moreover, in this population, many dimensions of emotion regulation are associated with different pathological eating patterns (with sometimes the same dimension associated with varying patterns of eating), which emphasizes the pleiotropic side of these dimensions. The same results are observed for the anxiety trait and impulsivity. Taken together, our results lead us to believe that patients with BED could benefit from the addition of an emotion regulation intervention, which could significantly improve their eating behaviors before surgery. It could also improve the outcomes of bariatric surgery [53]. Further research is needed to confirm our findings, to implement and evaluate emotion regulation interventions, and to characterize better neural correlates of emotion regulation in patients with BED.

Supplementary Materials: The following are available online at http://www.mdpi.com/2072-6643/12/10/3099/s1, Table S1: Results of regularized-method LASSO with inference, Table S2: Results of regularized-method LASSO with inference, Table S3: Bayesian approach: Population-Level Effects title.

Author Contributions: F.B., F.G. and S.B. designed the study and wrote the protocol. Z.D., F.B. and F.G. conducted the methodology and statistical analyses. F.B. and F.G. wrote the first version of the manuscript. Z.D., S.B., E.B. and A.K reviewed the different versions of the manuscript and all authors contributed to and have approved the final manuscript. All authors have read and agreed to the published version of the manuscript.

Funding: This research received no external funding.

Acknowledgments: The authors thank the volunteers who made the study possible through their participation and collaboration. This research constitutes part of the Ph.D. thesis of F.B. at the University of Reims.

Conflicts of Interest: All the authors declare they have no conflict of interest regarding the publication of this paper.

References

1. American Psychiatric Association. *Diagnostic and Statistical Manual of Mental Disorders: DSM-5*; American Psychiatric Association: Washington, DC, USA, 2013.
2. Brownley, K.A.; Berkman, N.D.; Peat, C.M.; Lohr, K.N.; Cullen, K.E.; Bann, C.M.; Bulik, C.M. Binge-eating disorder in adults: A systematic review and meta-analysis. *Ann. Intern. Med.* **2016**, *165*, 409–420. [CrossRef] [PubMed]
3. Bruce, B.; Wilfley, D. Binge eating among the overweight population: A serious and prevalent problem. *J. Am. Diet. Assoc.* **1996**, *96*, 58–61. [CrossRef]
4. Davis, C. The epidemiology and genetics of binge eating disorder (BED). *CNS Spectr.* **2015**, *20*, 522–529. [CrossRef]
5. Dawes, A.J.; Maggard-Gibbons, M., Maher, A.R.; Booth, M.J.; Miake-Lye, I.; Beroes, J.M.; Shekelle, P.G. Mental health conditions among patients seeking and undergoing bariatric surgery: A meta-analysis. *JAMA* **2016**, *315*, 150–163. [CrossRef] [PubMed]
6. Smith, K.E.; Orcutt, M.; Steffen, K.J.; Crosby, R.D.; Cao, L.; Garcia, L.; Mitchell, J.E. Loss of control eating and binge eating in the 7 years following bariatric surgery. *Obes. Surg.* **2019**, *29*, 1773–1780. [CrossRef] [PubMed]
7. Opolski, M.; Chur-Hansen, A.; Wittert, G. The eating-related behaviours, disorders and expectations of candidates for bariatric surgery. *Clin. Obes.* **2015**, *5*, 165–197. [CrossRef]
8. Castellini, G.; Godini, L.; Amedei, S.G.; Galli, V.; Alpigiano, G.; Mugnaini, E.; Veltri, M.; Rellini, A.H.; Rotella, C.M.; Faravelli, C.; et al. Psychopathological similarities and differences between obese patients seeking surgical and non-surgical overweight treatments. *Eat. Weight. Disord. EWD* **2014**, *19*, 95–102. [CrossRef]
9. Dingemans, A.; Danner, U.; Parks, M. Emotion regulation in binge eating disorder: A review. *Nutrients* **2017**, *9*, 1274. [CrossRef]
10. Polivy, J.; Herman, C.P. Etiology of binge eating: Psychological mechanisms. In *Binge Eating: Nature Assessment and Treatment*; Guilford Press: New York, NY, USA, 1993; pp. 173–205.
11. Gross, J.J. *Handbook of Emotion Regulation*, 2nd ed.; Guilford Press: New York, NY, USA, 2015.
12. Gratz, K.L.; Roemer, L. Multidimensional assessment of emotion regulation and dysregulation: Development, factor structure and initial validation of the difficulties in emotion regulation scale. *J. Psychopathol. Behav. Assess.* **2004**, *26*, 41–54. [CrossRef]

13. Prefit, A.-B.; Cândea, D.M.; Szentagotai-Tătar, A. Emotion regulation across eating pathology: A meta-analysis. *Appetite* **2019**, *143*, 104438. [CrossRef]
14. Willem, C.; Gandolphe, M.-C.; Roussel, M.; Verkindt, H.; Pattou, F.; Nandrino, J.-L. Difficulties in emotion regulation and deficits in interoceptive awareness in moderate and severe obesity. *Eat. Weight. Disord. EWD* **2019**, *24*, 633–644. [CrossRef] [PubMed]
15. Zijlstra, H.; van Middendorp, H.; Devaere, L.; Larsen, J.K.; van Ramshorst, B.; Geenen, R. Emotion processing and regulation in women with morbid obesity who apply for bariatric surgery. *Psychol. Health* **2012**, *27*, 1375–1387. [CrossRef] [PubMed]
16. Cella, S.; Fei, L.; D'Amico, R.; Giardiello, C.; Allaria, A.; Cotrufo, P. Binge eating disorder and related features in bariatric surgery candidates. *Open Med.* **2019**, *14*, 407–415. [CrossRef] [PubMed]
17. Giel, K.E.; Teufel, M.; Junne, F.; Zipfel, S.; Schag, K. Food-related impulsivity in obesity and binge eating disorder—A systematic update of the evidence. *Nutrients* **2017**, *9*, 1170. [CrossRef] [PubMed]
18. Lecrubier, Y.; Sheehan, D.; Weiller, E.; Amorim, P.; Bonora, I.; Harnett Sheehan, K.; Janavs, J.; Dunbar, G. The Mini International Neuropsychiatric Interview (MINI). A short diagnostic structured interview: Reliability and validity according to the CIDI. *Eur. Psychiatry* **1997**, *12*, 224–231. [CrossRef]
19. Sheehan, D.V.; Lecrubier, Y.; Sheehan, K.H.; Amorim, P.; Janavs, J.; Weiller, E.; Hergueta, T.; Baker, R.; Dunbar, G.C. The Mini-International Neuropsychiatric Interview (MINI.): The development and validation of a structured diagnostic psychiatric interview for DSM-IV and ICD-10. *J. Clin. Psychiatry* **1998**, *59*, 22–33. [PubMed]
20. Report of a WHO consultation. Obesity: Preventing and managing the global epidemic. *World Health Organ. Tech. Rep. Ser.* **2000**, *894*, 1–253.
21. World Medical Association Declaration of Helsinki. Ethical principles for medical research involving human subjects. *World Med. Assoc. J. Am. Med. Assoc.* **2013**, *310*, 2191–2194. [CrossRef]
22. Gormally, J.; Black, S.; Daston, S.; Rardin, D. The assessment of binge eating severity among obese persons. *Addict. Behav.* **1982**, *7*, 47–55. [CrossRef]
23. Marcus, M.D.; Wing, R.R.; Hopkins, J. Obese binge eaters: Affect, cognitions and response to behavioural weight control. *J. Consult. Clin. Psychol.* **1988**, *56*, 433–439. [CrossRef]
24. Brunault, P.; Gaillard, P.; Ballon, N.; Couet, C.; Isnard, P.; Cook, S.; Delbachian, I.; Réveillère, C.; Courtois, R. Validation of the French version of the binge eating scale: Examination of its factor structure, internal consistency and construct validity in a non-clinical and a clinical population. *L'Encephale* **2016**, *42*, 426–433. [CrossRef] [PubMed]
25. Henderson, M.; Freeman, C.P. A self-rating scale for bulimia. The "BITE". *Br. J. Psychiatry* **1987**, *150*, 18–24. [CrossRef] [PubMed]
26. Lluch, A.; Kahn, J.; Stricker-Krongrad, A.; Ziegler, O.; Drouin, P.; Méjean, L. Internal validation of a French version of the Dutch eating behaviour questionnaire. *Eur. Psychiatry J. Assoc. Eur. Psychiatr.* **1996**, *11*, 198–203. [CrossRef]
27. van Strien, T.; Frijters, J.E.; Bergers, J.P.A.; Defares, P.B. The Dutch Eating Behaviour Questionnaire (DEBQ) for assessment of restrained, emotional and external eating behaviour. *Int. J. Eat. Disord.* **1986**, *5*, 747–755. [CrossRef]
28. Dan-Glauser, E.; Scherer, K. The Difficulties in Emotion Regulation Scale (DERS): Factor structure and consistency of a French translation. *Swiss J. Psychol.* **2013**, *72*, 5–11. [CrossRef]
29. Billieux, J.; Rochat, L.; Ceschi, G.; Carré, A.; Offerlin-Meyer, I.; Defeldre, A.-C.; Khazaal, Y.; Besche-Richard, C.; Van der Linden, M. Validation of a short French version of the UPPS-P Impulsive Behavior Scale. *Compr. Psychiatry* **2012**, *53*, 609–615. [CrossRef]
30. Beck, A.T.; Steer, R.A.; Brown, G.K. *BDI-II, Beck Depression Inventory: Manual*; The Psychological Corporation: San Antonio, TX, USA, 1996.
31. Collet, L.; Cottraux, J. The shortened Beck depression inventory (13 items). Study of the concurrent validity with the Hamilton scale and Widlöcher's retardation scale. *L'Encephale* **1986**, *12*, 77–79.
32. Spielberger, C.D.; Gorsuch, R.L. *State-Trait Anxiety Inventory for Adults*; Mind Garden: Redwood, CA, USA, 1983; pp. 49–55. Available online: https://trove.nla.gov.au/version/39948405 (accessed on 21 March 2020).
33. Wils, J.; Djerada, Z.; Roca, F.; Duflot, T.; Iacob, M.; Remy-Jouet, I.; Joannides, R.; Bellien, J. Alteration in the availability of epoxyeicosatrienoic acids contributes with NO to the development of endothelial dysfunction in conduit arteries during aging. *Atherosclerosis* **2018**, *275*, 239–245. [CrossRef]

34. Akaike, H. A new look at the statistical model identification. *IEEE Trans. Autom. Control.* **1974**, *19*, 716–723. [CrossRef]
35. Corchia, A.; Wynckel, A.; Journet, J.; Moussi Frances, J.; Skandrani, N.; Lautrette, A.; Zafrani, L.; Lewandowski, E.; Reboul, P.; Vrigneaud, L.; et al. Metformin-related lactic acidosis with acute kidney injury: Results of a French observational multicenter study. *Clin. Toxicol.* **2020**, *58*, 375–382. [CrossRef]
36. Vinai, P.; Da Ros, A.; Speciale, M.; Gentile, N.; Tagliabue, A.; Vinai, P.; Bruno, C.; Vinai, L.; Studt, S.; Cardetti, S. Psychopathological characteristics of patients seeking for bariatric surgery, either affected or not by binge eating disorder following the criteria of the DSM IV TR and of the DSM 5. *Eat. Behav.* **2015**, *16*, 1–4. [CrossRef] [PubMed]
37. Tess, B.H.; Maximiano-Ferreira, L.; Pajecki, D.; Wang, Y.-P. Bariatric surgery and binge eating disorder: Should surgeons care about it? A literature review of prevalence and assessment tools. *Arq. Gastroenterol.* **2019**, *56*, 55–60. [CrossRef] [PubMed]
38. Svaldi, J.; Werle, D.; Naumann, E.; Eichler, E.; Berking, M. Prospective associations of negative mood and emotion regulation in the occurrence of binge eating in binge eating disorder. *J. Psychiatr. Res.* **2019**, *115*, 61–68. [CrossRef] [PubMed]
39. Gianini, L.M.; White, M.A.; Masheb, R.M. Eating pathology, emotion regulation, and emotional overeating in obese adults with binge eating disorder. *Eat. Behav.* **2013**, *14*, 309–313. [CrossRef] [PubMed]
40. Samuel, L.; Cohen, M. Expressive suppression and emotional eating in older and younger adults: An exploratory study. *Arch. Gerontol. Geriatr.* **2018**, *78*, 127–131. [CrossRef]
41. Lazarus, R.S.; Folkman, S. *Stress, Appraisal and Coping*; Springer Publishing: New York, NY, USA, 1984.
42. Whiteside, S.P.; Lynam, D.R. The Five Factor Model and impulsivity: Using a structural model of personality to understand impulsivity. *Personal. Individ. Differ.* **2001**, *30*, 669–689. [CrossRef]
43. Berg, J.M.; Latzman, R.D.; Bliwise, N.G.; Lilienfeld, S.O. Parsing the heterogeneity of impulsivity: A meta-analytic review of the behavioral implications of the UPPS for psychopathology. *Psychol. Assess.* **2015**, *27*, 1129–1146. [CrossRef]
44. Heeren, A.; Bernstein, E.E.; McNally, R.J. Deconstructing trait anxiety: A network perspective. *Anxiety Stress Coping* **2018**, *31*, 262–276. [CrossRef]
45. Meule, A. A history of "food addiction". In *Compulsive Eating Behavior and Food Addiction*; Cottone, P., Sabino, V., Moore, C.F., Koob, G.F., Eds.; Academic Press: Amsterdam, The Netherlands, 2019; pp. 1–13. [CrossRef]
46. Cyders, M.A.; Smith, G.T. Emotion-based dispositions to rash action: Positive and negative urgency. *Psychol. Bull.* **2008**, *134*, 807–828. [CrossRef]
47. Nicholls, W.; Devonport, T.J.; Blake, M. The association between emotions and eating behaviour in an obese population with binge eating disorder. *Obes. Rev. Off. J. Int. Assoc. Study Obes.* **2016**, *17*, 30–42. [CrossRef]
48. Tibshirani, R. Regression shrinkage and selection via the lasso. *J. R. Stat. Soc. Ser. B* **1996**, *58*, 267–288.
49. Paul-Christian Bürkner. brms: An R Package for Bayesian Multilevel Models Using Stan. *J. Stat. Softw.* **2017**, *80*, 1–28. [CrossRef]
50. *Harrell Franck Regression Modeling Strategies: With Applications to Linear Models, Logistic Regression, and Survival Analysis*; Springer Series in Statistics; Springer: New York, NY, USA, 2015.
51. Maugeri, A.; Barchitta, M. A systematic review of ecological momentary assessment of diet: Implications and perspectives for nutritional epidemiology. *Nutrients* **2019**, *11*, 2696. [CrossRef] [PubMed]
52. Yim, S.J.; Lui, L.; Lee, Y.; Rosenblat, J.D.; Ragguett, R.M.; Park, C.; Subramaniapillai, M.; Cao, B.; Zhou, A.; Rong, C.; et al. The utility of smartphone-based, ecological momentary assessment for depressive symptoms. *J. Affect. Disord.* **2020**, *274*, 602–609. [CrossRef] [PubMed]
53. Efferdinger, C.; König, D.; Klaus, A.; Jagsch, R. Emotion regulation and mental well-being before and six months after bariatric surgery. *Eat. Weight Disord. EWD* **2017**, *22*, 353–360. [CrossRef] [PubMed]

nutrients

MDPI

Article

Food Addiction and Cognitive Functioning: What Happens in Adolescents?

Christopher Rodrigue [1], Sylvain Iceta [2] and Catherine Bégin [1,*]

[1] School of Psychology, Laval University, Quebec, QC G1V 0A6, Canada; Christopher.Rodrigue.1@ulaval.ca
[2] Quebec Heart and Lung Institute, Laval University, Quebec, QC G1V 4G5, Canada; Sylvain.Iceta.1@ulaval.ca
* Correspondence: Catherine.Begin@psy.ulaval.ca; Tel.: +1-418-656-2131

Received: 29 September 2020; Accepted: 21 November 2020; Published: 26 November 2020

Abstract: This study aimed to examine cognitive factors associated to food addiction (FA) symptoms in a non-clinical sample of adolescents. A group of 25 adolescents (12–18 years; Mean age = 15.2 years) with a high level of FA symptoms (two and more) were compared to a control group without FA symptoms ($n = 25$), matched on sex and age, on four Cambridge Neuropsychological Test Automated Battery (CANTAB) neuropsychological tasks (MT: Multitasking Test; OTS: One Touch Stockings of Cambridge; SST: Stop Signal Task; RVP: Rapid Visual Information Processing). They were also compared on self-reported questionnaires assessing binge eating, depressive and anxiety symptoms, impulsivity levels, as well as executive functioning difficulties. Group comparisons did not show significant differences on neuropsychological tasks' performances. However, effect sizes' estimates showed small to medium effect sizes on three scores: adolescents with a high level of FA symptoms showed a higher probability of an error following an incorrect answer (OTS), a higher probability of false alarm, and a poorer target sensitivity (RVP). When referring to self-reported measurements, they reported significantly more executive functioning difficulties, more binge eating, depressive symptoms and higher impulsivity levels. Overall, results suggested that cognitive difficulties related to FA symptoms seem to manifest themselves more clearly when assessing daily activities with a self-reported questionnaire, which in turn are strongly related to overeating behaviors and psychological symptoms. Future longitudinal research is needed to examine the evolution of those variables, their relationships, and contribution in obesity onset. More precisely, the present findings highlighted the importance of affective difficulties related to this condition, as well as the need to take them into account in its assessment.

Keywords: food addiction; adolescents; executive functioning; CANTAB; Yale Food Addiction Scale

1. Introduction

Many authors have studied disordered eating behaviors in individuals with overweight and obesity through the lens of addictions. Indeed, they found similarities between addictive behaviors and compulsive overeating, namely behavioural and neurobiological [1–4]. These similarities led to the concept of food addiction (FA) [5]. Although there is no universal definition of FA, it is known as an excessive and abnormal intake of highly palatable foods [2,3]. Since then, the study of FA has multiplied, using the Yale Food Addiction Scale (YFAS), an instrument that consists in an adaptation of the diagnostic criteria for substance dependence, to food [2,6].

In order to deepen our understanding of the development of an addictive-like pattern of eating, it is necessary to target potential risk factors of this condition. A current hypothesis on the development of obesity suggests that deficits in executive functioning could contribute to problematic eating behaviors, attitudes towards food, and weight gain [7]. Similarly, deficits in executive functioning are considered as a central component in the development of addictive behaviors [8]. Executive function is an umbrella term including a series of cognitive processes needed to adapt to new situations, allowing to behave appropriately to the context and to produce future-oriented behaviors [9,10]. It includes central functions, like inhibition, working memory, and cognitive flexibility, which underlie many higher-order functions (e.g., reasoning, problem solving, and planning) [11,12]. Recently, a few studies have focused on the cognitive factors that could be involved in FA, mainly executive functions.

Up until now, some authors examined the cognitive factors underlying FA in adults, using neuropsychological tasks and questionnaires. In order to do so, they all compared groups of individuals endorsing higher levels of FA symptoms to individuals who endorsed fewer or none. Overall, studies conducted within adults from the general population revealed that individuals with more FA symptoms, showed faster reaction times in response to food cues among neutral pictures [13], as well as attentional biases to unhealthy food cues following a sad mood induction [14]; a poorer performance monitoring, and more difficulties to detect and process errors during the task [15]. Only one study failed to find differences between groups, on a specific task assessing inhibitory control [16]. Regarding studies in adults suffering from obesity, they essentially showed that FA symptoms were accompanied with a poorer performance on decision-making, significant deficits in sustained attention [17]; and more difficulties to detect and process errors, as well as more self-reported metacognitive difficulties [18]. Nevertheless, a very recent study failed to show specific neuropsychological impairments or difficulties in participants with FA [19]. Taken together, most of the previous studies tend to reveal that cognitive difficulties could be associated to FA symptoms. Thus, it is possible to think that these cognitive difficulties could represent risk factors and contribute to the development of FA.

Adolescence is also considered as a high-risk period to develop addictive-like behaviors, regarding the combination between less efficient emotional regulation processes and the immature impulse control, that characterize it [20]. A range of studies aimed to assess the prevalence of FA in adolescents, resulting in rates of 2 to 16% in the general population [21–25]; almost 17% in psychiatric inpatients and 10 to 38% in adolescents with overweight or obesity engaged in a weight-loss program [26–28]. These rates were similar to those observed in studies with adults [29]. Since then, a growing body of literature on the study of FA in adolescents has been observed, offering a broader understanding of this condition. For example, FA symptoms have already been found as highly correlated with more disrupted eating behaviors, as well as more impulsivity, depressive and anxiety symptoms in this population [21,22,25,26,30,31].

A recent large-scale study in adolescents aged from 12 to 18 years also showed that those with more FA symptoms also reported significantly more executive functioning difficulties on a self-reported scale (BRIEF). More precisely, they reported significantly more difficulties on both indexes, assessing behavioural regulation (inhibition, shifting, emotional control, monitoring) and metacognitive (working memory, planning/organize, organization of materials, and task completion) difficulties. It indicates that they reported more self-regulatory weakness in their everyday life, in comparison to adolescents with lower levels of FA symptoms [31]. So far, only one study has included neuropsychological tasks to assess executive functions in adolescents, according to FA symptoms. Hardee and her colleagues examined the relationships between FA symptoms and cerebral activity during an inhibitory task (Go/No-go), in a sample of adolescents [32]. Their results showed that, in participants with a higher level of FA symptoms, a hypoactivation in some brain areas was observed during the inhibitory phase of the task. They suggested that it could be associated to a poorer inhibitory control. However, no significant difference was observed when they were compared on their performances at the task. More studies are needed according to FA and cognitive functioning in adolescents, in order to identify if cognitive mechanisms are involved in addictive-like eating.

The main aim of the present study was to examine cognitive factors associated to FA symptoms in adolescents. More precisely, the objective was to assess sustained attention, as well as executive functions with neuropsychological tasks, in adolescents with a significant level of FA symptoms (two symptoms and more), and to compare them with a control group. Since very few studies have examined cognitive factors related to FA symptoms in adolescents, no formal hypothesis will be proposed.

2. Materials and Methods

2.1. Participants and Procedures

Participants were 50 adolescents (38 girls and 12 boys; Mean age = 15.20; *SD* = 1.62), selected from a larger non-clinical sample previously recruited to participate in a research study on eating behaviors in adolescents, as described in Rodrigue et al. [31]. To be included, participants were required to be aged 12–18, and have previously consent to be contacted to take part in the present study. No supplementary inclusion or exclusion criteria were added from the previous study. Within participants who accepted to be contacted subsequently for the actual study, those who reported a significant level of FA symptoms on the YFAS 2.0 (two criteria or more) were firstly contacted. This threshold represents the minimum number of criteria to endorse a FA diagnosis with a mild severity. In order to confirm the FA diagnosis, participants must additionally endorse a clinically significant impairment or distress criterion. Considering the exploratory nature of the present study and the study sample, this criterion was not taken into account in the recruitment of participants. Indeed, the objective of the study was to explore the cognitive functions of adolescents from a non-clinical sample, a population in which the early signs of FA could be observed without necessarily endorsing the diagnosis. As previously suggested in adults' studies, the inclusion of clinically significant impairment or distress criterion may not represent a valid marker of psychological distress as some people may endorse most of the criteria and not report distress [33]. Among the initial sample, a total of 25 participants reporting a significant level of FA symptoms on the YFAS 2.0 were recruited. Throughout the manuscript, this group has been categorized as "high FA group". Then, a comparison group of 25 adolescents who endorsed none of the FA symptoms was recruited in order to match the first group on sex and age (Control group). More precisely, for each participant from the high FA group, a control participant with the same sex and age was also recruited. Franken et al. previously used a similar procedure, in the study of the cognitive correlates of FA, in adults [15].

All participants were invited to come to our laboratory and completed, with a member of our research team trained in neuropsychological task administration, a demographic questionnaire as well as four cognitive tasks from the Cambridge Neuropsychological Test Automated Battery (CANTAB) in a quiet room without any background noises or distractions. Prior to the task completion, participants consented themselves to take part in the study and allowed us to use their previous answers to some of the questionnaires completed as part of the larger study, namely questionnaires assessing food addiction symptoms as well as binge eating, anxiety and depressive symptoms, impulsivity, and self-reported executive functioning difficulties (see Rodrigue et al.) [31]. A parental consent was required for those younger than 14 years of age. In order not to interfere in participants' performances, they were not informed of their group belonging; a general description of the study's objectives was given to each participant. They were also asked to self-report their height and weight, to calculate their Body Mass Index (BMI; kg/m^2). Self-reported height and weight has previously been shown as highly correlated with measured height and weight in adolescents [34,35]. According to the World Health Organization (WHO) body mass index-for-age percentile growth charts [36], 39 participants reported a healthy weight, seven reported being overweight, four reported suffering from obesity. Both groups were evenly balanced according to the BMI categories. The Laval University Research Ethics Committee approved the study. All participants received a 20$ monetary compensation for their participation.

2.2. Measures

2.2.1. Food Addiction

The French version of the Yale Food Addiction Scale 2.0 (YFAS 2.0) is a 35-item self-reported questionnaire used to assess food addiction symptoms over the previous year [6,37]. The YFAS 2.0 covers FA criteria based on the DSM-5 eleven diagnostic criteria for substance use disorders [38]. Items are answered on a 7-point Likert Scale ranging from 0 (Never) to 7 (Every day). To fulfill a criterion, participants must endorse at least one item related to the associated criterion. Two methods have been developed to interpret answers. First, it is possible to assess the presence/absence of the "FA diagnosis" if a participant has endorsed at least two criteria and reported functional impairment or clinical distress. The severity of the "FA diagnosis" can be categorized as "mild" (two or three criteria), "moderate" (four or five criteria), and "severe" (six or more criteria). The functional impairment/clinical distress criterion was not taken into account in the recruitment of participants. In the present study, an adapted version of the YFAS 2.0 was used to refer to age-appropriate activities, as in Rodrigue et al. (refer to Gearhardt et al. for adaptations of the YFAS for children or YFAS-C) [22,31]. To make sure that the YFAS 2.0 items were developmentally and culturally appropriate, minor adaptations have been made on the scale inspired by the YFAS-C and taking into account the specific use of the French language in the Province of Quebec. An inter-rating process was performed between two of the authors (C.R. and C.B) for each adaptation. Cronbach's alphas for this version of the YFAS 2.0 were 0.91 for the initial study, and 0.79 for the present sample.

2.2.2. Binge Eating Symptoms

The French version of the Binge Eating Scale (BES) is a 16-item self-reported questionnaire assessing symptoms related to behavioral, cognitive, and emotional manifestations of binge eating episodes [39,40]. Participants must choose, for each item, one of the four suggested sentences that most accurately described his or her situation. Each item has a designated weight on the total score, in proportion with its severity (between zero and three). The total score is obtained by summing up each item's score, and is ranging from 0 to 46. A score of 17 or less indicates a minimal, between 18 and 26 indicated a moderate, and 27 or more indicates severe binge eating symptomatology. In the present study, Cronbach's alpha of the BES was 0.91.

2.2.3. Anxiety Symptoms

The French version of the Multidimensional Anxiety Scale For Children- Self Report (MASC) is a 39-item self-reported questionnaire assessing anxiety-related emotional, cognitive, physical and behavioural symptoms in children and adolescents from 8 to 19 years old [41,42]. Participants must answer on a 4-point Likert scale for each item, ranging from "Rarely true about me" to "Often true about me". Items can be summed up in fours subscales: (1) Physical Symptoms, (2) Social Anxiety, (3) Harm Avoidance, and (4) Separation Anxiety. Summing subscales' scores can also produce a global score. T-scores are created for each score and should be interpreted using the following guidelines: <40 = low; 40–54 = Average; 55–59 = High average; 60–64 = Slightly elevated; 65–69 = Elevated; ≥70 = Very elevated. In the present study, only the global score was used and showed a good internal consistency ($\alpha = 0.91$).

2.2.4. Depressive Symptoms

The French version of the Beck Depression Inventory (BDI) is a 21-item self-reported questionnaire assessing depressive symptoms experienced during the last two weeks [43,44]. Each symptom is rated on a 4-point Likert scale from 0 (The symptom is not associated to any suffering) to 3 (The symptom is associated to intense suffering). The total score ranges from 0 to 63; A score from 0 to 13 indicates normal to minimal depressive symptoms, a score from 14 to 19 indicates mild to moderate depressive symptoms, a score from 20 to 28 indicates moderate depressive symptoms, and a score from 29 to 63 indicates severe depressive symptoms. The BDI has previously shown good psychometric properties for large-scale screening of depressive symptoms in adolescents [45–47]. For the present study, Cronbach's alpha of the questionnaire was 0.92.

2.2.5. Impulsivity

The French version of the UPPS Impulsive Behaviour Scale for adolescents [48,49] is a 45-item self-reported questionnaire used to assess impulsivity in adolescents aged between 12 and 19 years old. Items are rated on a 4-point Likert scale ranging from 1 (Disagree strongly) to 4 (Agree strongly). Items are divided in four mutually exclusive dimensions of impulsivity, namely urgency, lack of premeditation, lack of perseverance and sensation seeking. A higher score indicates a higher level of impulsivity. Cronbach's alphas for the subscales were good in the actual study, ranging from 0.82 to 0.89.

2.2.6. Cognitive Functioning

Cognitive functioning was assessed using the Connect Research Suite of the Cambridge Neuropsychological Test Automated Battery (CANTAB), a computerized test battery. The CANTAB is a well-validated standardized cognitive battery that provides rapid, sensitive and objective measures of multiple cognitive domains. These tasks have been validated to understand the role of specific brain functions among a wide range of disorders. Participants completed four of the CANTAB tasks with French voiceover instructions on an iPad Air 2, namely the Multitasking Test (MT), the One Touch Stockings of Cambridge (OTS), the Rapid Visual Information Processing (RVP), and the Stop Signal Task (SST), in that order. These tasks have been developed to assess two key domains, namely sustained attention and executive functioning. For more details on the following tasks and the CANTAB battery, see cantab.com.

2.2.7. Multitasking Test (MT)

MT is an eight-minute task of executive functioning that measures the participant's ability to use multiple sources of potentially conflicting information to guide behaviour, and to ignore task-irrelevant information. During the test, an arrow can appear on both sides of the screen and point in both directions. Before each trial, a cue is displayed at the top of the screen to indicate to the participant whether they must push the right or the left button, considering the instruction to identify the side or the direction of the arrow. The task includes some sections during which the rule is consistent across trials (single task) and other sections during which the rule may randomly change from trial to trial (multitasking). The multitasking sections require a higher cognitive demand than the single task sections. Moreover, some of the task trials display congruent stimuli (e.g., the arrow on the left side is pointing on the left side of the screen) and incongruent stimuli (e.g., the arrow on the left side is pointing on the right side of the screen). The incongruent trials require a higher cognitive demand than the congruent trials. MT includes practice blocks before each assessed block, to make sure that the participants understood the instructions. The main outcomes of this task are response latency and error scores, reflecting the ability to deal with multitasking as well as the interference of incongruent task-irrelevant information on the performance at the task. This task allows calculating the cost of using interchanged rules in opposition to consistent rules, and incongruent information in opposition to congruent information. More precisely, the outcomes of interest in the present study were: the number of trials for which the outcome was incorrect; the median reaction latency of response across all correct trials; Incongruency cost (a higher incongruency cost indicates that the subject takes longer to process conflicting information); Multitasking cost (a positive score indicates that the subject responds more slowly during multitasking blocks and a higher score indicates a higher cost of managing multiple sources of information).

2.2.8. One Touch Stockings of Cambridge (OTS)

OTS is a 10-min task, based upon the Tower of Hanoi test, measuring executive functioning and more precisely spatial planning and working memory. During the task, participants see two displays including three-color balls and the main instruction is to work out mentally the number of moves needed to make the lower display match the upper display. The OTS includes practice blocks to make sure that participants understood the instructions. Trials required one to six moves to complete. After an error, they are asked to rethink their solution and answer. Overall, there were 10 "easy" trials for which the upper display could be matched in one to three moves; and eight "hard" trials for which it could be matched in four to six moves. OTS outcome measures include the number of problems solved on the first choice, mean attempts required to obtain a correct solution, median latency to first choice (the median latency measured from the appearance of the stocking balls until the first choice was made, across all assessed trials where the participant's first choice was correct), median latency to obtain a correct solution. These measures are available for all problems and/or for problems with a specified number of moves (one to six). The probabilities of an error occurring when the previous trial was responded correctly (Error given correct; the probability of an error occurring when the previous trial was responded to correctly), or incorrectly (Error given incorrect; the probability of an error occurring when the previous trial was responded to correctly), were also documented.

2.2.9. Stop Signal Task (SST)

SST is a 20-min task measuring impulse control and response inhibition. It also allows measuring error monitoring after a response inhibition failure [50]. The first section of the test consists in a learning phase during which the participant must select the button associated to the direction in which the arrow points. During the second section of the test, the instruction remains the same, but when participants hear an audio tone (beep), they must hold back or inhibit their response. The SST uses a staircase design for the stop signal delay, meaning that the task allows adapting to the performance of the participant. More precisely, the stop signal delay increased by 50 ms following a successful inhibition and decreased by 50 milliseconds following a failed inhibition until the inhibition success rate stabilizes to 50%. It is suggested that the stop signal reaction time (SSRT) represents the time before which actions become ballistic and the participant is no longer able to cancel his action. The SST main outcome is the SSRT (a higher score indicates a worse stop signal reaction time). According to previous studies, performance monitoring was explored using the formula proposed by Bo, Aker, Billieux, and Landro [51]. More precisely, it allows calculating post-error slowing (PES; tendency to slow a response after a failure to stop), and post-success slowing (PSS; tendency to slow a response after a successful stop trial). PES was calculated by subtracting reaction times for "Go-after-go" trials to "Go-after-failure to stop trials", and post success slowing (PSS) by subtracting reaction time for "Go-after-go trials" to "Go-after-successful stop trials".

2.2.10. Rapid Visual Information Processing (RVP)

The RVP is a seven-minute test assessing sustained attention and working memory. During the test, digits from 2 to 9 appear in a white box in the middle of the screen, in a pseudo-random order at the rate of 100 digits/min. The task is divided in two phases. First, participants must detect one target sequence of digits and press the button at the centre of the screen when they see the last digit of the sequence (e.g., 3–5–7); in the second phase, they must detect three target sequences of digits (e.g., 3–5–7; 2–4–6; 4–6–8), and press the button at the centre when they see the last digit of one of the sequences. RVP outcome measures include the A′, which is a metric measure representing how good the participant is at detecting a targeted sequence. RVP outcome measures also include the median response latency on trials where the subject responded correctly across all trials, and the probability of a false alarm (False alarms ÷ (False alarms + Correct rejections)).

2.2.11. Behavior Rating Inventory of Executive Function–Self-Report Version (BRIEF-SR)

The French version of the BRIEF-SR is an 80-item self-reported questionnaire developed to assess adolescents' views (11–19 years old) of their own executive functions or self-regulatory strengths and weaknesses in their everyday life [52,53]. Items are scored on a 3-point Likert scale ranging from 1 (Never) to 3 (Often), and assess the frequency of some behaviors in the last six months. Items can be regrouped in a global score (Global Executive Composite; GEC) or in two indexes (Behavioural Regulation Index (BRI); and Metacognition Index (MI)). The Behavioural Regulation Index (BRI) includes subscales for Inhibition, Shift, Emotional Control and Monitoring, and the Metacognition Index (MI) includes subscales for Working Memory, Planning/Organize, Organization of Materials and Task Completion. T-scores are created for each subscale; a score between 60 and 64 indicates mildly elevated difficulties and a score of 65 and higher indicates clinical significant difficulties. Cronbach's alphas in the larger study were 0.83 for the BRI, 0.89 for the MI., and 0.81 for the GEC. For the present study, they were respectively of 0.80, 0.83, and 0.83.

2.3. Data Analysis

SPSS, version 24.0, was used for statistical analyses. Some of the variables of interest were transformed using windsorized transformation (MTT Total incorrect, MTT Incongruent cost, MTT Multitasking cost, and RVL Response latency), or using a combination of a windsorized and a logarithmic transformation (binge eating and depressive symptoms), considering their non-normal distributions [54]. Then, group comparisons were performed to compare both groups (high FA group and control group), on all variables of interest (MANOVAs and ANOVAs).

A one-way multivariate analysis of variance (MANOVA) was firstly performed to compare groups, considering the moderate correlations between the following variables: binge eating, depressive, and anxiety symptoms, as well as on impulsivity (Urgency, lack of premeditation, lack of perseverance), and self-reported executive functioning difficulties (Global Executive Composite), in order to corroborate the results from our previous study [31]. Thereafter, another one-way MANOVA was performed on all CANTAB outcomes previously described (see Measures section). More specifically, outcomes of interest were regrouped by task in the same analysis, considering their theoretical relationships under the same cognitive constructs. The latter were followed by univariate analyses of variance (ANOVAs) and descriptive analyses on all variables of interest, in order to document the severity of symptoms and difficulties in both groups. In order to quantify group comparisons' effect sizes, partial eta-squared values were calculated and interpreted following Cohen's guidelines (small ≥ 0.01; medium ≥ 0.06; large ≥ 0.14) [55].

3. Results

First, according to the YFAS 2.0 results, participants from the high FA group distributed as follows, according to the symptoms count: 10 participants reported two FA symptoms (40%), seven reported three symptoms (28%), four reported four symptoms (16%), one reported five symptoms (4%) and three participants reported six symptoms or more (12%). Within this group, four participants (16%) endorsed the FA diagnosis (i.e., two FA symptoms or more and the functional impairment or clinical distress criterion).

The initial one-way MANOVA first showed a significant difference between groups, when clustering binge eating, depressive, and anxiety symptoms, as well as impulsivity (UPPS Urgency, lack of premeditation, lack of perseverance), and self-reported executive functioning difficulties (BRIEF Global Executive Composite) ($F(7, 39) = 5.36$, $p < 0.001$, Wilks' $\Lambda = 0.51$). Following that, One-way ANOVAs between groups showed significant differences with a medium effect size for UPPS- Lack of premeditation, and strong effect sizes for binge eating and depressive symptoms, UPPS- Urgency, UPPS- Lack of perseverance, and executive functioning difficulties. A non-significant difference with a medium effect size was found for anxiety symptoms.

A second set of one-way MANOVAs was performed to compare both groups on each CANTAB task separately, regrouping key scores previously presented. Results showed no significant differences between groups on the Multitasking Test (MTT) ($F(4,45) = 0.45$, $p = 0.77$, Wilks' $\Lambda = 0.96$), the One Touch Stockings of Cambridge (OTS) ($F(6,39) = 1.47$, $p = 0.21$, Wilks' $\Lambda = 0.82$), the Stop Signal Task (SST) ($F(3,46) = 0.25$, $p = 0.86$, Wilks' $\Lambda = 0.98$), and the Rapid Visual Information Processing (RVP) ($F(3,46) = 1.26$, $p = 0.30$, Wilks' $\Lambda = 0.92$). Interactions between groups and, sex, age (12–14 and 15–18), and BMI categories were tested independently, and were not statistically significant. However, based on independent group comparisons for each task's outcomes, it is possible to observe three effects sizes ranging from small (OTS–Probability of error given incorrect = 0.04; RVP–Probability of false alarm = 0.03) to medium (RVP–A′ = 0.07). Descriptive statistics and ANOVAs outcomes for each dependent variable are shown in Tables 1 and 2.

Table 1. Group comparisons–Self-reported variables (high FA group vs. control group).

	High FA Group (n = 25)		Control Group (n = 25)		Contrasts		
	M	SD	M	SD	F	P	Partial Eta-Squared
Binge eating symptoms (BES) [a]	14 (1.10)	10.15 (0.26)	5.29 (0.70)	4.10 (0.33)	21.49	< 0.01	0.32
Depressive symptoms (BDI) [a]	18.6 (1.20)	11.55 (0.31)	6.60 (0.69)	6.06 (0.45)	20.86	<0.01	0.32
Anxiety symptoms (MASC)	52.04	18.54	42.54	20.25	2.81	0.10	0.06
Urgency (UPPS)	30.22	6.62	23.83	6.52	11.10	<0.01	0.20
Lack of premeditation (UPPS)	24.96	5.41	21.16	5.04	6.17	<0.05	0.12
Lack of perseverance (UPPS)	22.57	4.99	17.83	4.40	11.92	<0.01	0.16
Sensation seeking (UPPS)	33.35	8.03	32.72	8.14	0.07	0.79	<0.01
Executive functioning difficulties-Global Executive Composite (GEC)	54.30	8.83	46.11	10	8.76	<0.01	0.16

Note: BES = Binge Eating Scale; BDI = Beck Depression Inventory; MASC = Multidimensional Anxiety Scale for Children; UPPS = Urgency, Premeditation (lack of), Perseverance (lack of), Sensation Seeking – Impulsive Behavior Scale. [a] Windsorized and logarithmic transformations were applied. Transformed scores are indicated in brackets.

Table 2. Group comparisons–CANTAB neuropsychological tasks (high FA group vs. control group).

	High FA Group (n = 25)		Control Group (n = 25)		Contrasts		
	M	SD	M	SD	F	P	Partial Eta-Squared
Executive functioning							
MTT [a]							
Total incorrect [a]	7.08	6.83	6.84	6.71	0.02	0.90	<0.01
Reaction latency (ms) [b]	550.00	83.35	565.42	83.48	0.43	0.52	<0.01
Incongruency cost [ab]	35.86	43.92	33.84	25.07	0.04	0.84	<0.01
Multitasking cost [ab]	199.04	96.43	181.49	107.21	0.37	0.55	<0.01
OTS							
Problems solved on first choice	10.71	2.24	10.86	1.75	0.07	0.80	<0.01
Mean choice to correct [a]	1.48	0.36	1.45	0.25	0.12	0.73	<0.01
Latency to first choice (ms) [b]	9172.50	4208.17	8944.32	2832.11	0.05	0.83	<0.01
Latency to correct (ms) [b]	9827.42	4304.16	10,478.77	3773.67	0.30	0.59	<0.01
Probability of error given correct	0.31	0.17	0.32	0.13	0.11	0.75	<0.01
Probability of error given incorrect	0.26	0.22	0.19	0.18	1.69	0.20	0.04
SST							
Stop Signal Reaction Time (SSRT)	208.04	28.22	202.82	39.85	0.29	0.60	<0.01
Post-error slowing	18.14	92.27	34.02	73.81	0.45	0.51	<0.01
Post-success slowing	7.18	73.01	13.58	64.67	0.11	0.74	<0.01
Sustained attention							
RVP							
A′	0.88	0.06	0.91	0.05	3.41	0.07	0.07
Response latency (ms) [ab]	430.78	63.17	428.62	49.46	0.02	0.89	<0.01
Probability of false alarm	0.03	0.10	0.02	0.07	1.28	0.26	0.03

Note: ms = milliseconds. MMT = Multitasking Test; OTS = One Touch Stockings of Cambridge; SST = Stop Signal Task; RVP = Rapid Visual Information Processing; A′ = Sensitivity to target. [a] Windsorized transformations were applied. [b] Median score.

Finally, it should be noted that correlation analyses showed moderate to strong significant associations between the following self-reported variables: FA symptoms, binge eating, depressive and anxiety symptoms, UPPS Urgency scale and the BRIEF Global Executive Composite. Correlations between self-reported data and neuropsychological tasks' outcomes were much smaller, revealing few significant associations. Among all CANTAB scores, only the Rapid Visual Information Processing (RVP) outcomes, A′ (sensitivity to the target) and the probability of false alarm, were significantly and moderately correlated with FA symptoms.

4. Discussion

Addictive-like eating behaviors are associated with many negative outcomes, namely weight gain, obesity and related medical conditions, a poorer quality of life, as well as more psychiatric comorbidities. Beyond this bleak picture, little is known about the risk factors that could possibly contribute to the emergence of those eating behaviors in adulthood. Thus, the present study aimed to investigate cognitive factors associated to FA symptoms in adolescents, in order to identify potential cognitive risk factors in the development of FA at this developmental stage. More specifically, the aim of this study was to compare a group of adolescents with high levels of FA symptoms (two symptoms and more), to a control group without FA symptoms, on neuropsychological tasks assessing key cognitive domains (sustained attention and executive functioning). To the best of our knowledge, our study is among the first to use neuropsychological tasks in the study of FA in adolescents.

First of all, group comparisons on self-reported data showed that participants with a high level of FA symptoms reported a significantly more severe profile on almost all self-reported assessed variables compared to adolescents from the control group. However, our results showed no statistically significant differences between groups, among key scores for all four tasks of the CANTAB neuropsychological. It means that participants from both groups showed similar performances on tasks assessing sustained attention and executive functioning. Since CANTAB cognitive tests have been highly validated as sensitive measures to examine cognitive functions related to brain networks, it is possible to think that FA symptoms in adolescents are not clearly accompanied with a specific pattern of cognitive difficulties that can be captured by neuropsychological tasks in a context of neutral stimuli. These results are supported by Hardee et al., who also noted an absence of difference on neuropsychological tasks (inhibitory control), according to FA symptoms [32]. This pattern also seems to be observed in related conditions, like binge-eating disorder (BED), which is mainly characterized by frequent overeating episodes in a discrete period of time, and a feeling of lack of control over the food intake during those episodes [38]. For example, Kittel, Schmidt, and Hilbert, only found small differences between adolescents with BED and obesity and control adolescents with obesity only, on well-validated neuropsychological tasks assessing multiple cognitive functions (sustained attention, inhibition, cognitive flexibility and decision-making) [56]. Our results as well as those of recently obtained from Hardee and Kittel's studies may indicate that compulsive overeating behaviors are not typically associated to specific cognitive dysfunctions in this developmental stage, or that the associated cognitive difficulties cannot be captured by standard neuropsychological tasks. It is however important to note that neuroimaging findings from Hardee et al., revealed cerebral activity that could be associated with a poorer inhibitory control in participant with a higher level of FA symptoms [32]. According to this result, we cannot yet rule out the presence of cognitive difficulties or impairments in adolescents with FA symptoms.

Nevertheless, our results showed slightly poorer performances from participants within the high FA group according to effect sizes estimates, on three specific scores. Since these differences between groups were not statistically significant and effect sizes were small, the following hypotheses aim to stimulate reflection about potential cognitive factors associated to FA symptoms, according to previous findings in adults. Firstly, compared to the control group, participants from the high FA group showed a slightly higher probability of producing an error on the OTS –assessing planning and working memory – only when the previous trial was responded incorrectly. No difference between groups was observed when the previous trial was responded correctly (i.e., when the participant previously gave the expected response). It is possible to think that, after an error, it was harder for those with significant FA symptoms to adjust their behaviors in order to avoid the same error on the next trial. Previous findings, from adult studies, suggested that individuals with more FA symptoms showed more difficulties in error or performance monitoring [15,18]. Finally, both groups slightly distinguished themselves on two of the RVP scores, used to measure sustained attention. More specifically, compared to the control group, subjects from the high FA group showed a poorer target sensitivity with a medium effect size, suggesting that they were a little less efficient in detecting the sequences they were asked to; and a slightly higher probability of false alarm with a small effect size, suggesting that they tended a little bit more often to identify the target as present, when it was absent or incomplete. These outcomes could be associated with one's difficulties to focus his attention on the ongoing task [57]. Accordingly, Steward et al. suggested deficits in sustained attention, in adults with obesity and FA [17]. Thus, FA symptoms could be associated with a poorer ability to sustain attention, as soon as in adolescence. However, these results should be replicated in order to support these hypotheses. More precisely, it would be interesting to reproduce this study in a clinical sample of adolescents endorsing the FA diagnosis, as a way to examine if the same patterns can be observed, but with larger effect sizes.

Moreover, an interesting finding is the discrepancy between group comparisons on self-reported questionnaires, and those on neuropsychological tasks. Our results showed a significant difference between groups according to self-reported executive functioning difficulties whereas no clear distinction was found on the neuropsychological tasks. Self-reported measures aim to assess the participant's view of his own self-regulatory strengths and weaknesses in his everyday life, whereas neuropsychological tasks consist in a performance-based task designed to objectively assess cognitive functioning. Accordingly, it is possible to think that a self-reported measure of executive functioning is a more sensitive way to assess self-regulation or executive difficulties in adolescents with an addictive-like pattern of eating, while neuropsychological tasks seem not to offer a convergent source of information. Recently, Demidenko, Huntley, Martz, and Keating have argued that self-reported measures of cognitive constructs would represent a more consistent predictor of risky behaviors in adolescents (for example substance use), than cognitive tasks [58]. Moreover, they found that associations between risky behaviors and self-reported measures were stronger than with cognitive tasks outcomes. The same observation can be made from our data, as the magnitude of the associations between FA symptoms and the self-reported measure of executive functioning was clearly larger than with the neuropsychological tasks' outcomes. Demidenko and his collaborators proposed that even though cognitive tasks predicted some risky behaviors, they seemed to require a greater power to detect a small effect, in opposition to the self-reported measures [58]. Accordingly, they reconsidered the use of neuropsychological tasks in laboratory settings as a way to infer real-world risky behaviors in adolescents. A similar pattern, in which self-reported measures of executive functioning showed clearer differences between groups than cognitive tasks, has been observed in a previous study in adults suffering from severe obesity [18].

It is possible to think that classical neuropsychological tasks do not allow capturing the affective components of executive functioning or hot executive functions. The latter are considered as major factors in the development of an addiction, involved in the weighting of pros and cons in the decision-making process [8,59]. Although the self-reported measure does not directly activate the affective components of executive functioning either, it includes items referring to emotional control in everyday life, assessing the impact of executive function difficulties on the emotional expression, as well as one's ability to regulate or control his/her responses to emotions (e.g., "I overreact to small problems"). Even though adolescents from the high FA group did not reported clinically significant difficulties on the self-reported measure, it nevertheless allows to significantly discriminate participants according to FA symptoms. The present study highlights the relationship between self-reported executive functioning and negative affect, as depressive and anxiety symptoms, as well as with urgency. Considering the previous statements and the absence of significant difference on neuropsychological tasks, FA seems to be mainly characterized by a tendency to experiment more distress and negative affects and a tendency to act on them impulsively as well as a difficulty to regulate them satisfactorily.

Besides, those could represent vulnerability factors in the development of addictive-like eating. Still, the subjective nature of self-reported measures naturally comes with personal biases that could be associated with many factors hardly measurable; it could also affect study results and come with an over or under evaluation of one's own difficulties varying from one individual to another. Even though self-reported measures are known to provide valid findings, their subjective nature limits the associated conclusions. Thus, it would be interesting to measure the cognitive correlated of FA in adolescents in a more objective way, but in a setting with better ecological validity in order to capture the affective components of those functions in a real-life setting. We also think that a structured interview to assess eating behaviors, psychological symptoms, and impulsivity would be a great addition, to objectively assess those variables as well.

Our results should be considered in light of some limitations. First, the present sample was recruited within the general population, with a limited access to adolescents with overweight and obesity. The narrowed BMI range of the present sample has limited the inclusion of BMI categories in the analyses, as well as in the discussion of our results. In addition, the use of a subthreshold of the FA diagnosis in the recruitment of the high FA group could also represent a limitation to the present findings and conclusions. Indeed, it could have contributed to the small and statistically non-significant differences between both groups, which could have been clearer with the FA diagnosis as a threshold. Participants' selection for the present study could also have induced a selection bias. More precisely, participants consented to be contacted for the present study, after completing a battery of questionnaires on their eating behaviors and psychological condition. It is possible that, those who presented a more severe profile, were not interested in participating in this project. A majority of the participants came from private schools, which could also have an impact on the representativeness of the sample. Moreover, since participants were received in time slots that were more convenient for them, we did not control for the moment of the testing and it could have varied across participants; it is also possible that it influenced slightly their performances, even though we asked them to pick a moment when they felt rested and awakened. Finally, the relatively small sample size, and the cross-sectional design also represented limitations in the interpretation of the outcomes. Even though the actual sample size was adequate to test our hypotheses, a larger sample size could have allowed clearer results and conclusions, especially in the examination of cognitive factors associated with FA symptoms. In order to support the present findings, future studies in this field should also include longitudinal study designs, in order to clarify the direction of the associations between psychological symptoms, cognitive functioning and FA symptoms. Beyond self-reported measures and neuropsychological tasks, neuroimaging technic would also represent an interesting addition in future studies. Finally, considering the clear relationships between FA symptoms and variables surrounding negative affect, it would also be interesting to examine neuropsychological substrates of FA, in a context involving a greater affective load.

An important strength of the present study was the use of the CANTAB, a well-validated and sensitive measure cognitive functions. The use of a computerized cognitive battery reduced the potential sources of errors across the whole process, including testing and data entry. Moreover, the use of an alternative measure of executive functioning (BRIEF-SR), offered an interesting look to the participants' perception of their own cognitive difficulties, as compared to their objective performance on cognitive tasks. Although participants' selection in the general population has limitations, it also represented strength of the present study. More precisely, we accessed to a group of adolescents with potentially emerging addictive-like eating behaviors accompanied with preclinical psychological symptoms. The majority of this subgroup could represent an understudied category of adolescents who seems to be in a critical period in the development of FA, during which a concerning distress is reported without being associated with important consequences. Internalized symptoms related to FA could also suggest a greater difficulty to detect and manage this condition before the development of more obvious health problems (e.g., overweight and obesity, type 2 diabetes, cardiovascular problems).

Finally, the selection of a healthy control group, according to FA symptoms, also represented strength of the study design.

5. Conclusions

The main objective of the present study was to explore cognitive factors associated to FA symptoms in adolescents, by comparing adolescents with a significant number of symptoms to a control group, on multiple neuropsychological tasks. Even though our results did not show any significant differences or impairments on the key scores of the tasks, outcomes allowed some interesting comments and hypotheses on the potential vulnerability factors of FA in adolescents. Firstly, it highlighted the absence of clear cognitive impairment or difficulties in adolescents with a high level of FA symptoms, as assessed with neuropsychological tasks. Even though some subtle cognitive distinctions have been observed and discussed, we cannot actually conclude that cognitive difficulties are implied in the emergence of FA symptoms. Then, a prominent finding was the distinction between self-reported questionnaires and neuropsychological tasks, in the discrimination of adolescents with and without FA symptoms. Regarding this finding, we suggested that self-reported measures assessing depressive symptoms (or negative affect), impulsivity, and executive functioning difficulties were more sensitive than the computerized tasks to assess the severity of this condition. It suggests that the presence of FA symptoms in adolescents could be accompanied with a more impaired self-reported condition, including internalized symptoms, in a developmental period of major changes and distress. At this stage, the propensity to experiment negative affect and difficulties in regulating them could represent a central vulnerability factor in the development of an addictive-like pattern of eating, and should be a focal point in the future research, the clinical assessment and treatment of this condition. More studies are needed to clarify the neurocognitive mechanisms underlying FA in adolescents.

Author Contributions: Conceptualization, C.R. and C.B.; methodology, C.R. and C.B.; software, C.R. and C.B.; validation, C.R., C.B. and S.I.; formal analysis, C.R.; investigation, C.R.; resources, C.B.; data curation, C.R.; writing—original draft preparation, C.R. and C.B.; writing—review and editing, C.R., C.B. and S.I.; visualization, C.R., C.B. and S.I.; supervision, C.B. and S.I.; project administration, C.R.; funding acquisition, C.B. All authors have read and agreed to the published version of the manuscript.

Funding: This research received no external funding. Mr. Christopher Rodrigue reports a doctoral training grant from the Fonds de recherche du Québec-Santé (FRQS) during his doctoral studies.

Acknowledgments: The authors thank the participants of the study, for their collaboration. This research constitutes part of the Ph.D. thesis of C.R. at the Laval University.

Conflicts of Interest: The authors declare no conflict of interest.

References

1. Corsica, J.A.; Pelchat, M.L. Food addiction: True or false? *Curr. Opin. Gastroenterol.* **2010**, *26*, 165–169. [CrossRef] [PubMed]
2. Gearhardt, A.N.; Corbin, W.R.; Brownell, K.D. Preliminary validation of the Yale Food Addiction Scale. *Appetite* **2009**, *52*, 430–436. [CrossRef] [PubMed]
3. Gearhardt, A.N.; Corbin, W.R.; Brownell, K.D. Food addiction: An examination of the diagnostic criteria for dependence. *J. Addict. Med.* **2009**, *3*, 1–7. [CrossRef] [PubMed]
4. Gearhardt, A.N.; Yokum, S.; Orr, P.T.; Stice, W.R.; Corbin, W.R.; Brownell, K.D. Neural correlates of food addiction. *Arch. Gen. Psychiatry* **2011**, *68*, 808–816. [CrossRef] [PubMed]
5. Davis, C.; Carter, J.C. Compulsive overeating as an addiction disorder. A review of theory and evidence. *Appetite* **2009**, *53*, 1–8. [CrossRef]
6. Gearhardt, A.N.; Corbin, W.R.; Brownell, K.D. Development of the Yale Food Addiction Scale 2.0. *Psychol. Addict. Behav.* **2016**, *30*, 113–121. [CrossRef]
7. Dohle, S.; Diel, K.; Hofmann, W. Executive functions and self-regulation of eating behaviors: A review. *Appetite* **2018**, *124*, 4–9. [CrossRef]
8. Noël, X.; Brevers, D.; Bechara, A. A neurocognitive approach to understanding the neurobiology of addiction. *Curr. Opin. Neurobiol.* **2013**, *23*, 632–638. [CrossRef]

9. Luria, A.R. *The Working Brain: An Introduction to Neuropsychology*; Basic Books: New York, NY, USA, 1973; Basic Books.
10. Norman, D.A.; Shallice, T. Attention to action: Willed and automatic control of behavior. In *Consciousness and Self-Regulation*; Davidson, R.J., Ed.; Plenum Press: New York, NY, USA, 1986; pp. 1–14.
11. Collins, A.; Koechlin, E. Reasoning, learning, and creativity: Frontal lobe function and human decision-making. *PLoS Biol.* **2012**, *10*, 1–16. [CrossRef]
12. Diamond, A. Executive functions. *Annu. Rev. Psychol.* **2013**, *64*, 135–168. [CrossRef]
13. Meule, A.; Lutz, A.; Vögele, C.; Kübler, A. Women with elevated food addiction symptoms show accelerated reactions, but no impaired inhibitory control, in response to pictures of high-calories food-cues. *Eat. Behav.* **2012**, *13*, 423–428. [CrossRef]
14. Frayn, S.; von Ranson, K.M. A sad mood increases attention to unhealthy food images in women with food addiction. *Appetite* **2016**, *100*, 55–63. [CrossRef]
15. Franken, I.H.A.; Nijs, I.M.T.; Toes, A.; van der Veen, F.M. Food addiction is associated with impaired monitoring. *Biol. Psychol.* **2018**, *131*, 49–53. [CrossRef]
16. VanderBroek-Stice, L.; Stojek, M.K.; Beach, S.R.H.; vanDellen, M.R.; Mackillop, J. Multidimensional assessment of impulsivity in relation to obesity and food addiction. *Appetite* **2017**, *112*, 59–68. [CrossRef]
17. Steward, T.; Mestre-Bach, G.; Vintro-Alcaraz, C.; Lozana-Madrid, M.; Aguera, Z.; Fernández-Formoso, J.A.; Granero, R.; Jiménez-Murcia, S.; Vilarrasa, N.; García-Ruiz-de-Gordejuela, A.; et al. Food addiction and impaired executive functions in women with obesity. *Eur. Eat. Disord. Rev.* **2018**, *26*, 574–584. [CrossRef] [PubMed]
18. Rodrigue, C.; Ouellette, A.S.; Lemieux, S.; Tchernof, A.; Biertho, L.; Bégin, C. Executive functioning and psychological symptoms in food addiction: A study among individuals with severe obesity. *Eat. Weight Disord. Stud. Anorex. Bulim. Obes.* **2018**, *23*, 469–478. [CrossRef] [PubMed]
19. Blume, M.; Schmidt, R.; Hilbert, A. Executive functioning in obesity, food addiction, and binge-eating disorder. *Nutrients* **2019**, *11*, 54. [CrossRef] [PubMed]
20. Noël, X. Why adolescents are at risk of misusing alcohol and gambling. *Alcohol Alcohol.* **2014**, *49*, 165–172. [CrossRef]
21. Ahmed, A.Y.; Sayed, A.M.; Mostafa, M.K.; Abdelaziz, E.A. Food addiction relations to depression and anxiety in Egyptian adolescents. *Gaz. Egypt. Paediatr. Assoc.* **2016**, *64*, 149–153. [CrossRef]
22. Gearhardt, A.N.; Roberto, C.A.; Seamans, M.J.; Corbin, W.R.; Brownell, K.D. Preliminary validation of the Yale Food Addiction Scale for children. *Eat. Behav.* **2013**, *14*, 508–512. [CrossRef]
23. Laurent, J.S.; Sibold, J. Addictive-like eating, body mass index, and psychological correlates in a community sample of preadolescents. *J. Pediatric Health Care* **2015**, *30*, 216–223. [CrossRef] [PubMed]
24. Mies, G.W.; Treur, J.L.; Larsen, J.K.; Halberstadt, J.; Pasman, J.A.; Vink, J.M. The prevalence of food addiction in a large sample of adolescents and its association with addictive substances. *Appetite* **2017**, *118*, 97–105. [CrossRef]
25. Zhao, Z.; Yanan, M.; Han, Y.; Liu, Y.; Yang, K.; Zhen, S.; Wen, D. Psychosocial correlates of food addiction and its association with quality of life in a non-clinical adolescent sample. *Nutrients* **2018**, *10*, 837. [CrossRef] [PubMed]
26. Meule, A.; Hermann, T.; Kübler, A. Food addiction in overweight and obese adolescents seeking weight-loss treatment. *Eur. Eat. Disord. Rev.* **2015**, *23*, 193–198. [CrossRef] [PubMed]
27. Schulte, E.M.; Jacques-Tiura, A.J.; Gearhardt, A.N.; Naar, S. Food addiction prevalence and concurrent validity in African American adolescents with obesity. *Psychol. Addict. Behav.* **2017**, *32*, 187–196. [CrossRef]
28. Tompkins, C.L.; Laurent, J.; Brock, D.W. Food addiction: A barrier for effective weight management for obese adolescents. *Child. Obes.* **2017**, *13*, 1–8. [CrossRef]
29. Hauck, C.; Weib, A.; Schulte, E.M.; Meule, A.; Ellrott, T. Prevalence of 'Food Addiction' as Measured with the Yale Food Addiction Scale 2.0 in a representative German sample and its association with sex, age and weight categories. *Obes. Facts* **2017**, *10*, 12–24. [CrossRef]
30. Albayrak, O.; Föcker, M.; Kliewer, J.; Esber, S.; Peters, T.; de Zwaan, M.; Hebebrand, J. Eating-related psychopathology and food addiction in adolescent psychiatric inpatients. *Eur. Eat. Disord. Rev.* **2017**, *25*, 214–220. [CrossRef]

31. Rodrigue, C.; Gearhardt, A.N.; Bégin, C. Food addiction in adolescents: Exploration of psychological symptoms and executive functioning difficulties in a non-clinical sample. *Appetite* **2019**, *141*, 104303. [CrossRef]
32. Hardee, J.E.; Phaneuf, C.; Cope, L.; Zucker, R.; Gearhardt, A.N.; Heitzeg, M. Neural correlates of inhibitory control in youth with symptoms of food addiction. *Appetite* **2020**, *148*, 104578. [CrossRef]
33. Ouellette, A.S.; Rodrigue, C.; Lemieux, S.; Tchernof, A.; Biertho, L.; Bégin, C. Establishing a food addiction diagnosis using the Yale food addiction scale: A closer look at the clinically significant distress/functional impairment criterion. *Appetite* **2018**, *129*, 55–61. [CrossRef] [PubMed]
34. Elgar, F.J.; Roberts, C.; Tudor-Smith, C.; Moore, L. Validity of self-reported height and weight and predictors of bias in adolescents. *J. Adolesc. Health* **2005**, *37*, 371–375. [CrossRef] [PubMed]
35. Foncesa, H.; Silva, A.M.; Matos, M.G.; Esteves, I.; Costa, P.; Guerra, A.; Gomes-Pedro, J. Validity of BMI based on selfreported weight and height in adolescents. *Acta Paediatr.* **2009**, *99*, 83–88. [CrossRef]
36. Who Growth Charts. WHO Growth Charts. Diettitians of Canada Web Site. Available online: https://www.dietitians.ca/Dietitians-Views/Prenatal-and-Infant/WHO-Growth-Charts/WHO-Growth-Charts-Set-1.aspxUpdatedMarch2014 (accessed on 1 May 2018).
37. Brunault, P.; Courtois, R.; Gearhardt, A.N.; Gaillard, P.; Journiac, K.; Cathelain, S.; Réveillère, C.; Ballon, N. Validation of the French version of the DSM-5 Yale Food Addiction Scale (YFAS 2.0) in a nonclinical sample. *Can. J. Psychiatry* **2016**, *62*, 1–12. [CrossRef]
38. American Psychiatric Association. *Diagnostic and Statistical Manual of Mental Disorders*, 5th ed.; American Psychiatric Association: Washington, DC, USA, 2013.
39. Brunault, P.; Gaillard, P.; Ballon, N.; Couet, C.; Isnard, P.; Cook, S.; Delbachian, I.; Reveillere, C.; Courtois, R. Validation de la version française de la Binge Eating Scale: étude de sa structure factorielle, de sa consistance interne et de sa validité de construit en population clinique et non clinique. *L'Encéphale* **2016**, *42*, 426–433. [CrossRef]
40. Gormally, J.; Black, S.; Daston, S.; Rardin, D. The assessment of binge-eating severity among obese persons. *Addict. Behav.* **1982**, *7*, 47–55. [CrossRef]
41. Turgeon, L.; Chartrand, É.; Robaey, P.; Gauthier, A.-K. Qualités psychométriques de la version québécoise du Multidimensional Anxiety Scale for Children (MASC) [Psychometric properties of a French version of the Muldidimentional Aniety Scale for Children (MASC)]. *Rev. Fr. Clin. Comport. Cogn.* **2006**, *11*, 1–8.
42. March, J.S.; Parker, J.D.; Sullivan, K.; Stallings, P.; Conners, C.K. The Multidimensional Anxiety Scale for Children (MASC): Factor structure, reliability, and validity. *J. Am. Acad. Child Adolesc. Psychiatry* **1997**, *36*, 554–565. [CrossRef]
43. Beck, A.T.; Steer, R.A.; Brown, G.K. *Inventaire de Dépression de Beck (BDI-II): Manuel de Notation [BDI-II, Beck Depression Inventory: Manual]*, 2nd ed.; Pearson Canada Assessment: Toronto, ON, Canada, 1998.
44. Beck, A.T.; Steer, R.A.; Brown, G.K. *BDI-II, Beck Depression Inventory: Manual*, 2nd ed.; The Psychological Corporation: San Antonio, TX, USA, 1996.
45. Byrne, B.M.; Baron, P.; Campbell, T.L. The Beck Depression Inventory (French version): Testing for gender-invariant factorial structure for nonclinical adolescents. *J. Adolesc. Res.* **1994**, *9*, 166–179. [CrossRef]
46. Osman, A.; Barrios, F.X.; Gutierrez, P.M.; Williams, J.E.; Balley, J. Psychometric properties of the Beck Depression Inventory-II in nonclinical adolescent samples. *J. Clin. Psychol.* **2008**, *64*, 83–102. [CrossRef]
47. Wu, P.C. Longitudinal measurement invariance of Beck Depression Inventory-II in early adolescents. *Assessment* **2017**, *24*, 337–345. [CrossRef] [PubMed]
48. D'Acremont, M.; van der Linden, M. Adolescent impulsivity: Findings from a community sample. *J. Youth Adolesc.* **2005**, *34*, 427–435. [CrossRef]
49. Whiteside, S.P.; Lynam, D.R. The 5-Factors Model and impulsivity: Using a structural model of personality to understand impulsivity. *Personal. Individ. Differ.* **2001**, *30*, 669–689. [CrossRef]
50. Verbruggen, F.; Logan, G.D. Response inhibition in the stop-signal paradigm. *Trends Cogn. Sci.* **2008**, *12*, 418–424. [CrossRef] [PubMed]
51. Bo, R.; Aker, M.; Billieux, J.; Landro, N.I. Binge drinkers are fast, able to stop—But they fail to adjust. *J. Int. Neuropsychol. Soc.* **2015**, *22*, 38–46. [CrossRef] [PubMed]
52. Roy, A.; Fournet, N.; Roulin, J.-L.; Le Gall, D. *BRIEF: Inventaire D'évaluation Comportementale des Fonctions Executives*; Hogref: Paris, France, 2012.

53. Isquith, P.K.; Gioia, G.A.; Guy, S.C. *Behavior Rating Inventory of Executive Function (BRIEF–SR)*; PAR: Florida, FL, USA, 2008. [CrossRef]
54. Tabachnik, B.G.; Fidell, L.S. *Using Multivariate Statistics*, 5th ed.; Allyn & Bacon: New York, NY, USA, 2007.
55. Cohen, J. A power primer. *Psychol. Bull.* **1992**, *112*, 155–159. [CrossRef]
56. Kittel, R.; Schmidt, R.; Hilbert, A. Executive functions in adolescents with binge-eating disorder and obesity. *Int. J. Eat. Disord.* **2017**, *50*, 933–941. [CrossRef]
57. Conners, C.K. *Conners' Continuous Performance Test II (CPT-II)*; Multi-Health Systems: North Tonwanda, NY, USA, 2004.
58. Demidenko, M.I.; Huntley, E.D.; Martz, E.M.; Keating, D.P. Adolescent health risk behaviors: Convergent, discriminant and predictive validity of self-report and cognitive measures. *J. Youth Adolesc.* **2019**, *48*, 1765–1785. [CrossRef]
59. Barkley, R.A. *Executive Functions: What They Are, How They Work, and Why They Evolved*; Guilford Press: New York, NY, USA, 2012.

nutrients MDPI

Article

The Influence of Response Inhibition Training on Food Consumption and Implicit Attitudes toward Food among Female Restrained Eaters

Noam Weinbach [1,*], Eldad Keha [2,3], Hila Leib [2] and Eyal Kalanthroff [2,*]

1 School of Psychological Sciences, University of Haifa, Haifa 3498838, Israel
2 Department of Psychology, The Hebrew University of Jerusalem, Jerusalem 9190501, Israel; eldadkeha@gmail.com (E.K.); hilaleib326@gmail.com (H.L.)
3 Department of Psychology, Achva Academic College, Arugot 7980400, Israel
* Correspondence: nweinbach@psy.haifa.ac.il (N.W.); eyal.kalanthroff@mail.huji.ac.il (E.K.)

Received: 29 September 2020; Accepted: 20 November 2020; Published: 24 November 2020

Abstract: Restrained eaters display difficulties engaging in self-control in the presence of food. Undergoing cognitive training to form associations between palatable food and response inhibition was found to improve self-control and influence eating behaviors. The present study assessed the impact of two such response inhibition trainings on food consumption, food-related anxiety, and implicit attitudes toward food among female restrained eaters (Dutch Eating Behavior Questionnaire-restrained eating subscale ≥ 2.5). In Experiment 1, 64 restrained eaters completed either one of two training procedures in which they were asked to classify food vs. non-food images: a food-response training, in which stop cues were always associated with non-food images, or a balanced food-response/inhibition training, in which participants inhibited motor actions to food and non-food stimuli equally. The results revealed reduced snack consumption following the food-response/inhibition training compared to the food-response training. The food-response training was associated with increased levels of food-related anxiety. In Experiment 2, the same training procedures were administered to 47 restrained eaters, and implicit attitudes toward palatable foods were assessed. The results revealed an increase in positive implicit attitudes toward palatable foods in the food-response/inhibition group but not in the food-response training group. The results suggest that balancing response inhibition and execution across food and non-food stimuli may reduce overeating while retaining positive attitudes toward food among female restrained eaters.

Keywords: restrained eating; response inhibition; stop-signal task; implicit associations; cognitive training; inhibitory control

1. Introduction

Restrained eaters, or "chronic dieters", are individuals who restrict food intake in an attempt to promote weight loss or avoid gaining weight [1]. Restrained eaters are generally highly motivated to restrict their food intake in order to control body weight. Paradoxically, several studies have shown that restrained eaters consume larger amounts of food compared to unrestrained eaters [2,3]. Additionally, severe restrained eating is a significant risk factor for eating disorders and is associated with increased levels of depression and anxiety [4]. As such, it is important to better understand the underlying factors that affect restrained eating and how to promote more balanced eating patterns.

Exposure to palatable high-calorie foods is one of the common risk factors for overeating among restrained eaters (for review see [5]). Difficulty engaging in self-control while being exposed to palatable foods may be a consequence of a failure to activate neurocognitive abilities, such as response inhibition. Response inhibition is an executive function which allows one to pursue goal-directed behavior by overriding actions or thoughts based on a strong internal predisposition or external lure [6]. Indeed, multiple studies have demonstrated an association between response inhibition and eating behaviors (for review see [7]). Several researchers have suggested that inefficient activation of response inhibition, especially following exposure to food stimuli, may reflect difficulties in self-control and maintain binge eating episodes in individuals with bulimia nervosa, binge eating disorder, and obesity [8–12]. In contrast, superior response inhibition triggered following the presentation of high-calorie foods has been demonstrated in patients with anorexia nervosa [13], a disorder characterized by dangerous weight loss due severe dietary restraint, fear of gaining weight, and body image disturbance [14]. Specifically, over-activation of response inhibition was hypothesized to allow patients with anorexia nervosa to endure prolonged periods of self-starvation [13,15]. Taken together, it seems that imbalanced activation of response inhibition (i.e., inefficient/under-activation or superior/over-activation) may underlie various disordered eating styles (i.e., overeating and restricted eating, respectively).

Restrained eating is not a psychiatric disorder as eating disorders are. However, dietary restraint is a core feature in eating disorders such as bulimia and anorexia nervosa [14]. Thus, studying response inhibition among healthy individuals with restrained eating may shed light on the phenomenon, independently of comorbid psychopathologies and physical complications that are commonly associated with eating disorders. Imbalanced activation of response inhibition was also reported in nonclinical samples of restrained eaters [16–18]. For example, in a previous study, we showed that restrained eaters were better at inhibiting a response following exposure to palatable food images compared to non-food images [18]. However, when being exposed to neutral non-food stimuli, restrained eaters' response inhibition abilities were poorer compared to that of unrestrained eaters [18]. This pattern suggests that a strong activation of response inhibition following exposure to food stimuli may support restrained eaters' goal to reduce food consumption. However, in the long run, due to a general deficit in inhibitory resources, such restriction may lead to a paradoxical breakdown of control over eating behaviors [3]. Again, this pattern strengthens the notion that imbalanced activation of response inhibition abilities may be involved in disordered eating among restrained eaters. Taken together with the evidence reviewed above, it is not surprising that studies found that training response inhibition can directly influence eating behaviors (for reviews see [19,20]).

Computerized training procedures that train response inhibition to food usually involve associating images of palatable foods with stopping by presenting images of food along with task cues that instruct stopping an action. The association formed between palatable foods and stopping was shown to reduce food consumption among healthy individuals and those with obesity [19,20]. Similar training interventions were used with other clinical populations in which a stimulus that commonly triggers unwanted maladaptive behavior (e.g., compulsions in obsessive–compulsive disorder) was associated with response inhibition in order to extinguish compulsive behaviors [21]. Interestingly, associating stop cues with palatable food images influences not only eating but also attitudes toward palatable foods among healthy individuals and those with obesity. For example, in a series of studies, Chen and colleagues have shown that palatable food stimuli are rated as less attractive after a training task that associated palatable food images with stop cues [22–24].

To date, most studies have attempted to reduce food consumption and create negative attitudes toward food among different populations by associating response inhibition with food stimuli. However, it seems that a more therapeutic goal for restrained eaters would be to achieve greater balance between response inhibition and response execution in the presence of food, rather than training them to constantly stop their responses in the presence of palatable foods. In other words, self-control in the presence of food should reflect flexibility between food consumption and restriction—an ability that seems to be lacking in restrained eaters. Improving self-control in the presence of food in such a way may also reduce food-related anxiety and increase positive attitudes toward food among individuals who chronically restrict food intake.

The goal of the present study was to assess the impact of two response inhibition training procedures on food consumption, food-related anxiety, and implicit attitudes toward palatable foods among female restrained eaters. In one training group, restrained eaters completed a behavioral task in which palatable food images were always associated with response execution and non-food images with response inhibition. That is, restrained eaters never had to inhibit their response upon seeing food and always had to inhibit their response when seeing non-food stimuli (i.e., food-response group). In a second training group, the task was modified so that restrained eaters had to inhibit their response to food and non-food images in an equal proportion (i.e., food-response/inhibition group). The primary outcome measures were snack consumption in a bogus taste test, changes in food-related anxiety (Experiment 1) and implicit attitudes toward palatable foods in the food–valence compatibility task (Experiment 2) following the training. We expected that snack consumption will be smaller in the balanced food-response/inhibition training group compared to that in the food-response training group. Additionally, we expect that the food-response/inhibition training will result in reduced food-related anxiety and an increase in positive implicit attitudes toward high-calorie foods.

2. Experiment 1

2.1. Method

2.1.1. Participants

Sixty-eight restrained eaters participated in this experiment in return for a small monetary reward. Because restrained eating as a means to control weight is far more common in females than in males, only female restrained eaters were recruited for the current study. Demographics and characteristics of the sample are shown in Table 1. All participants had normal or corrected-to-normal vision, had no history of attention deficits or dyslexia, and were unaware of the purposes of the experiment. All participants were recruited from an undergraduate university sample. Potential participants were initially screened using an online survey, which included the Restrained Eating subscale of the Dutch Eating Behavior Questionnaire (DEBQ-R) [25]. Participants who scored >2.5 were invited to the lab by email to participate in the study. Body mass index (BMI) lower than 18.5 or greater than 35 was used as an exclusion criterion. Four participants did not complete the bogus taste test, and therefore, their results were not further analyzed (three refused to taste at least one of each snack and one was fasting due to a religious holiday). The final sample included 64 female participants. The participants were randomly assigned to either the food-response training group or the balanced food-response/inhibition training group (Table 1).

Table 1. Characteristics of training and control groups.

		Experiment 1		
Factor	**Food response (n = 32)**	**Food response/inhibition (n = 32)**	t	**p-value**
Age	24.06 (2.4) [19–32]	24.13 (2.2) [19–31]	$t(62) = 0.11$	0.914
BMI	23.85 (3.3) [18.4–30.1]	24.16 (4.1) [15.2–35]	$t(62) = 0.34$	0.736
DEBQ-R	3.07 (2.88) [2.6–3.6]	3.08 (0.3) [2.6–3.6]	$t(62) = 0.13$	0.898
		Experiment 2		
	Food response (n = 23)	**Food response/inhibition (n = 24)**		**p-value**
Age	25.04 (4.1) [20–40]	26.54 (6.6) [20–49]	$t(45) = 0.94$	0.354
BMI	24.08 (3.7) [8.8–34.2]	25.33 (3.5) [20.3–35]	$t(45) = 1.18$	0.243
DEBQ-R	3.81 (0.5) [3.1–4.6]	3.78 (0.5) [3.1–5]	$t(45) = -0.18$	0.859

Mean, (SD), [Range] of sample characteristics. Independent t-test analyses for age, BMI (body mass index), and DEBQ-R (Restrained Eating subscale of the Dutch Eating Behavior Questionnaire reveal no differences between the groups.

2.1.2. Procedure

The study was approved by the ethical committee of the Hebrew University of Jerusalem and followed the American Psychological Association (APA) ethical standards. To reduce differences in a priori states of hunger, participants were asked to refrain from eating or drinking anything but water 3 h prior to the experiment. After signing an informed consent form, participants completed the following steps (a) participants completed self-report measures, including the DEBQ-R and questions regarding their hunger ("How hungry are you at the moment from 1—Not hungry at all to 10—Extremely hungry"), weight, and height, and were asked to assess their current level of food-related anxiety using a visual analog scale (VAS; "How anxious are you right now about issues related to food and eating"). (b) Participants were randomly assigned into one of the two groups: the food-response or the food-response/inhibition training group. Based on the training group, each group completed a different version of the food stop-signal task (F-SST) [18] (further details below). (c) Following the task, participants completed a bogus taste task to measure food consumption. (d) Finally, participants were asked to answer self-report questions, identical to those asked at baseline, regarding their hunger and current level of food-related anxiety (using a VAS).

In order to ensure that participants were unaware of the purpose of the experiment, the taste test and the F-SST were presented as two separate studies. Participants were told that they were recruited for a laboratory taste test but were told that they will have to wait 15 minutes after completing the questionnaires and before the taste test. Then, they were offered to "use this time" to participate in "a different study in our lab" (the F-SST) for additional payment. All participants agreed. Participants received a total of ~10 USD (5 USD for the "original" taste test and another 5 USD for the "additional" computerized task).

2.1.3. Measures

The Restrained Eating Subscale of the Dutch Eating Behavior Questionnaire (DEBQ-R) [25]. The restrained eating subscale includes 10 questions regarding one's tendency to restrict food consumption. Ratings are made on a five-point Likert scale. Participants completed the Hebrew version of the questionnaire. The questionnaire was translated to Hebrew in the following way: Two independent translators translated the questionnaire to Hebrew. Inconsistencies were then discussed. Next, a third translator reverse-translated the questionnaire back to the English to ensure clear understanding of all items. Cronbach's alpha value in the original study was 0.95 [25] and in the current study it was 0.89.

The Food Stop-Signal Task (F-SST; Figure 1). The two versions of the F-SST were administered to associate food/no-food stimuli with either response or stopping behaviors. Each trial in the task started with a black fixation point presented at the center of a white screen for 1000 ms. Next, participants were shown an image of either a food or non-food item in the center of the screen (i.e., a go signal). Participants were instructed to press the "z" key on the keyboard for food stimuli or the "?" key for non-food stimuli as fast as possible. Forty images of food (18 sweet and 22 savory) and 40 non-food images (household items) were selected from the "food pics" database [26]. On a random selection of 25% of the trials, a stop signal (i.e., a blue frame that appeared for 50 ms) was presented after a "stop-signal delay" (SSD) of 300 ms. Participants were instructed to withhold their response upon seeing the stop signal. Each trial ended with a 500 ms inter-trial interval. The task started with 32 practice trials that included feedback on accuracy and response times (RTs). The experimental task included 240 trials.

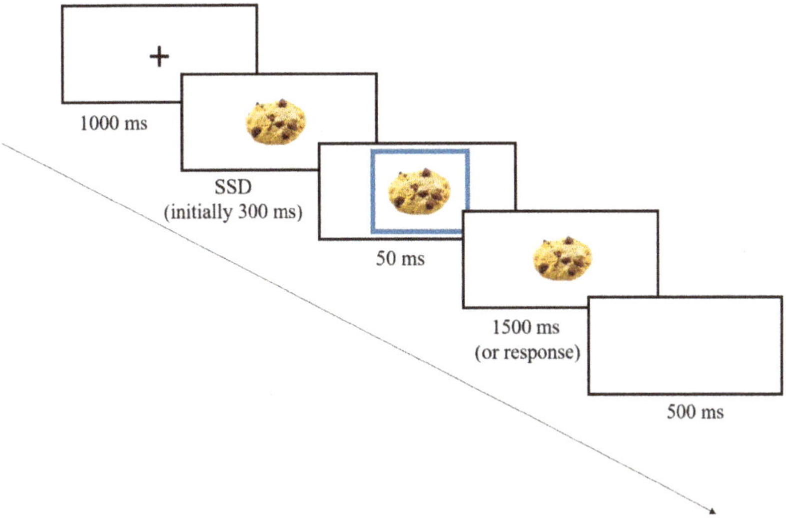

Figure 1. An example of a stop-food trial in the F-SST (food stop-signal task).

In order to create two training groups, the proportion of trials in which a stop signal followed food or non-food stimuli was manipulated. In the food-response training group, all 60 trials that included a stop signal were trials in which the stop signal appeared after non-food images and never after food items. That is, this task always involved executing a behavioral motor response when being exposed to food images. In the food-response/inhibition training group, the stop signals were distributed equally across trials, which included food and non-food images as go-signals (i.e., 30 stop signals on food trials and 30 on non-food trials), creating a balance between response inhibition following exposure to food and non-food items.

The bogus taste test [27] was used to measure food consumption. Three bowls of palatable snacks containing chocolate-covered peanuts (M&Ms; 3.4 kcal each), hazelnut biscuits (Loacker; 17.4 kcal each), and pretzel sticks (5 kcal each) were presented to each participant (all snacks were about the same size). Note that none of these snacks were used in the F-SST task. Participants were asked to "taste the snack from each bowl" (in a consistent order) and were asked to rate the taste of each snack on a scale of 1 to 10. No instruction was given regarding the amount food that the participant needed to taste (tasting at least 1 snack from each of the three bowls was mandatory in order to participate in the study). After each participant, the bowls were collected, and the total amount of snacks eaten was recorded.

2.2. Results

As presented in Table 1, there were no differences between the two groups of restrained eaters in age, BMI, and DEBQ-R. Before assessing differences in food consumption between the groups, we conducted a t-test to assess differences in hunger level between the groups before the training and found no differences in hunger level between the food-response training group (mean = 4.47, SD = 2.23) and the food-response/inhibition training group (mean = 3.97, SD = 1.77) ($t(62) = 0.99$, $p = 0.324$).

To ensure proper task engagement, we calculated two measures for the F-SST task: the nsRT (RT in no-stop-signal trials) and the nsACC (accuracy in no-stop-signal trials). There were no significant differences between the groups in both nsRT (food-response training: mean = 540, food-response/inhibition training: mean = 527; $t(62) = 0.71$, $p = 0.480$) and nsACC (food-response training: mean = 0.96, food-response/inhibition training: mean = 0.95; $t(62) = 0.55$, $p = 0.585$), indicating similar task engagement in both groups, on measures that are not affected by the specific version of the task.

Next, food consumption was calculated for each participant as the sum of snacks tasted from all three bowls. To test our primary hypothesis, independent t-tests were carried out to assess differences between the training groups in food consumption. As was hypothesized, the results showed reduced food consumption in the food-response/inhibition training group compared to the food-response training group. This was evident both in number of snacks eaten (5.53 (SE = 0.39) in the food-response group vs. 4.34 (SE = 0.29) in the food-response/inhibition group; $t(62) = 2.45$, $p = 0.017$, Cohen's d = 0.61; Figure 2A) and in the total amount of kcal consumed (48.88 (SE = 4.23) in the food-response group vs. 37.34 (SE = 3.27) in the food-response/inhibition group; $t(62) = 2.16$, $p = 0.034$, Cohen's d = 0.54).

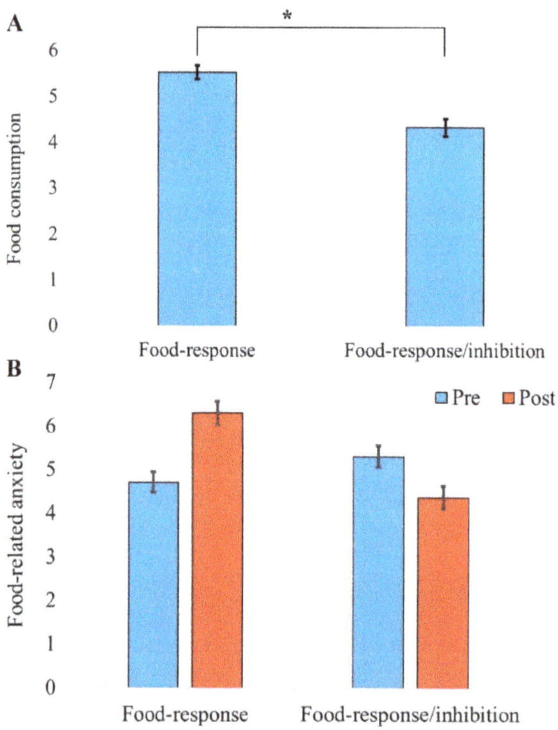

Figure 2. (**A**) Differences in snack consumption between the food-response and the food-response/inhibition training groups. The y-axis represents the total number of snacks eaten from the three bowls. (**B**) Changes in food-related anxiety as a function of training group and time. The y-axis shows the food-related anxiety score on the visual analog scale (VAS). The x-axis represents group. Error bars represent 1 standard error from the mean. * indicates $p < 0.05$.

In order to assess changes in food-related anxiety following the trainings, we conducted a two-way mixed model (ANOVA) on food-related anxiety score on the VAS with time (pre vs. post) as a within-subject factor and group (food-response training vs. food-response/inhibition training) as a between-subject factor. There were no main effects for time ($F(1, 62) = 0.37$, $p = 0.547$) or group ($F(1, 62) = 2.08$, $p = 0.154$). Importantly, the group × time interaction was significant ($F(1, 62) = 5.46$, $p = 0.023$, $\eta_p^2 = 0.08$). Planned comparisons revealed a trend toward increased food-related anxiety from pre- to post-training in the food-response training group ($t(31) = 1.91$, $p = 0.06$, Cohen's d = 0.56; Figure 2B), whilst in the food-response/inhibition group, there was a reduction in food-related anxiety from pre- to post-training, but this did not reach significance level ($t(31) = 1.36$, $p = 0.18$; Figure 2B).

2.3. Discussion Experiment 1

In line with the a priori hypothesis, the results showed reduced food consumption in the food-response/inhibition training group compared to that in the food-response training group. This supports the notion that a response inhibition training aimed to balance response execution and inhibition to food stimuli can influence actual food consumption among female restrained eaters. In addition, differences between the training groups were observed in food-related anxiety. Specifically, a trend showing an increase in food-related anxiety following the food-response training and reduced food-related anxiety following the food-response/inhibition training resulted in a significant interaction between time and training group. That said, the simple effects testing pre- and post-training changes in each group separately did not reach a significance level. Nevertheless, the interaction between time and group showed that food-specific response inhibition training can influence not only eating behaviors but also the emotional response to food. This is important because previous studies have shown that restrained eaters tend to overeat in response to negative emotions such as stress [28]. High-calorie food stimuli, such as those used in the training task, are considered threat-provoking stimuli for restrained eaters because high-calorie foods are associated with weight gain. Thus, the fact that the food-response/inhibition training exposed participants to palatable high-calorie foods yet reduced food-related anxiety is clinically meaningful. While Experiment 1 implies that a balanced food-response/inhibition training can help prevent overeating and change food-related anxiety, it does not inform us regarding the participants' attitudes toward food. Previous studies have shown that following response inhibition trainings, palatable foods are often rated as less attractive [24]. Therapeutically, encouraging negative attitudes toward food among restrained eaters may not be ideal since these individuals already hold more negative attitudes toward food compared to unrestrained eaters [29]. Therefore, Experiment 2 was conducted to assess whether and how a balanced food-response/inhibition training influences restrained eaters' implicit attitudes toward food stimuli.

3. Experiment 2

3.1. Method

3.1.1. Participants

Fifty-three female restrained eaters, who did not participate in Experiment 1, participated in this experiment in return for small monetary compensation (~5 USD). Inclusion and exclusion criteria were identical to Experiment 1. The study was administered online. Six participants were excluded from further analyses due to low task engagement: five due to less than 70% accuracy in the food–valence compatibility task (FVCT; either before or after training) and one due to an extremely high number (> 2.5 SD) of no response trials. The final sample therefore included 47 female participants. The participants were randomly assigned to either the food-response training (n = 23) or the food-response/inhibition training (n = 24) group. Demographic and clinical characteristics of the two groups are presented in Table 1.

3.1.2. Procedure

The study was approved by the Institutional Review Board of the Hebrew University of Jerusalem and followed APA ethical standards. Procedures in Experiment 2 were almost identical to those of Experiment 1. In Experiment 2, the SSD in the food-stop-signal task was initially set to 300 ms and a tracking procedure was then applied (see [30]). The only differences were that participants completed the FVCT (Figure 3) before and after completing the F-SST in order to assess changes in their implicit attitudes toward palatable foods, and that the taste test was not administered.

3.1.3. Measures

The food–valence compatibility task (FVCT; Figure 3) was used to assess implicit attitudes toward food by examining response interference caused by associating palatable food images with positive and negative words. Each trial of the task began with a 1000 ms fixation followed by the target word that was presented for 1000 ms or until response. In each trial, one of four clearly positive words (i.e., excellent, wonderful, great, or pleasurable) or one of four negative words (i.e., gross, disgusting, terrible, or horrifying) was randomly selected and presented at the upper center panel of the screen. Under the target words, prime-pictures were presented. The prime-pictures were randomly selected out of 10 palatable food and 10 non-food pictures. Participants were asked to categorize the words to positive vs. negative-valence words by pressing the "Z" key (with their left hand) or the "?" key (with their right hand), respectively. Instructions emphasized the need to respond to the target word as quickly and as accurately as possible. Prior to the experimental block, a training block was administered. This training block was used in order to train participants to associate a left response with positive words and a right response with negative words. The training block was identical to the experimental block but consisted of the word "good," that appeared in the upper left side of the screen, and the word "bad," that appeared at the upper right side of the screen. The prime-pictures were not presented in the training block, and participants received feedback for response time (RT) and accuracy. The task therefore started with a block of 12 random training trials followed by an experimental block of 80 trials.

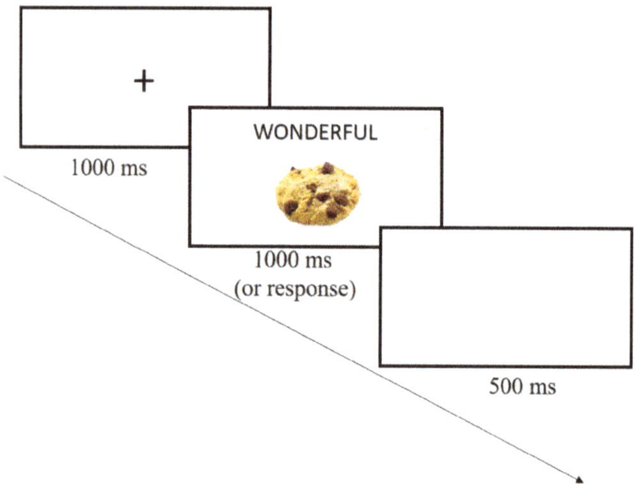

Figure 3. An example of a positive food trial of the food–valence compatibility task (FVCT) used in Experiment 2. Participants are asked to classify pleasant words to a "positive" category and unpleasant words to a "negative" category.

The main dependent measure of the FVCT was the food–valence association effect, which is calculated as RT for non-food trials minus RT for food trials. The food–valence association effect is calculated separately for each valence condition (positive vs. negative). In the negative-valence condition, a larger effect (non-food RT > food RT) indicates that food and negative words are associated and thus suggests more negative attitudes toward palatable food, whilst in the positive-valence condition, a larger effect (non-food RT > food RT) indicates that food and positive words are associated and thus suggests more positive attitudes toward palatable foods.

3.2. Results

As presented in Table 1, there were no differences between the two groups of restrained eaters in age, BMI, and DEBQ-R. Mean reaction time (RT) in the FVCT was calculated for each participant. The food–valence association effect was calculated for each participant in each valence (negative vs. neutral) and time (pre- vs. post-training) condition.

To ensure proper task engagement, we calculated two measures for the F-SST task: the nsRT (RT in no-stop-signal trials) and the nsACC (accuracy in no-stop-signal trials). There were no significant differences between the groups in both nsRT (food-response training: mean = 592, food-response/inhibition training: mean = 583; $t(45) = 0.33$, $p = 0.740$) and nsACC (food-response training: mean = 0.93, food-response/inhibition training: mean = 0.94; $t(45) = 1.40$, $p = 0.166$), indicating similar task engagement in both groups.

In order to validate our measurement of implicit association toward food (food–valence association effect), we tested whether this effect at baseline (prior to practice) correlated with the DEBQ-R scores. To that end, we subtracted the food–valence association effect for the negative valence from the food–valence association effect for the positive valence (i.e., a more positive attitude toward food or a less negative attitude toward food will result in a larger estimate). Results yielded a significant correlation between this estimate and DEBQ-R scores at baseline ($r(46) = -0.31$, $p = 0.035$), indicating that highly restrained eaters will exhibited more negative and less positive attitudes toward food.

A three-way mixed model ANOVA was carried out on the food–valence association effect, with group (food-response vs. food-response/inhibition training) as a between-subject factor, and valence (negative vs. positive) and time (pre-training vs. post-training) as within-subject factors (Figure 4 and Table 2). The results revealed a significant main effect for valence, ($F(1, 45) = 58.57$, $p < 0.001$, $\eta_p^2 = 0.57$), indicating a larger food–valence association effect for the positive compared to negative valence condition. There was also a significant main effect for time ($F(1, 45) = 5.40$, $p = 0.025$, $\eta_p^2 = 0.11$), indicating a larger food–valence association effect pre-training (mean effect = −7.3, SD = 34.3) compared to post-training (mean effect = 6.0, SD = 22.9). The main effect for group was not significant ($F(1, 45) = 2.58$, $p = 0.116$). Importantly, the three-way interaction between group, valence, and time was significant, ($F(1, 45) = 5.12$, $p = 0.029$, $\eta_p^2 = 0.10$). To further investigate this interaction, planned comparisons were carried out to examine the effects of time and group for each valence condition separately. In the positive associations condition, the results showed a significant increase in the food–valence association effect (i.e., an increase in implicit positive attitudes toward food) pre- to post-training in the food-response/inhibition training group ($F(1,44) = 13.052$, $p < 0.001$, $\eta_p^2 = 0.22$), but not in the food-response training group ($F(1,44) = 1.094$, $p = 0.301$, $\eta_p^2 = 0.02$). In the negative association condition, there were no pre to post differences in the food–valence association effect in the food-response/inhibition group ($F(1,44) = 2.257$, $p = 0.14$, $\eta_p^2 = 0.04$), nor in the food-response training group ($F(1,44) = 1.21$, $p = 0.276$, $\eta_p^2 = 0.02$) (Figure 4 and Table 2).

Table 2. Results of Experiment 2—food–valence compatibility task (FVCT).

Factor	Positive Valence				Negative Valence			
	Pre-Training		Post-Training		Pre-Training		Post-Training	
	Non-food	Food	Non-food	Food	Non-food	Food	Non-food	Food
Food response/inhibition	646 (97) [0.91]	623 (93) [0.95]	645 (92) [0.84]	573 (79) [0.96]	614 (96) [0.95]	647 (111) [0.86]	560 (80) [0.97]	612 (98) [0.87]
Food response	621 (89) [0.87]	592 (76) [0.93]	613 (85) [0.87]	569 (78) [0.95]	569 (71) [0.97]	625 (84) [0.87]	558 (67) [0.97]	598 (86) [0.89]

Note: Mean reaction time, (SD), and [accuracy] for the different conditions of Experiment 2.

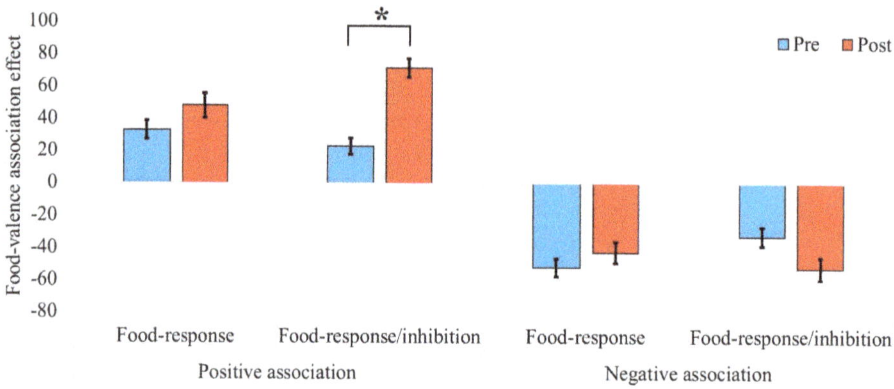

Figure 4. Differences in the food–valence association effect as a function of group, time, and valence. The food–valence association effect was calculated as mean response time (RT) in the non-food condition minus that in the food conditions. On the left panel (positive-valence condition) higher scores indicate greater positive associations whilst on the right panel (negative-valence condition) higher scores indicate greater negative associations. Error bars represent 1 standard error from the mean. * indicates $p < 0.05$.

3.3. *Discussion Experiment 2*

The results of Experiment 2 revealed an increase in positive implicit attitudes toward palatable foods following the food-response/inhibition training, but not following the food-response training. There were no differences between the training groups in negative implicit attitudes toward food. Previous studies have shown that inhibiting motor responses while being exposed to food stimuli leads to a food devaluation effect. Specifically, inhibited food stimuli were rated as less attractive following such training procedures [23,24]. Our results showed that balanced response inhibition training can increase, rather than reduce, positive implicit associations regarding food stimuli among female restrained eaters. This is especially important considering that restrained eaters already hold negative attitudes toward food [29]. Therefore, finding ways to increase positive attitudes toward food among these individuals, while retaining self-control, is clinically important.

4. Discussion

The current study compared two response inhibition trainings on food consumption, food-related anxiety (Experiment 1), and implicit attitudes toward food (Experiment 2) among female restrained eaters. Results yielded that a food-response/inhibition training that balanced the requirement to inhibit and respond to food and non-food stimuli reduced food consumption and increased positive implicit attitudes toward food. On the other hand, a food-response training procedure that encouraged a response to food stimuli and inhibition to non-food stimuli increased food-related anxiety and had no influence on implicit attitudes toward food.

The current study contributes to the existing literature in several important ways. First, response inhibition trainings are currently being investigated as means for influencing food consumption. However, most studies form consistent associations between food and stopping a response [19,20]. Besides reduction of food consumption, such training procedures also elicit negative attitudes toward food among participants [22–24]. In contrast to those studies, the present study showed that a response inhibition training procedure that balances the requirement to stop a response to food and non-food stimuli reduces food consumption but also improves implicit attitudes toward food. This finding is clinically important because therapeutically, we would like to help restrained eaters achieve more flexible eating behaviors while retaining positive attitudes toward food (i.e., not experiencing food as a threat). Second, previous studies have shown that restrained eaters have a general deficit in response inhibition to non-food stimuli [16–18] and exert more inhibitory resources in order to inhibit their response to food stimuli compared to non-restrained eaters [18,31]. It has been postulated that in the long run, the exhaustion of inhibitory resources may subsequently lead to disinhibited eating behaviors such as overeating [3]. These findings suggest that overeating among restrained eaters may be the result of a lack of balance between over-activation and under-activation of inhibitory reactions to food and non-food stimuli. The present study supports this theory by showing that self-control over eating may be achieved by training restrained eaters to balance response inhibition and execution between food and non-food stimuli. Our results indicate that such balancing may improve restrained eaters' self-control in the presence of food.

In a broader perceptive, the current study adds to the mounting evidence suggesting that response inhibition is a modular process that can be trained and result in real-life behavioral changes. Specifically, individuals differ in their ability to use response inhibition as was previously indicated in behavioral and imaging studies [32–34]. Individual differences in response inhibition mean that some individuals experience marked difficulties recruiting inhibitory resources in everyday situations. As noted earlier, general response inhibition failures have been documented among restrained eaters, which may result in difficulty engaging in self-control in the presence of palatable foods [16,18]. Nevertheless, as the current study and others demonstrate, the modular nature of response inhibition allows training an individual to stop an automatic response in the presence of specific environmental cues. Indeed, previous studies have shown that conditioning response inhibition to activate during exposure to specific-environmental cues such as beer, cigarettes, and chocolate in lab-based experiments can subsequently alter how the trained individual behave when presented with these type of cues in real-life situations [35–37].

Increasing the knowledge on how response inhibition interacts with environmental cues and how it can be trained may have implications on various psychological disorders that are characterized by stimulus-driven behaviors or impulse-control problems. In fact, recent studies have provided preliminary reports that such conditioning can be beneficial in augmenting treatments for clinical populations (e.g., [38]). For example, in a recent study, a modified version of the stop-signal task was used to condition automatic inhibition in treatment of refractory patients with obsessive–compulsive disorder [21]. This study demonstrated that associating disorder-specific stimuli with stopping can not only change behaviors but also reduce unwanted intrusive negative cognitions. Similarly, the training procedures used in the current study, not only influenced food intake but also had an effect on food-related anxiety and attitudes toward food. Other studies have also shown that response inhibition can result in attitudinal changes regarding food cues that were associated with stopping a response [22–24]. Taken together, converging evidence suggest that response inhibition trainings can act to regulate emotions, thoughts, and behaviors.

With respect to clinical implications, although restrained eating is not considered a psychiatric disorder, it is associated with elevated levels of depression and anxiety [4]. Moreover, studies have reported that dietary restraint is a proxy for developing eating disorders such as binge eating disorder and bulimia nervosa (for review see [39]). Restrained eating is also associated with weight gain and obesity [40]. Therefore, there is great importance in identifying ways to improve restrained eaters' control over eating while retaining positive attitudes toward food. Response inhibition training such as that tested in the current study may, in the future, show promise as a means to regulate disordered eating patterns among individuals who are at high risk for developing eating disorders. Nevertheless, the field of cognitive training using appetitive food cues is relatively new. There are still many questions that require answers before such training can be offered as clinical interventions. For example, it is still not clear whether response inhibition trainings using appetitive cues have a lasting effect. Additionally, the exact amount of training sufficient for eliciting long-term effects is yet to be determined. Lastly, most response-inhibition trainings using appetitive food cues focus on changing eating behaviors. As such, there is still much to learn regarding the impact of such trainings on emotional and attitudinal factors that can be clinically meaningful and influence one's emotional experience during exposure to various foods. In the future, it would also be interesting to assess such training procedures as potential add-on treatments for eating disorders in which over- and under-activation of response inhibition in the presence of food represent core clinical symptoms of the disorders such as self-starvation in anorexia nervosa and binge eating in bulimia nervosa and binge eating disorder [8,13].

Several limitations of the current study should be addressed. Experiment 1 did not include a baseline measurement of food consumption so that the purpose of the training would not be revealed. However, a lack of baseline food consumption measurement makes it difficult to determine whether the difference found in snack consumption between the training groups is because of a reduction of food intake in the food-response/inhibition group or an increase in food intake in the food-response training group. Future studies should include a baseline measurement of food intake or include a third control group that does not perform any training. Nevertheless, the results showed that there were no baseline differences between the groups in hunger level. A second limitation is that Experiment 2 was run online using a modest sample. Our initial plan was to replicate the results on food consumption and add the implicit attitude measures, but due to COVID-19, we could not conduct a lab-based experiment with a taste test. Thus, future studies will need to replicate the results with a larger sample in order to affirm the beneficial role of the food-response/inhibition task on food consumption and implicit attitudes toward food. Finally, the DEBQ-R does not have a standardized threshold for defining high-restrained eating. Therefore, it could be that other thresholds than that used in the current study would yield different results. However, it is important to note that the same threshold was used in both training groups as the study only tested high-restrained eaters.

To conclude, the current study revealed that response inhibition training that balances the requirement to stop a response to food and non-food stimuli can reduce food consumption and improve positive attitudes toward food among restrained eaters. This study adds to the existing knowledge regarding how eating behaviors can be modulated using cognitive training procedures that target neurocognitive mechanisms suggested to underlie disordered eating. Future studies should investigate the utility of such training procedures as intervention programs with a goal to achieving long-term effects on eating-related thoughts, emotions, and behaviors.

Author Contributions: Conceptualization, E.K. (Eyal Kalanthroff), and N.W; data collection, E.K. (Eldad Keha), and H.L.; formal analysis, E.K. (Eyal Kalanthroff), E.K. (Eldad Keha), and H.L.; writing—original draft preparation, N.W., E.K. (Eyal Kalanthroff), E.K. (Eldad Keha), and H.L.; writing—review and editing, N.W. and E.K. (Eyal Kalanthroff). All authors have read and agreed to the published version of the manuscript.

Funding: This research was funded by the Israel Science Foundation (1341/18, (PI: Eyal Kalanthroff and the National Institute for Psychobiology, Israel (215-17-18b, (PI: Eyal Kalanthroff)).

Acknowledgments: We thank Hadar Naftalovich for her useful input on this article. We thank Ashely Ackerling and Noa Bar for their assistance in data collection.

Conflicts of Interest: The authors declare no conflict of interest.

References

1. Tuschl, R.J. From dietary restraint to binge eating: Some theoretical considerations. *Appetite* **1990**, *14*, 105–109. [CrossRef]
2. Houben, K.; Jansen, A. Lacking skills to improve self-control: Reward-induced loss of inhibitory control and overeating in restrained eaters. *J. Exp. Psychopathol.* **2014**, *5*, 29–37. [CrossRef]
3. Jansen, A.; van den Hout, M. On being led into temptation: "Counterregulation" of dieters after smelling a "preload". *Addict. Behav.* **1991**, *16*, 247–253. [CrossRef]
4. Hagan, K.E.; Forbush, K.T.; Chen, P.Y. Is dietary restraint a unitary or multi-faceted construct? *Psychol. Assess.* **2017**, *29*, 1249–1260. [CrossRef]
5. Polivy, J.; Herman, C.P. Restrained eating and food cues: Recent findings and conclusions. *Curr. Obes. Rep.* **2017**, *6*, 79–85. [CrossRef]
6. Diamond, A. Executive functions. *Annu. Rev. Psychol.* **2013**, *64*, 135–168. [CrossRef]

7. Bartholdy, S.; Dalton, B.; O'Daly, O.G.; Campbell, I.C.; Schmidt, U. A systematic review of the relationship between eating, weight and inhibitory control using the stop signal task. *Neurosci. Biobehav. Rev.* **2016**, *64*, 35–62. [CrossRef]
8. Wu, M.; Hartmann, M.; Skunde, M.; Herzog, W.; Friederich, H.C. Inhibitory control in bulimic-type eating disorders: A systematic review and meta-analysis. *PLoS ONE* **2013**, *8*, e83412. [CrossRef]
9. Houben, K.; Nederkoorn, C.; Jansen, A. Eating on impulse: The relation between overweight and food-specific inhibitory control. *Obesity* **2014**, *22*, 6–8. [CrossRef]
10. Lavagnino, L.; Arnone, D.; Cao, B.; Soares, J.C.; Selvaraj, S.; Brockmeyer, T.; Sinno, M.H.; Skunde, M.; Wu, M.; Woehning, A.; et al. Inhibitory control in obesity and binge eating disorder: A systematic review and meta-analysis of neurocognitive and neuroimaging studies. *Neurosci. Biobehav. Rev.* **2016**, *9*, 299–309. [CrossRef]
11. Svaldi, J.; Naumann, E.; Trentowska, M.; Schmitz, F. General and food-specific inhibitory deficits in binge eating disorder. *Int. J. Eat. Disord.* **2014**, *47*, 534–542. [CrossRef] [PubMed]
12. Smith, K.E.; Mason, T.B.; Schaefer, L.M.; Juarascio, A.; Dvorak, R.; Weinbach, N.; Crosby, R.D.; Wonderlich, S.A. Examining intra-individual variability in food-related inhibitory control and negative affect as predictors of binge eating using ecological momentary assessment. *J. Psychiatr. Res.* **2019**. [CrossRef] [PubMed]
13. Weinbach, N.; Lock, J.; Bohon, C. Superior response inhibition to high-calorie foods in adolescents with anorexia nervosa. *Behav. Res. Ther.* **2020**, *124*, 103441. [CrossRef]
14. American Psychiatric Association. *Diagnostic and Statistical Manual of Mental Disorders*, 5th ed.; American Psychiatric Association: Washington, DC, USA, 2013.
15. Hill, L.; Peck, S.K.; Wierenga, C.E.; Kaye, W.H. Applying neurobiology to the treatment of adults with anorexia nervosa. *J. Eat. Disord.* **2016**, *4*, 31. [CrossRef] [PubMed]
16. Nederkoorn, C.; Van Eijs, Y.; Jansen, A. Restrained eaters act on impulse. *Pers. Individ. Dif.* **2004**, *37*, 1651–1658. [CrossRef]
17. Dong, D.; Lei, X.; Jackson, T.; Wang, Y.; Su, Y.; Chen, H. Altered regional homogeneity and efficient response inhibition in restrained eaters. *Neuroscience* **2014**, *266*, 116–126. [CrossRef]
18. Ganor-Moscovitz, N.; Weinbach, N.; Canetti, L.; Kalanthroff, E. The effect of food-related stimuli on inhibition in high vs. low restrained eaters. *Appetite* **2018**, *131*, 53–58. [CrossRef]
19. Jones, A.; Hardman, C.A.; Lawrence, N.; Field, M. Cognitive training as a potential treatment for overweight and obesity: A critical review of the evidence. *Appetite* **2018**, *124*, 50–67. [CrossRef]
20. Jones, A.; Di Lemma, L.C.G.; Robinson, E.; Christiansen, P.; Nolan, S.; Tudur-Smith, C.; Field, M. Inhibitory control training for appetitive behaviour change: A meta-analytic investigation of mechanisms of action and moderators of effectiveness. *Appetite* **2016**, *97*, 16–28. [CrossRef]
21. Kalanthroff, E.; Steinman, S.A.; Schmidt, A.B.; Campeas, R.; Simpson, H.B. Piloting a personalized computerized inhibitory training program for individuals with obsessive-compulsive disorder. *Psychother. Psychosom.* **2018**, *87*, 52–55. [CrossRef]
22. Chen, Z.; Veling, H.; Dijksterhuis, A.; Holland, R.W. Do impulsive individuals benefit more from food go/no-go training? Testing the role of inhibition capacity in the no-go devaluation effect. *Appetite* **2018**, *124*, 99–110. [CrossRef]
23. Chen, Z.; Veling, H.; De Vries, S.P.; Bijvank, B.O.; Janssen, I.M.C.; Dijksterhuis, A.; Holland, R.W. Go/No-Go training changes food evaluation in both morbidly obese and normal-weight individuals. *J. Consult. Clin. Psychol.* **2018**, *86*, 980–990. [CrossRef]
24. Chen, Z.; Veling, H.; Dijksterhuis, A.; Holland, R.W. How does not responding to appetitive stimuli cause devaluation: Evaluative conditioning or response inhibition? *J. Exp. Psychol. Gen.* **2016**, *145*, 1687–1701. [CrossRef]
25. Van Strien, T.; Frijters, J.E.R.; Bergers, G.P.A.; Defares, P.B. The Dutch Eating Behavior Questionnaire (DEBQ) for assessment of restrained, emotional, and external eating behavior. *Int. J. Eat. Disord.* **1986**, *5*, 295–315. [CrossRef]
26. Blechert, J.; Meule, A.; Busch, N.A.; Ohla, K. Food-pics: An image database for experimental research on eating and appetite. *Front. Psychol.* **2014**, *5*, 617. [CrossRef]
27. Robinson, E.; Haynes, A.; Hardman, C.A.; Kemps, E.; Higgs, S.; Jones, A. The bogus taste test: Validity as a measure of laboratory food intake. *Appetite* **2017**, *116*, 223–231. [CrossRef]

28. Evers, C.; Dingemans, A.; Junghans, A.F.; Boevé, A. Feeling bad or feeling good, does emotion affect your consumption of food? A meta-analysis of the experimental evidence. *Neurosci. Biobehav. Rev.* **2018**, *92*, 195–208. [CrossRef]
29. Papies, E.K.; Stroebe, W.; Aarts, H. Who likes it more? Restrained eaters' implicit attitudes towards food. *Appetite* **2009**, *53*, 279–287. [CrossRef]
30. Verbruggen, F.; Logan, G.D.; Stevens, M.A. STOP-IT: Windows executable software for the stop-signal paradigm. *Behav. Res. Methods* **2008**, *40*, 479–483. [CrossRef]
31. Zhou, Y.; Liu, Y.; Du, J.; Chen, H. Effects of food exposure on food-related inhibitory control in restrained eaters: An ERP study. *Neurosci. Lett.* **2018**, *672*, 130–135. [CrossRef]
32. Forstmann, B.U.; Jahfari, S.; Scholte, H.S.; Wolfensteller, U.; van den Wildenberg, W.P.M.; Ridderinkhof, K.R. Function and structure of the right inferior frontal cortex predict individual differences in response inhibition: A model-based approach. *J. Neurosci.* **2008**, *28*, 9790–9796. [CrossRef]
33. White, C.N.; Congdon, E.; Mumford, J.A.; Karlsgodt, K.H.; Sabb, F.W.; Freimer, N.B.; London, E.D.; Cannon, T.D.; Bilder, R.M.; Poldrack, R.A. Decomposing decision components in the stop-signal task: A model-based approach to individual differences in inhibitory control. *J. Cogn. Neurosci.* **2014**, *26*, 1601–1614. [CrossRef]
34. Congdon, E.; Mumford, J.A.; Cohen, J.R.; Galvan, A.; Aron, A.R.; Xue, G.; Miller, E.; Poldrack, R.A. Engagement of large-scale networks is related to individual differences in inhibitory control. *Neuroimage* **2010**, *53*, 653–663. [CrossRef] [PubMed]
35. Adams, S.; Mokrysz, C.; Attwood, A.S.; Munafò, M.R. Resisting the urge to smoke: Inhibitory control training in cigarette smokers. *R. Soc. Open Sci.* **2017**, *4*, 170045. [CrossRef] [PubMed]
36. Houben, K.; Havermans, R.C.; Nederkoorn, C.; Jansen, A. Beer à no-go: Learning to stop responding to alcohol cues reduces alcohol intake via reduced affective associations rather than increased response inhibition. *Addiction* **2012**, *107*, 1280–1287. [CrossRef]
37. Houben, K.; Jansen, A. Chocolate equals stop. Chocolate-specific inhibition training reduces chocolate intake and go associations with chocolate. *Appetite* **2015**, *87*, 318–323. [CrossRef]
38. Allom, V.; Mullan, B.; Hagger, M. Does inhibitory control training improve health behaviour? A meta-analysis. *Health Psychol. Rev.* **2016**, *10*, 168–186. [CrossRef]
39. Schaumberg, K.; Anderson, D.A.; Anderson, L.M.; Reilly, E.E.; Gorrell, S. Dietary restraint: What's the harm? A review of the relationship between dietary restraint, weight trajectory and the development of eating pathology. *Clin. Obes.* **2016**, *6*, 89–100. [CrossRef]
40. Polivy, J.; Herman, C.P.; Mills, J.S. What is restrained eating and how do we identify it? *Appetite* **2020**, *155*, 104820. [CrossRef]

nutrients

MDPI

Article

The Psycho-Affective Roots of Obesity: Results from a French Study in the General Population

Lena Bourdier [1,*], Melina Fatseas [1,2], Anne-Solène Maria [3,4], Arnaud Carre [5] and Sylvie Berthoz [2,6]

1 Addictology Department, University Hospital of Bordeaux, 33000 Bordeaux, France; melina.fatseas@u-bordeaux.fr
2 CNRS, EPHE, INCIA, UMR 5287, Université de Bordeaux, F-33000 Bordeaux, France; sylvie.berthoz-landron@inserm.fr
3 Private Practice, Stimulus Consulting Ltd., Coastal Road, Cascavelle 90203, Mauritius; annesolenemaria@gmail.com
4 UVSQ, CESP, INSERM, Université Paris-Sud and Université Paris Descartes, 75014 Paris, France
5 LIP/PC2S, Grenoble Alpes University and Savoie Mont Blanc University, F-38000 Grenoble, France; arnaud.carre@univ-smb.fr
6 Department of Psychiatry for Adolescents and Young Adults, Institut Mutualiste Montsouris, 75014 Paris, France
* Correspondence: bourdier.lena@gmail.com

Received: 31 August 2020; Accepted: 24 September 2020; Published: 28 September 2020

Abstract: The aim of the study was to examine the extent to which obese people differ in their emotionally driven and addictive-like eating behaviors from normal-weight and overweight people. A total of 1142 participants were recruited from a general population, by a web-based cross-sectional survey assessing anxiety/depression (Hospital Anxiety and Depression Scale), emotional eating (Emotional Appetite Questionnaire), food addiction (modified Yale Food Addiction Scale), and intuitive eating (Intuitive Eating Scale-2). The statistical design was based on analyses of (co)variance, correlograms, and mediations. A set of Body Mass Index (BMI) group comparisons showed that obese people reported higher levels of depression and emotional eating and that they experienced more severe and frequent food addiction symptoms than overweight and normal-weight people. Associations between anxiety, depression, food addiction symptoms' count, and the difficulties to rely on hunger and satiety cues were found across all weight classes, suggesting that addictive-like eating may represent a unique phenotype of problematic eating behavior that is not synonymous with high BMI or obesity. Conversely, the interrelation between anxiety/depression, emotional eating, and the difficulties to rely on hunger and satiety cues was found only among obese participants, and negative emotional eating mediated the association between depression and anxiety and the difficulties to rely on hunger and satiety cues. This study emphasizes the necessity to develop more comprehensive approaches integrating emotional dysregulation and addictive-like eating behaviors to improve weight management and quality of life of obese people.

Keywords: obesity; food addiction; emotional eating; intuitive eating; depression; anxiety

1. Introduction

Obesity is a multifactorial disease involving an interplay between environmental, genetic, biological, and psychological factors [1]. Among them, both homeostatic dysregulation, which results in poor interoceptive awareness and low sensitivity to the physiological hunger and satiety signals [2–4], and emotional dysregulation [5–7] are increasingly being discussed as possible factors involved in non-nutritional eating. Failures in weight management may be partly explained by an incomplete understanding of the psychological obesity risk and maintaining factors [6].

Indeed, while a decrease or suppression of food intake in response to stress and negative mood has been conceived as the natural, typical, distress response because of physiological changes that mimic satiety [8], it is now acknowledged that important individual differences modulate the way people intake food in the same conditions, with as many as 30 to 50% of people who report eating more during stressful periods [9]. Consistently with seminal descriptions of the psychological aspects of hyperphagia and obesity made in the 1950s by Bruch [10], Hamburger [11], and Stunkard [12], and as conceptualized in the Emotionally Driven Eating Model [6,13,14], some individuals appear to be susceptible to unhealthy shifts towards energy-dense and highly palatable (HP) food items when being emotional [15–17]. Both experimental and epidemiological studies on this issue have consistently identified overweight and obese people as being particularly prone to these shifts, and these findings are part of the conceptual framework of the recently proposed Clinical Obesity Maintenance Model [6].

From a neurobiological perspective, it is now established that the regulation of food intake originates from the orchestration of the activity of neural circuits involved in both somatic and affective (or emotional) homeostatic processes. These links have been viewed as the basis for the development of undercontrolled, nonhomeostatic, and addictive-like consumption of high-energy-dense and HP foods [18,19]. While there exists no consensual definition of what should be considered addictive-like eating patterns [20], the concept of food addiction (FA) has been recently identified as a potential underlying mechanism of overeating and unsuccessful attempts to reduce calorie intake [21,22]. Such an eating behavior triggers the neurobiological cascade associated with the brain reward pathways, in a similar way as the association between stress (either intrinsic or extrinsic) and drug addiction [19,23,24].

No standardized definition of emotionally driven eating behaviors exists, but the concept of Emotional eating (EE) is generally defined as the overconsumption of food in response to negative effects rather than in response to feelings of hunger, which places the individual at risk for overweight and obesity [9,25]. Emotional eating has been viewed as a potential precursor of compulsive overeating and addictive-like eating behaviors [26–28] and accumulating evidence suggests that (i) individuals with high levels of negative affectivity are prone to use food for self-medication purposes and to adopt addictive-like eating behaviors [27–30], and (ii) that psychological distress has differential effects on anthropometric indices (BMI, waist circumference and weight gain) as a function of the level of EE or FA (i.e., that emotionally driven and addictive-like eating act as mediators between low mood and high body weight) [28,31,32]. In addition, a diagnosis of FA, as measured by the Yale Food Addiction Scales (YFAS, mYFAS, YFAS2.0, mYFAS2.0), has been found to be positively associated with depression and EE, and FA and EE are prevalent among high BMI populations [33–36]. However, the extent to which these patterns of association are specific to obesity or concerns all weight classes remains largely unexplored.

Different studies have examined emotionally driven and addictive-like eating behaviors in high BMI populations, but the majority of them either included patients seeking bariatric surgery or they did not clearly differentiate obese people from overweight people [33,34]. Of note, besides the prevalence of FA, the question of whether obese and overweight people differ in the type of FA symptoms they endorse has been overlooked. Nonetheless, we suggest that a better understanding of these issues should help to tailor additional therapeutic options.

Accordingly, the aim of the present study was to examine, in a sample from the general population, the extent to which obese people differ in their emotionally driven and addictive-like eating behaviors not only from normal-weight people but also from overweight people. We expected the obese group would present the highest levels of these behaviors and symptoms. Moreover, we expected to observe stronger positive associations between the level of psychological distress and both the EE score and the FA symptoms score and stronger negative associations between the level of psychological distress and the level of sensitivity to the physiological hunger and satiety signals among the obese group than the other two groups.

2. Materials and Methods

2.1. Participants and Procedures

Participants were recruited from a larger web-based cross-sectional survey data-set on eating behaviors [28]. All participants were adults and engaged freely in the study for no financial compensation. The web survey link was sent to participants using online social media and platforms and via institutional mailing lists. The first page of the online survey included information regarding the purposes of the study and a note about the fundamental principles of ethical scientific research and the French Code of Ethics of Psychologists. Information about anonymity, confidentiality, and data protection was given. In addition, it was explained that all the provided and collected information would only be used to meet the objectives of the research. Participants were then asked to provide their electronic, informed consent prior to their participation in the study. The survey demanded between 25 and 30 min to complete. For the present study, we included participants with a BMI of at least 18.5 kg/m^2 and with no missing data for our variables of interest, reducing the initial sample size from 1349 participants to 1142 participants.

2.2. Ethical Considerations

This study was conducted in accordance with the ethical standards described in the Declaration of Helsinki. The study was approved by the Ethics Committee of the University Savoie Mont Blanc (CEREUS_2016_4).

2.3. Measurements

Self-reported sociodemographic information was collected (age, gender, and level of education). Participants also provided self-reported height and weight to calculate Body Mass Index (BMI) as weight (kg)/height (m)^2. Standard categories of BMI were constituted according to the World Health Organization: 18.5–24.9 (normal-weight), 25–29.9 (overweight), and 30 or more (obesity).

2.3.1. Anxiety and Depression

The Hospital Anxiety and Depression Scale (*HAD*) is a 14-item self-report questionnaire that assesses the level of anxious and depressive symptoms during the past week [37,38]. The HAD includes two subscales: Anxiety (7 items) and Depression (7 items). Participants were asked to rate the extent to which they agreed with each statement on a 4-point scale rating from 0 to 3. In this study, Cronbach's alphas for the *HAD Anxiety* and *Depression* subscales were 0.79 and 0.75, respectively.

2.3.2. Emotional Eating

The Emotional Appetite Questionnaire (*EMAQ*) is a 22-item self-report questionnaire assessing variations of food intake in response to different emotional states and situations [39,40]. The scale contains 9 items assessing negative emotions, 5 items assessing positive emotions, 5 items assessing negative situations, and 3 items assessing positive situations. For each item, participants were asked to rate on a 9-point Likert-type scale whether they ate less (from 1 to 4), the same (5), or more (from 6 to 9) food compared to usual. In the present study, we used the EMAQ global positive score (obtained by averaging the EMAQ-positive emotions and positive situations scores) and the EMAQ global negative score (obtained by averaging negative emotions and negative situations scores). In this sample, Cronbach's alphas were 0.88 for the *EMAQ-Positive* subscale and 0.83 for the *EMAQ-Negative* subscale.

2.3.3. Intuitive Eating

The Intuitive Eating Scale-2 (*IES-2*) is a self-report questionnaire designed to assess attitudes and behaviors towards eating in response to physiological cues [41,42]. The IES-2 encompasses 18 items divided into three subscales: *Eating for Physical rather than Emotional Reasons* (EPR: 8 items), *Reliance on*

Hunger and Satiety Cues (RHSC: 4 items), and *Unconditional Permission to Eat* (UPE: 6 items). Items were answered using a 5-point response format ranging from 1 ("*Strongly disagree*") to 5 ("*Strongly agree*"). In this sample, Cronbach's alpha was 0.90 for the EPR subscale, 0.87 for the RHSC subscale. and 0.70 for the UPE subscale.

2.3.4. Food Addiction

The modified Yale Food Addiction Scale (*mYFAS*) is a short version of the original YFAS [43,44] designed to assess the behavioral indices of addictive-like eating [45]. This 9-item self-report questionnaire was developed for epidemiologic studies. Seven items are based on DSM-IV-TR symptoms of addiction: *Loss of control* (substance taken in larger amount and for a longer period than intended); *Cut down* (persistent desire or repeated unsuccessful attempt to quit); *Time spent* (much time/activity to obtain, use, recover); *Impact activities* (important social, occupational, or recreational activities given up or reduced); *Withdrawal* (characteristic withdrawal symptoms; substance taken to relieve withdrawal); *Despite problems* (use continues despite knowledge of adverse consequences) and *Tolerance* (marked increase in amount; marked decrease in effect). Two additional items (*Clinical distress* and *Clinical impairments*) are used to assess *Clinical significance*. This questionnaire includes five frequency response options that range from 0 ("*Never*") to 4 ("*More than 4 times/week*"). The mYFAS provides two scoring options: a "Symptom Count" scoring option (i.e., a count of food addiction symptoms, ranging from 0 to 7) and a "Diagnostic" scoring option (presence of 3 or more symptoms in addition to the presence of *Clinical significance*) [45]. In addition, a severity score above the cut-off was calculated for each item of the mYFAS. In this sample, Cronbach's alpha for the mYFAS was 0.73.

2.4. Statistical Analyses

Descriptive statistics were computed using means, standard deviations (SD), and ranges for continuous variables, and using counts and percentages for categorical variables. The main effects of BMI groups for age and gender were tested using one-way analysis of variance (ANOVAs) and Chi-square tests (χ^2), respectively. Age differed significantly between the three BMI groups. As is known to affect BMI, EE, and FA [46,47], the main effects of BMI groups and comparisons between pairs of BMI groups for the mood and eating variables were performed using separate analyses of covariance (ANCOVAs) with age as the covariate. Effect sizes were estimated using partial eta-squares ($\eta_p{}^2$) and Cramers' V. Value of $\eta_p{}^2$ around 0.01 was associated with a small effect, value around 0.06 was associated with a medium effect, and value around 0.14 was associated with a large effect [48]. A value of Cramer's V can be interpreted as negligible (0–0.10), weak (0.10–0.20), moderate (0.20–0.30), relatively strong (0.40–0.60), strong (0.60–0.80), or very strong (0.80–1) [49].

To examine if the associations between the mood and eating variables vary by a group of BMI, correlation matrix using Spearman correlation coefficients, and corresponding correlograms were performed in each BMI group separately. A correlogram is a graphical representation of the correlations for all pairs of variables. The color legend of the correlogram shows the correlation coefficients and the corresponding colors [50]. The intensity of the color is proportional to the correlation coefficient (*r*), so strong correlations (i.e., the closest to −1 or 1) are displayed in dark boxes. No significant correlations are displayed in white, positive correlations are displayed in blue and negative correlations are displayed in red.

Finally, based on the finding among the obese group that the level of depression or anxiety, negative emotional eating, and capacity to rely on internal cues to regulate food intake were interrelated, we examined if negative emotional eating (*EMAQ Negative* score) mediated the association between the level of psychological distress (*HAD Depression* or *Anxiety* score) and the reliance on internal cues (*IES-2 Reliance on Hunger and Satiety Cues* score) (see Supplementary Figure S1). We followed the basic steps for mediation analysis [51]:

- Step 1: To show that the predictor was significantly associated with the outcome variable, we estimated the unmediated effects of the *HAD Depression* and *HAD Anxiety* scores on the *IES-2 Reliance on Hunger and Satiety Cues* score (i.e., total effect);
- Step 2: To verify that the predictor was associated with the mediator, we estimated the direct effects of the *HAD Depression* and *HAD Anxiety* scores on the *EMAQ Negative* score (i.e., *a* paths),
- Step 3: To verify that the mediator was associated with the outcome, we estimated the direct effect of the *EMAQ Negative* score on the *IES-2 Reliance on Hunger and Satiety Cues* score (i.e., *b* paths);
- Step 4: To establish that the mediator affects the predictor–outcome relationship, we estimated the direct effects (i.e., *c* paths, adjusted for the mediator) and indirect effects (i.e., $a \times b$ paths) of the *HAD Depression* and *HAD Anxiety* scores on the *IES-2 Reliance on Hunger and Satiety Cues* score.

We used the bootstrapping resampling technique (with a 1000 sample) and reported the estimates (B) and their respective standard errors and confidence intervals as well as the percentage of mediation.

Analyses of variance, covariance, χ^2 tests, and mediation models were performed using Jamovi version 1.1, Jamovi, Sydney, Australia [52]. The correlograms were carried out using R 2.15.2, R Core Team, Vienna, Austria [53]. An alpha of 0.05 was retained as a significant threshold for all statistical tests.

3. Results

3.1. Descriptive Statistics of the Sample

Participant characteristics and scale scores are presented in Table 1. Of the 1142 participants, based on their BMI, 82.1% of them (n = 938) reported being normal-weight (NW), 12.9% of them (n = 147) reported being overweight (OW) and 5% of them (n = 57) reported being obese (OB). Among the obese participants, 63.2% reported moderate obesity (Class 1: BMI of 30 to 34.9), 26.3% reported severe obesity (Class 2: BMI of 35 to 39.9), and 10.5% reported morbid obesity (Class 3: BMI of 40 or higher). The mean ages were 22.7 years (±6.6) for normal-weight participants (75.6% women), 25.3 years (±10.1) for overweight participants (68.7% women), and 28.6 years (±10.3) for obese participants (80.7% women).

Table 1. Descriptive statistics of the sample.

	n	%	
Gender			
Men	286	25	
Women	856	85	
Level of education			
High School degree	26	2.3	
Bachelor's degree	760	66.8	
Master's degree	321	28.2	
Doctorate degree	30	2.6	
mYFAS			
Diagnosis	117	10.2	
	M	*SD*	*Min–Max*
Age	23.4	7.5	18–68
BMI	22.7	3.8	18.5–57.8
HAD			
Anxiety	8.0	3.8	0–19
Depression	4.1	3.2	0–17
EMAQ			
Positive	4.9	0.9	1–8.6
Negative	4.4	1.3	1–8.8

Table 1. *Cont.*

	M	*SD*	*Min–Max*
IES-2			
EPR	3.3	1.1	1–5
RHSC	3.3	0.9	1–5
UPE	3.5	1.0	1–5
mYFAS			
Symptoms Count	1.6	1.4	0–7

M: Mean. SD: Standard Deviation. BMI: Body Mass Index. HAD: Hospital Anxiety and Depression Scale. EMAQ: Emotional Appetite Questionnaire. IES-2: Intuitive Eating Scale 2; EPR: Eating for physical rather than emotional reasons; RHSC: Reliance on Hunger and Satiety Cues; UPE: Unconditional Permission to Eat. mYFAS: modified Yale Food Addiction Scale.

3.2. BMI Group Comparisons

The main effect of gender was not significant (χ^2 (2, n = 1142) = 4.3; p = 0.119). The results showed a main effect for age (F(2,1139) = 23.0; $p < 0.001$; $\eta^2 = 0.04$), and the post-hoc tests (Bonferroni-corrected) highlighted that obese participants were older than the overweight participants, who were themselves older than the normal-weight participants (respectively: OB/OW mean difference = 3.25, SD = 1.2, $p < 0.05$; OB/NW mean difference = 5.86, SD = 1.0, $p < 0.001$; OW/NW mean difference = 2.60, SD = 0.7, $p < 0.001$). In view of this result, all the remaining BMI group comparisons were adjusted for age (ANCOVAs). Table 2 summarizes the BMI group comparisons for the mood and eating behaviors variables.

Concerning mood measures (HAD), there was a main effect for *Depression* ($p < 0.001$) and pairwise comparisons adjusted for age showed that scores were significantly higher among the obese group than the overweight and normal-weight groups (OB/OW mean difference = 1.61, SD = 0.5; OB/NW mean difference = 1.96, SD = 0.4). There was no main effect for *Anxiety* (p = 0.220).

The analyses indicated a main effect for positive emotional eating (*EMAQ Positive*: $p < 0.001$) and pairwise comparisons adjusted for age showed that the obese and overweight participants reported lower scores than the normal-weight participants did (OB/NW mean difference = 0.50, SD = 0.1; OW/NW mean difference = 0.32, SD = 0.8). There was also a main effect for negative emotional eating (*EMAQ Negative; p* < 0.001), and pairwise comparisons adjusted for age indicated that the obese participants reported higher scores than the overweight participants, who themselves reported higher scores than the normal-weight participants (OB/OW mean difference = 0.63, SD = 0.2; OB/NW mean difference = 1.20, SD = 0.2; OW/NW mean difference = 0.58, SD = 0.1).

Regarding intuitive eating (*IES-2*), the main effect of BMI groups emerged for the *Eating for physical rather than emotional reasons* subscale (*EPR*: $p < 0.001$) and the *Reliance on Hunger and Satiety Cues* subscale (*RHSC*: $p < 0.001$). Pairwise comparisons adjusted for age highlighted that the obese participants had lower scores than the overweight participants, who themselves reported lower scores than the normal-weight participants for *EPR* (OB/OW mean difference = 0.36, SD = 0.16; OB/NW mean difference = 0.83, SD = 0.1; OW/NW mean difference = 0.47, SD = 0.9). For *RHSC*, the obese and overweight participants reported lower scores than the normal-weight participants (OB/NW mean difference = 0.69, SD = 0.1; OW/NW mean difference = 0.49, SD = 0.8). There were no significant differences between the BMI groups for the *Unconditional Permission to Eat* subscale (*UPE*: p = 0.445).

Table 2. BMI group comparisons.

	Obese (OB)		Overweight (OW)		Normal Weight (NW)										
n	57		147		938										
%	5.0		12.9		82.1										
											Pairwise comparisons adjusted for age				
Measure	M	SD	M	SD	M	SD	F	df	p	η_p^2	Groups' comparison	F	df	p	η_p^2
HAD															
Anxiety	8.8	4.3	8.1	3.9	7.9	3.7	1.9	2.1138	NS	—	—	—	—	—	—
											OB > OW	8.9	1.201	0.003	0.04
Depression	5.8	4.4	4.2	3.0	3.9	3.1	9.4	2.1138	<0.001	0.02	OB > NW	16.4	1.992	<0.001	0.02
											OW = NW	—	—	—	—
EMAQ															
											OB = OW	—	—	—	—
Positive	4.3	1.2	4.5	1.0	4.8	0.9	14.3	2.1127	<0.001	0.02	OB < NW	11.6	1.982	<0.001	0.01
											OW < NW	13.6	1.1073	<0.001	0.01
											OB > OW	6.8	1.199	0.01	0.03
Negative	5.5	1.6	4.9	1.3	4.3	1.2	34.5	2.1131	<0.001	0.06	OB > NW	49.9	1.986	<0.001	0.05
											OW > NW	29.7	1.1076	<0.001	0.03
IES-2															
											OB < OW	4.5	1.201	0.038	0.02
EPR	2.6	1.1	2.9	1.0	3.4	1.1	28.1	2.1138	<0.001	0.05	OB < NW	35.4	1.992	<0.001	0.03
											OW < NW	27.9	1.1082	<0.001	0.03
											OB = OW	—	—	—	—
RHSC	2.7	1.0	2.9	1.0	3.4	1.0	30.3	2.1138	<0.001	0.05	OB < NW	31.0	1.992	<0.001	0.03
											OW < NW	38.2	1.1082	<0.001	0.03
UPE	3.5	0.9	3.3	0.9	3.5	1.0	2.5	2.1138	NS	—	—	—	—	—	—
mYFAS Symptom Severity															
											OB = OW	—	—	—	—
1—Loss of control	0.3	0.4	0.2	0.4	0.1	0.3	5.7	2.1138	<0.001	0.01	OB > NW	17.1	1.989	<0.001	0.02
											OW = NW	—	—	—	—
											OB = OW	—	—	—	—
2—Cut down	0.2	0.4	0.2	0.4	0.1	0.2	10.9	2.1138	<0.001	0.02	OB > NW	12.9	1.992	<0.001	0.01
											OW > NW	14.6	1.1082	0.001	0.01

Table 2. *Cont.*

Measure	Obese (OB) M	Obese (OB) SD	Overweight (OW) M	Overweight (OW) SD	Normal Weight (NW) M	Normal Weight (NW) SD	F	df	p	η_p^2	Pairwise comparisons adjusted for age: Groups' comparison	F	df	p	η_p^2
n	57		147		938										
%	5.0		12.9		82.1										
3—Time spent	0.4	0.7	0.4	0.7	0.2	0.5	4.1	2.1138	<0.05	0.01	OB = OW	—	—	—	—
											OB=NW	—	—	—	—
											OW > NW	6.4	1.1082	0.012	0.01
4—Impact activities	0.3	0.6	0.1	0.4	0.1	0.4	4.2	2.1138	<0.05	0.01	OB > OW	4.3	1.201	0.04	0.02
											OB > NW	8.1	1.992	0.004	0.01
											OW = NW	—	—	—	—
5—Withdrawal	0.2	0.6	0.1	0.3	0.1	0.3	6.6	2.1138	<0.001	0.01	OB > OW	7.1	1.201	0.008	0.03
											OB > NW	12.9	1.992	0.001	0.01
											OW = NW	—	—	—	—
6—Despite problems	1.4	1.4	1.6	1.4	1.5	1.5	0.4	2.1138	NS	—	—	—	—	—	—
7—Tolerance	1.2	1.5	1.0	1.3	0.7	1.2	7.4	2.1138	<0.001	0.01	OB = OW	—	—	—	—
											OB > NW	9.4	1.990	0.002	0.01
											OW > NW	7.6	1.1081	0.006	0.01
8—Clinical distress	0.6	0.9	0.4	0.7	0.2	0.5	27.3	2.1138	<0.001	0.05	OB > OW	6.8	1.200	0.01	0.03
											OB > NW	45.1	1.992	<0.001	0.04
											OW > NW	18.3	1.1082	<0.001	0.02
9—Clinical impairments	0.4	0.8	0.2	0.5	0.1	0.3	20.5	2.1138	<0.001	0.03	OB > OW	7.7	1.200	0.006	0.04
											OB > NW	40.1	1.992	<0.001	0.04
											OW > NW	6.8	1.1082	0.009	0.01
mYFAS Symptom Count	2.2	1.8	1.8	1.6	1.5	1.4	10.8	2.1138	<0.001	0.02	OB = OW	—	—	—	—
											OB > NW	17.1	1.989	<0.001	0.02
											OW > NW	8.2	1.1079	0.004	0.01

df: degrees of freedom. η_p^2: partial eta-squares. OB: Obese; OW: Overweight; NW: Normal Weight. HAD: Hospital Anxiety and Depression Scale. EMAQ: Emotional Appetite Questionnaire. IES-2: Intuitive Eating Scale 2; EPR: Eating for physical rather than emotional reasons; RHSC: Reliance on Hunger and Satiety Cues; UPE: Unconditional Permission to Eat. mYFAS: modified Yale Food Addiction Scale. NS: not significant.

Concerning the measure of food addiction (*mYFAS*), comparisons were conducted on the symptom count and the symptom severity as well as on the symptom and diagnosis prevalence. The results showed a main effect of BMI groups for the *Symptom Count* ($p < 0.001$) and pairwise comparisons (adjusted for age) highlighted that obese and overweight participants reported higher scores than normal-weight participants (OB/NW mean difference = 0.73, SD = 0.2; OW/NW mean difference = 0.33, SD = 0.1).

Concerning symptom severity, the analyses indicated a main effect for *Loss of control* ($p < 0.001$), *Cut down* ($p < 0.001$), *Time spent* ($p < 0.05$), *Impact activities* ($p < 0.05$), *Withdrawal* ($p < 0.001$), *Tolerance* ($p < 0.001$), *Clinical distress* ($p < 0.001$), and *Clinical impairments* ($p < 0.001$), while the groups did not significantly differ from each other for *Despite problems* ($p = 0.840$). The pairwise comparisons (see Table 2) indicated that the obese participants differed significantly from the normal-weight participants for all the symptoms' severity except for *Time Spent* ($p = 0.078$). In addition, the obese participants differed significantly from the overweight participants for *Impact activities* (OB/OW mean difference = 0.16, SD = 0.7), *Withdrawal* (OB/OW mean difference = 0.17, SD = 0.1), *Clinical distress* (OB/OW mean difference = 0.28, SD = 0.1), and *Clinical impairments* (OB/OW mean difference = 0.25, SD = 0.1), but these two groups did not differ significantly for *Loss of control*, *Cut down*, *Time spent*, and *Tolerance* symptoms' severity.

Regarding symptoms' prevalence (see Figure 1), significant differences between the BMI groups emerged for *Loss of control* (χ^2 (2, n = 1142) = 14.0; $p < 0.001$; Cramers'V = 0.11), with a higher proportion among the obese group than the normal-weight group only. The results also showed a main effect for *Impact activities* (χ^2 (2, n = 1142) = 8.0; $p < 0.05$; Cramers'V = 0.08) and *Withdrawal* (χ^2 (2, n = 1142) = 11.1; $p < 0.005$; Cramers'V = 0.10) with a higher proportion among the obese group than among both the overweight and normal-weight groups. In addition, the results highlighted significant differences between the BMI groups for *Cut down* (χ^2 (2, n = 1142) = 22.9; $p < 0.001$; Cramers'V = 0.14), and *Clinical significance* (χ^2 (2, n = 1142) = 41.1; $p < 0.001$; Cramers'V = 0.19), with a higher proportion among both the obese and overweight groups than the normal-weight group. There was no significant main effect of BMI groups for *Time spent* (χ^2 (2, n = 1142) = 5.6; $p = 0.800$), *Despite problems* (χ^2 (2, n = 1142) = 2.7; $p = 0.259$) and *Tolerance* (χ^2 (2, n = 1142) = 5.7; $p = 0.570$).

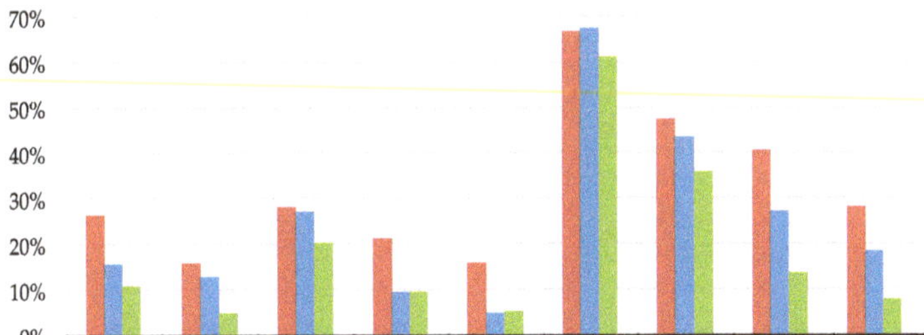

	Loss of Control	Cut Down	Time Spent	Impact Activities	Withdrawal	Despite Problem	Tolerance	Clinical Significance	Diagnostic
Obese	26.3% ^{a}	15.8% ^{a}	28.1% ^{a}	21.1% ^{a}	15.8% ^{a}	66.7% ^{a}	47.4% ^{a}	40.4% ^{a}	28.1% ^{a}
Overweight	15.6% ^{a,b}	12.9% ^{a}	27.2% ^{a}	9.5% ^{b}	4.8% ^{b}	67.3% ^{a}	43.5% ^{a}	27.2% ^{a}	18.4% ^{a}
Normal Weight	10.8% ^{b}	4.8% ^{b}	20.3% ^{a}	9.5% ^{b}	5.3% ^{b}	61.0% ^{a}	35.9% ^{a}	13.6% ^{b}	7.9% ^{b}

Figure 1. Prevalence of modified Yale Food Addiction Scale (*mYFAS*) symptoms and food addiction diagnosis by BMI group. For each pair of BMI groups, the proportions are compared using a z-test. If a pair of values is significantly different, the values have different subscript letters assigned to them.

Finally, the analyses indicated a main effect of BMI groups for the mYFAS *Diagnosis* prevalence (χ^2 (2, n = 1142) = 35.9; $p < 0.001$; Cramers'V = 0.18) and binary logistic regressions showed that relative to the normal-weight group, the odds ratio of meeting the FA diagnosis was 4.56 (95% CI (2.44–8.51),

$p < 0.001$; Cramers'V = 0.16) for the obese and 2.63 (95% CI (1.62–4.25), $p < 0.001$; Cramers'V = 0.12) for the overweight participants.

3.3. Correlograms

Figure 2 presents the correlograms of the correlation matrix between the variables of interest for the obese, overweight, and normal-weight groups separately. The results showed that *HAD Anxiety* and *Depression* scores were significantly positively correlated with the majority of mYFAS symptoms severity among all BMI groups. However, mood and symptom severity scores were more strongly correlated among the obese group than among the other two groups, particularly for *Loss of control* (i.e., *mYFAS_1*) and *Clinical impairments* (i.e., *YFAS_9*), with coefficient values around 0.5 for *HAD Anxiety* and 0.6 for *HAD Depression*.

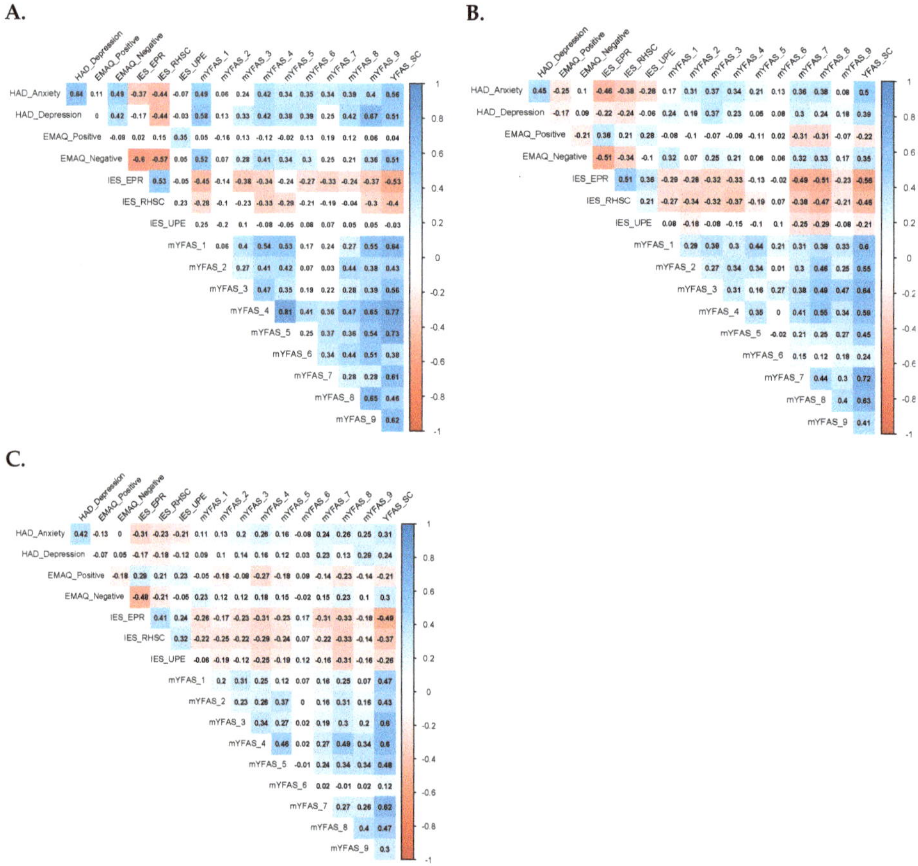

Figure 2. Correlograms for each BMI group. (**A**), Correlogram for Obese group (n = 57). (**B**), Correlogram for Overweight group (n = 147). (**C**), Correlogram for Normal Weight group (n = 938). Positive correlations are displayed in blue and negative correlations are displayed in red. The darkness of the color is proportional to the correlation coefficient, such that the strong correlations (i.e., the closest to −1 or 1) are represented in dark boxes. Nonsignificant correlations are displayed in white. HAD Anxiety: Hospital Anxiety and Depression Scale—Anxiety subscale. HAD Depression: Hospital Anxiety and Depression Scale—Depression subscale. EMAQ Positive: Emotional Appetite Questionnaire Positive subscale. EMAQ Negative: Emotional Appetite Questionnaire Negative subscale.

IES-2: Intuitive Eating Scale 2; -EPR: Eating for physical rather than emotional reasons; -RHSC: Reliance on Hunger and Satiety Cues; -UPE: Unconditional Permission to Eat. mYFAS_1: Loss of control; mYFAS_2: Cut down; mYFAS_3: Time spent; mYFAS_4: Impact activities; mYFAS_5: Withdrawal; mYFAS_6: Despite problems; mYFAS_7: Tolerance; mYFAS_8: Clinical distress; mYFAS_9: Clinical impairments.

Unlike the findings among the overweight or normal-weight groups, *EMAQ Negative* scores were significantly positively correlated with the *HAD Anxiety* ($r = 0.49$) and *HAD Depression* scores ($r = 0.42$) among the obese group. Regarding intuitive eating (IES-2), the *Reliance on Hunger and Satiety Cues (RHSC)* subscale scores were significantly and negatively correlated with *HAD Anxiety*, *HAD Depression*, and *EMAQ Negative* scores among all BMI groups, but the correlation values were the highest among the obese group (*IES-2 RHSC* and *HAD Anxiety* or *HAD Depression*: $r = -0.44$; *IES-2RHSC* and *EMAQ Negative*: $r = -0.57$).

3.4. Mediation Analyses

Based on these results, we tested if negative emotional eating (*EMAQ Negative* scores) in the obese group mediated the observed positive association between psychological distress (*HAD Anxiety* and *Depression* scores) and the lack of reliance on internal cues to regulate food intake (*IES-2 RHSC* scores). For each mediation model, path estimates, indirect and total effect estimates, as well as the percentage of mediation, are presented in Table 3.

For both *HAD* subscales, high scores were associated with low *IES-2 RHSC* scores (Model 1: *HAD Dep → IES-2 RHSC*; Model 2: *HAD Anx → IES-2 RHSC*) and high *EMAQ Negative* scores predicted low *IES-2 RHSC* scores independently from *HAD* scores (Model 1: *EMAQ Neg → IES-2 RHSC*; Model 2: *EMAQ Neg → IES-2 RHSC*). Moreover, for both models, the indirect effects were significant (Model 1: *HAD Dep → EMAQ Neg → IES-2 RHSC*; Model 2: *HAD Anx → EMAQ Neg → IES-2 RHSC*), indicating that for both models *EMAQ Negative* scores did act as mediators in the association between high *HAD Depression* or *Anxiety* scores and low *IES-2 RHSC* scores. The proportion of the total effect explained by the indirect effect was 47.6% and 54.7% for Models 1 and 2, respectively.

Table 3. Direct, indirect, and total effects of the two mediation models among the obese group.

Models Tested	% Mediation	B	SE	p	95% CI Lower	95% CI Upper
Model 1: HAD Dep → EMAQ Neg → IES-2 RHSC						
Direct effects	52.4					
Path a: HAD Dep → EMAQ Neg		0.138	0.05	0.004	0.035	0.225
Path b: EMAQ Neg → IES-2 RHSC		−0.309	0.07	<0.001	−0.448	−0.165
Path c: HAD Dep → IES-2 RHSC		−0.047	0.03	0.109	−0.100	0.013
Indirect effect (a X b)	47.6					
HAD Dep → EMAQ Neg → IES-2 RHSC		−0.043	0.018	0.020	−0.082	−0.0096
Total effect (c + a X b)	100					
HAD Dep → IES-2 RHSC + HAD Dep → EMAQ Neg → IES-2 RHSC		−0.089	0.027	<0.001	−0.133	−0.029
Model 2: HAD Anx → EMAQ Neg → IES-2 RHSC						
Direct effects	45.3					
Path a: HAD Anx → EMAQ Neg		0.171	0.04	<0.001	0.085	0.2446
Path b: EMAQ Neg → IES-2 RHSC		−0.303	0.08	<0.001	−0.453	−0.1446
Path c: HAD Anx → IES-2 RHSC		−0.043	0.027	0.116	−0.0966	0.0112
Indirect effect (a X b)	54.7					
HAD Anx → EMAQ Neg → IES-2 RHSC		−0.052	0.018	0.004	−0.089	−0.018
Total effect (c + a X b)	100					
HAD Anx → IES-2 RHSC + HAD Anx → EMAQ Neg → IES-2 RHSC		−0.094	0.023	<0.001	−0.138	−0.047

B: Standardized estimate. SE: Standard Error. 95% CI: 95%Confidence Interval. See supplementary Figure S1 for an illustration of Paths a, b, and c as well as the indirect and total effects.

4. Discussion

We examined the extent to which obese people differ in their emotionally driven and addictive-like eating behaviors not only from normal-weight but also overweight people in a sample from the French general population. We confirmed previous findings that have been reported in high BMI population, by showing that the two high BMI groups reported higher levels of depressed mood, eating less intuitively but more in response to their negative emotions, and that they presented more severe and/or frequent symptoms of addictive-like eating behaviors than normal-weight people [34,54–56]. In addition, we found an increase in FA diagnosis prevalence (as defined by the mYFAS), with the odds for presenting the condition being more than four times higher among the obese group and more than two times higher among the overweight group than among the normal-weight people. The prevalence of FA diagnosis in the obese participants was comparable to the prevalence of FA diagnosis reported in studies using the longer version of the scale (i.e., the YFAS: 15–25% [57]). In all BMI groups, the most often endorsed symptom by the participants was «*Use despite aversive emotional/physical problem*», with comparable high prevalence in the three groups (on average 65%). Although this symptom is commonly reported [33,57], this high rate among the normal-weight group was unexpected as it is much closer to the rates described in clinical samples e.g., bariatric surgery candidates, binge eating disorder: 40-75% [58,59]) than in community samples (9-23% [47,59,60]) using the YFAS and YFAS 2.0.

Further, we found an increased frequency of the *Loss of control* and *Inability to Cut Down* symptoms by weight classes, but with comparable prevalence between the Obese and Overweight participants. They are both core components and characteristic behavioral features of addiction that have been critically incriminated in the « downwardly escalating dimension » along the continuum of overeating in C. Davis' psychobiological model of eating behaviors [27]. Interestingly, the same pattern of association between indicators of anxiety or depression, FA, and a lack of intuitive eating was found across all weight classes, suggesting that addictive-like eating may represent a unique phenotype of problematic eating behavior that is not synonymous with BMI and obesity, including a complex pattern of interaction between psychological distress, emotion regulation and addictive process. Such findings suggest that individuals prone to FA may turn to excessive food consumption as a coping strategy for heightened emotional distress, similar to individuals with a substance use disorder [23].

Moreover, besides these findings, we believe the present study also adds to the field by providing a more fine-grained distinction between Obese and Overweight people and highlighting individual characteristics that appeared more specifically associated with the obese phenotype. Indeed, Obese participants reported more severe depressive symptoms than the Overweight participants, which is in line with the well-known depression-obesity association and co-occurrence [54]. Combined with the fact that Obese individuals also reported eating even more than the Overweight participants when facing negative emotions or situations, our study further supports the suggestion of a bidirectional link between obesity and depression, more particularly, with the atypical depression subtype [54,61,62]. Emotional eating has been shown to be (i) exacerbated in obese women, (ii) associated with both consumption of highly palatable food and weight gain [9,55] and (iii) it is a negative factor for post-bariatric surgery weight management outcomes [63]. Moreover, an emerging line of evidence points out that negative EE acts as a mediator between depression and obesity and that it may be a marker of atypical depression [28,31,32]. Here, we found a mediation effect of negative EE on the association between psychological distress (for both depression and anxiety) and the difficulties to rely on hunger and satiety cues, difficulties that are, in turn, known to place the person at risk for increased weight [56]. The present data, thus, complement these observations and suggest that obese individuals get caught in a downward spiral and vicious circle leading to an 'interoceptive blindness' due to a specific interplay between their negative affect and their eating patterns. Of important note, it seems this dynamic is not so much an issue of the perceived intensity of the negative affective states as an issue of the obese individual's negative emotional experience per se, because the Obese group admittedly reported higher levels of depressed mood, but similar levels of anxiety, than the other two groups. Our results are in line with previous studies in non-clinical [7] and clinical samples with obesity or eating disorders [36,64] and point out the role of emotion regulation on eating behavior across different weight classes. While the present findings suggest higher alterations in emotional regulation among individuals with obesity, our study also highlights the role of EE in depression and altered interoception of satiety signals, that is a well-known crucial component for regulating food intake. Our study adds a piece of knowledge on this topic, by showing that individuals with obesity could be more vulnerable to such effects, and offers interesting perspectives for improving intervention approaches aimed at reducing compulsive eating behaviors and body weight. These results also seem to support the *Emotionally Driven Eating Model* [65] considering alterations in emotional regulation and cognitive processing as a key mechanism of inappropriate eating behaviors and overeating. Further studies should address in daily life emotion trajectories, emotional regulation strategies, satiety signals and eating behaviors using Ecological Momentary Assessment to confirm the real time temporal dynamics and relationships between these variables among obese patients.

Further, in addition to replicating the observed association between FA, EE and depression, the present study is, to the best of our knowledge, the first one to statistically compare if the prevalence and severity of FA symptoms vary across high BMI classes. Besides the finding that Obese participants reported more severe levels of *Clinical distress* and *Impairments* than the Overweight participants, *Impact Activities* and *Withdrawal* were found to distinguish these two groups as well. In the mYFAS, the wording of the symptom *Impact activities* clearly refers to the negative emotional experience associated with the overconsumption (i.e., « *I have spent time dealing with negative feelings from overeating certain food*») and the fact that it is frequently endorsed by the obese group is consistent with their high levels of depression. This symptom may be related to ruminative thinking, which is a cognitive process that has been associated with the severity of eating disorders symptomatology in both clinical and non-clinical populations [66] and may lead to EE [67]. Moreover, ruminative thinking has been found to impair cognitive flexibility and decision making, which are processes that have been found to be impaired in obese individuals [6,68]. Additional studies are needed to confirm our suggestion and provide further arguments for incorporating anti-rumination therapy for people with comorbid obesity and depression.

The prevalence of *Withdrawal* symptom was three times higher in the Obese group than the Overweight group. Although the suggestion that withdrawal syndromes occur to certain food items has been subject to heavy criticism in the early days of the FA construct, a growing line of experimental evidence has emerged in animal and human studies, and showing notably psychological signs of withdrawal in humans [69]. The mYFAS was based on DSM-IV-TR criteria of the SUD, so it does not evaluate *Craving*, a symptom that is tightly associated with *Withdrawal*. Therefore, we could not ascertain if its absence biased the results. Nonetheless, the frequency of withdrawal symptom endorsement remains high in obese people even when items on *Craving* are considered using the DMS-5 version of the scale (i.e YFAS 2.0 [47,59]). To gain knowledge on this issue, a recently developed self-report, the Highly Processed Withdrawal Food Scales [70], might prove beneficial in future research.

Although the current study provides important information about emotionally-driven and addictive-like eating behaviors by weight class, some limitations should be considered. First, researchers should know that women are more prone than men to (i) show symptoms of psychological distress, (ii) report EE, and (iii) to be affected by obesity [15,61]. Therefore, the number of women in our sample could have influenced our results. Another limitation concerns the use of self-reports that raises the question of the ability for introspection, the gap between the participant's perceptions and realities, or the social desirability bias in the areas of weight and eating behaviors. Furthermore, although some authors highlighted the role of the nutritional and/or chemical composition of HP food in emotionally-driven and addictive-like eating behaviors [71], the type of food consumed was not considered in this study. Finally, personal and psychiatric risk factors for EE, FA and obesity, such as traumatic experiences/PTSD or binge eating disorder [58,72], were not assessed in the study, and these factors may have affected the findings.

Despite these limitations, the present study has important clinical implications. The hypothesis that a distinct mechanism drives excessive weight gain among obese individuals involving EE, psychological distress, and intuitive eating points to the need for specific and integrated interventions in this population. In view of the high level of clinically significant impairments and distress of FA among obese participants, assessment of symptoms and/or diagnosis of food addiction should be systematically considered in this population. A more comprehensive approach integrating emotional dysregulation and addictive-like eating behaviors could improve weight management and quality of life. The key role of EE in this group highlights the need to promote emotion regulation skills in the treatment of obesity. The efficacy of such interventions should be further investigated in randomized controlled trials.

This study confirms a complex pattern of interaction between psychological distress, emotion regulation and addictive process. Such findings suggest that individuals prone to FA may turn to excessive food consumption as a coping strategy to relieve negative affects, similar to individuals with a substance use disorder. More importantly, this study showed that for the obese individuals emotional eating plays a mediation effect between psychological distress and the difficulties to rely on hunger and satiety cues. This emphasizes the role of emotional dysregulation in obesity risk and addiction vulnerability with a potential significant impact on the perception of satiety signals. In summary, this study highlighted the central role of emotional eating and negative affectivity in the maintenance of non-homeostatic eating behaviors among obese individuals. By showing a specific pathway between psychological distress, emotional eating, and a lack of intuitive eating in obese people, our findings support the hypothesis of a distinct mechanism buffering weight management in this population. It also paves the way for designing interventions that aim to reduce compulsive eating behaviors or body weight in this population. In view of the food addiction prevalence and symptoms' severity among the obese people, this study suggests that therapeutic approaches of addictive disorders should be proposed in the presence of FA. To progress in this domain, Ecological Momentary Assessments and mobile applications could offer a paradigm shift, first in the way ecologically valid data can be collected in daily life, and then, in turn, in the way personalized care could be offered depending on the individual's needs.

Supplementary Materials: The following are available online at http://www.mdpi.com/2072-6643/12/10/2962/s1, Figure S1: Mediation analyses path diagram.

Author Contributions: Conceptualization, L.B., M.F., and S.B.; formal analysis, L.B., A.-S.M., and S.B.; methodology, L.B. and S.B.; writing—original draft preparation, L.B., M.F., A.-S.M., A.C., and S.B. All authors significantly participated in interpreting the results and revising the manuscript.

Funding: This research received no external funding.

Acknowledgments: We thank Cecile Dantzer for her help in setting up the online survey and collecting data.

Conflicts of Interest: The authors declare no conflict of interest.

References

1. Hruby, A.; Hu, F.B. The Epidemiology of Obesity: A Big Picture. *Pharmacoeconomics* **2015**, *33*, 673–689. [CrossRef] [PubMed]
2. Herbert, C.; Pollatos, O. Attenuated interoceptive sensitivity in overweight and obese individuals. *Eat. Behav.* **2014**, *15*, 445–448. [CrossRef] [PubMed]
3. Herman, C.P.; Polivy, J. A boundary model for the regulation of eating. *Res. Publ. Assoc. Res. Nerv. Ment. Dis.* **1984**, *62*, 141–156. [CrossRef] [PubMed]
4. Kaplan, H.I.; Kaplan, H.S. The psychosomatic concept of obesity. *J. Nerv. Ment. Dis.* **1957**, *125*, 181–201. [CrossRef]
5. Bruch, H. *Eating Disorders: Obesity, Anorexia Nervosa, and the Person Within*; Basic Books: New York, NY, USA, 1973; ISBN 978-0-465-01782-9.

6. Raman, J.; Smith, E.; Hay, P. The Clinical Obesity Maintenance Model: An Integration of Psychological Constructs including Mood, Emotional Regulation, Disordered Overeating, Habitual Cluster Behaviours, Health Literacy and Cognitive Function. *J. Obes.* **2013**, *2013*, 1–9. [CrossRef]
7. Casagrande, M.; Boncompagni, I.; Forte, G.; Guarino, A.; Favieri, F. Emotion and overeating behavior: Effects of alexithymia and emotional regulation on overweight and obesity. *Eat. Weight. Disord. Stud. Anorexia Bulim. Obes.* **2020**, *25*, 1333–1345. [CrossRef]
8. Gold, P.W.; Chrousos, G.P. Organization of the stress system and its dysregulation in melancholic and atypical depression: High vs low CRH/NE states. *Mol. Psychiatry* **2002**, *7*, 254–275. [CrossRef]
9. Gibson, E.L. The psychobiology of comfort eating: Implications for neuropharmacological interventions. *Behav. Pharmacol.* **2012**, *23*, 442–460. [CrossRef]
10. Bruch, H. Psychological Aspects of Overeating and Obesity. *J. Psychosom. Res.* **1964**, *5*, 269–274. [CrossRef]
11. Hamburger, W.W. Psychological Aspects of Obesity. *Bull. N. Y. Acad. Med.* **1957**, *33*, 771–782.
12. Stunkard, A.; McLaren-Hume, M. The results of treatment for obesity: A review of the literature and report of a series. *AMA Arch. Intern. Med.* **1959**, *103*, 79–85. [CrossRef] [PubMed]
13. Macht, M. How emotions affect eating: A five-way model. *Appetite* **2008**, *50*, 1–11. [CrossRef] [PubMed]
14. Anversa, R.G.; Campbell, E.J.; Ch'Ng, S.S.; Gogos, A.; Lawrence, A.J.; Brown, R. A model of emotional stress-induced binge eating in female mice with no history of food restriction. *Genes Brain Behav.* **2019**, *19*, e12613. [CrossRef] [PubMed]
15. Camilleri, G.M.; Méjean, C.; Kesse-Guyot, E.; Andreeva, V.A.; Bellisle, F.; Hercberg, S.; Péneau, S. The Associations between Emotional Eating and Consumption of Energy-Dense Snack Foods Are Modified by Sex and Depressive Symptomatology. *J. Nutr.* **2014**, *144*, 1264–1273. [CrossRef]
16. Oliver, G.; Wardle, J.; Gibson, E.L. Stress and Food Choice: A Laboratory Study. *Psychosom. Med.* **2000**, *62*, 853–865. [CrossRef]
17. Nguyen-Rodriguez, S.T.; Unger, J.B.; Spruijt-Metz, D. Dietary correlates of emotional eating in adolescence. *Appetite* **2007**, *49*, 494–499. [CrossRef]
18. Bellisle, F.; Drewnowski, A.; Anderson, G.H.; Westerterp-Plantenga, M.; Martin, C.K. Sweetness, Satiation, and Satiety. *J. Nutr.* **2012**, *142*, 1149S–1154S. [CrossRef]
19. Volkow, N.D.; Wang, G.-J.; Tomasi, D.; Baler, R.D. Obesity and addiction: Neurobiological overlaps. *Obes. Rev. Off. J. Int. Assoc. Study Obes.* **2013**, *14*, 2–18. [CrossRef]
20. Paterson, C.; Lacroix, E.; Von Ranson, K.M. Conceptualizing addictive-like eating: A qualitative analysis. *Appetite* **2019**, *141*, 104326. [CrossRef]
21. Lerma-Cabrera, J.M.; Carvajal, F.; Lopez-Legarrea, P. Food addiction as a new piece of the obesity framework. *Nutr. J.* **2015**, *15*, 5. [CrossRef]
22. Davis, C.; Curtis, C.; Levitan, R.D.; Carter, J.C.; Kaplan, A.S.; Kennedy, J.L. Evidence that 'food addiction' is a valid phenotype of obesity. *Appetite* **2011**, *57*, 711–717. [CrossRef] [PubMed]
23. Sinha, R. Role of addiction and stress neurobiology on food intake and obesity. *Biol. Psychol.* **2017**, *131*, 5–13. [CrossRef]
24. Sinha, R.; Jastreboff, A.M. Stress as a common risk factor for obesity and addiction. *Biol. Psychiatry* **2013**, *73*, 827–835. [CrossRef]
25. Frayn, M.; Knäuper, B. Emotional Eating and Weight in Adults: A Review. *Curr. Psychol.* **2017**, *37*, 924–933. [CrossRef]
26. Piccinni, A.; Marazziti, D.; Vanelli, F.; Franceschini, C.; Baroni, S.; Costanzo, D.; Cremone, I.M.; Veltri, A.; Dell'Osso, L. Food addiction spectrum: A theoretical model from normality to eating and overeating disorders. *Curr. Med. Chem.* **2015**, *22*, 1631–1638. [CrossRef]
27. Davis, C. From Passive Overeating to "Food Addiction": A Spectrum of Compulsion and Severity. *ISRN Obes.* **2013**, *2013*, 1–20. [CrossRef] [PubMed]
28. Bourdier, L.; Orri, M.; Carre, A.; Gearhardt, A.; Romo, L.; Dantzer, C.; Berthoz, S. Are emotionally driven and addictive-like eating behaviors the missing links between psychological distress and greater body weight? *Appetite* **2018**, *120*, 536–546. [CrossRef] [PubMed]
29. Parylak, S.L.; Koob, G.F.; Zorrilla, E.P. The dark side of food addiction. *Physiol. Behav.* **2011**, *104*, 149–156. [CrossRef]
30. Singh, M. Mood, food, and obesity. *Front. Psychol.* **2014**, *5*, 5. [CrossRef]

31. Konttinen, H.; Männistö, S.; Sarlio-Lähteenkorva, S.; Silventoinen, K.; Haukkala, A. Emotional eating, depressive symptoms and self-reported food consumption. A population-based study. *Appetite* **2010**, *54*, 473–479. [CrossRef]
32. Van Strien, T.; Konttinen, H.; Homberg, J.R.; Engels, R.C.; Winkens, L.H. Emotional eating as a mediator between depression and weight gain. *Appetite* **2016**, *100*, 216–224. [CrossRef] [PubMed]
33. Pursey, K.M.; Stanwell, P.; Gearhardt, A.N.; Collins, C.E.; Burrows, T.L. The Prevalence of Food Addiction as Assessed by the Yale Food Addiction Scale: A Systematic Review. *Nutrients* **2014**, *6*, 4552–4590. [CrossRef] [PubMed]
34. Meule, A.; Gearhardt, A.N. Ten Years of the Yale Food Addiction Scale: A Review of Version 2.0. *Curr. Addict. Rep.* **2019**, *6*, 218–228. [CrossRef]
35. Koski, M.; Naukkarinen, H. Severe obesity, emotions and eating habits: A case-control study. *BMC Obes.* **2017**, *4*, 2. [CrossRef] [PubMed]
36. Nicholls, W.; Devonport, T.J.; Blake, M. The association between emotions and eating behaviour in an obese population with binge eating disorder. *Obes. Rev. Off. J. Int. Assoc. Study Obes.* **2016**, *17*, 30–42. [CrossRef]
37. Lépine, J.P.; Godchau, M.; Brun, P.; Lempérière, T.H. Évaluation de l'anxiété et de la dépression chez des patients hospitalisés dans un service de médecine interne. *Ann. Med. Psychol.* **1985**, *143*, 175–189.
38. Zigmond, A.S.; Snaith, R.P. The Hospital Anxiety and Depression Scale. *Acta Psychiatr. Scand.* **1983**, *67*, 361–370. [CrossRef]
39. Nolan, L.J.; Halperin, L.B.; Geliebter, A. Emotional Appetite Questionnaire. Construct validity and relationship with BMI. *Appetite* **2009**, *54*, 314–319. [CrossRef]
40. Bourdier, L.; Lalanne, C.; Morvan, Y.; Kern, L.; Romo, L.; Berthoz, S. Validation and Factor Structure of the French-Language Version of the Emotional Appetite Questionnaire (EMAQ). *Front. Psychol.* **2017**, *8*, 8. [CrossRef]
41. Camilleri, G.M.; Méjean, C.; Bellisle, F.; Andreeva, V.A.; Sautron, V.; Hercberg, S.; Péneau, S. Cross-cultural validity of the Intuitive Eating Scale-2. Psychometric evaluation in a sample of the general French population. *Appetite* **2015**, *84*, 34–42. [CrossRef]
42. Tylka, T.L.; Van Diest, A.M.K. The Intuitive Eating Scale–2: Item refinement and psychometric evaluation with college women and men. *J. Couns. Psychol.* **2013**, *60*, 137–153. [CrossRef] [PubMed]
43. Gearhardt, A.N.; Corbin, W.R.; Brownell, K.D. Food addiction: An examination of the diagnostic criteria for dependence. *J. Addict. Med.* **2009**, *3*, 1–7. [CrossRef] [PubMed]
44. Brunault, P.; Ballon, N.; Gaillard, P.; Réveillère, C.; Courtois, R. Validation of the French version of the yale food addiction scale: An examination of its factor structure, reliability, and construct validity in a nonclinical sample. *Can. J. Psychiatry Rev. Can. Psychiatr.* **2014**, *59*, 276–284. [CrossRef] [PubMed]
45. Flint, A.J.; Gearhardt, A.N.; Corbin, W.R.; Brownell, K.D.; Field, A.E.; Rimm, E.B. Food-addiction scale measurement in 2 cohorts of middle-aged and older women. *Am. J. Clin. Nutr.* **2014**, *99*, 578–586. [CrossRef] [PubMed]
46. Abdella, H.M.; El Farssi, H.O.; Broom, D.R.; Hadden, D.A.; Dalton, C. Eating Behaviours and Food Cravings; Influence of Age, Sex, BMI and FTO Genotype. *Nutrients* **2019**, *11*, 377. [CrossRef]
47. Hauck, C.; Weiß, A.; Schulte, E.M.; Meule, A.; Ellrott, T. Prevalence of 'Food Addiction' as Measured with the Yale Food Addiction Scale 2.0 in a Representative German Sample and Its Association with Sex, Age and Weight Categories. *Obes. Facts* **2017**, *10*, 12–24. [CrossRef]
48. Cohen, J. *Statistical Power Analysis for the Behavioral Sciences*, 2nd ed.; Routledge: Hillsdale, NJ, USA, 1988; ISBN 978-0-8058-0283-2.
49. Rea, L.M.; Parker, R.A. *Designing and Conducting Survey Research: A Comprehensive Guide*; John Wiley & Sons: Hoboken, NJ, USA, 2014.
50. Holland, M. *Understanding & Applying Basic Statistical Methods Using R*; Scientific e-Resources: Waltham Abbey, UK, 2019.
51. Baron, R.M.; Kenny, D.A. The moderator–mediator variable distinction in social psychological research: Conceptual, strategic, and statistical considerations. *J. Pers. Soc. Psychol.* **1986**, *51*, 1173–1182. [CrossRef]
52. The Jamovi Project. Jamovi. (Version 1.1). 2019. Available online: https://www.jamovi.org (accessed on 21 February 2020).
53. Team, R.C. *R: A Language and Environment for Statistical Computing*; R Foundation for Statistical Computing: Vienna, Austria, 2014.

54. Luppino, F.S.; de Wit, L.M.; Bouvy, P.F.; Stijnen, T.; Cuijpers, P.; Penninx, B.W.J.H.; Zitman, F.G. Overweight, obesity, and depression: A systematic review and meta-analysis of longitudinal studies. *Arch. Gen. Psychiatry* **2010**, *67*, 220–229. [CrossRef]
55. Jáuregui-Lobera, I.; Montes-Martínez, M. Emotional Eating and Obesity. In *Psychosomatic Medicine [Online First]*; IntechOpen: London, UK, 2020.
56. Camilleri, G.M.; Méjean, C.; Bellisle, F.; Andreeva, V.A.; Kesse-Guyot, E.; Hercberg, S.; Péneau, S. Intuitive eating is inversely associated with body weight status in the general population-based NutriNet-Santé study. *Obesity* **2016**, *24*, 1154–1161. [CrossRef]
57. Meule, A.; Gearhardt, A.N. Five years of the Yale Food Addiction Scale: Taking stock and moving forward. *Curr. Addict. Rep.* **2014**, *1*, 193–205. [CrossRef]
58. Gearhardt, A.N.; White, M.A.; Masheb, R.M.; Morgan, P.T.; Crosby, R.D.; Grilo, C.M. An examination of the food addiction construct in obese patients with binge eating disorder. *Int. J. Eat. Disord.* **2011**, *45*, 657–663. [CrossRef] [PubMed]
59. Meule, A.; Müller, A.; Gearhardt, A.N.; Blechert, J. German version of the Yale Food Addiction Scale 2.0: Prevalence and correlates of 'food addiction' in students and obese individuals. *Appetite* **2017**, *115*, 54–61. [CrossRef] [PubMed]
60. Gearhardt, A.N.; Corbin, W.R.; Brownell, K.D. Development of the Yale Food Addiction Scale Version 2.0. *Psychol. Addict. Behav.* **2016**, *30*, 113–121. [CrossRef] [PubMed]
61. Mannan, M.; Mamun, A.A.; Doi, S.A.; Clavarino, A. Is there a bi-directional relationship between depression and obesity among adult men and women? Systematic review and bias-adjusted meta analysis. *Asian J. Psychiatry* **2016**, *21*, 51–66. [CrossRef]
62. Van Strien, T. Causes of distress-induced emotional eating. In *Genes, Brain, and Emotions: Interdisciplinary and Translational Perspectives*; Oxford University Press: New York, NY, USA, 2019.
63. Castellini, G.; Godini, L.; Amedei, S.G.; Faravelli, C.; Lucchese, M.; Ricca, V. Psychological effects and outcome predictors of three bariatric surgery interventions: A 1-year follow-up study. *Eat. Weight. Disord. Stud. Anorexia, Bulim. Obes.* **2014**, *19*, 217–224. [CrossRef]
64. Dingemans, A.; Danner, U.; Parks, M. Emotion Regulation in Binge Eating Disorder: A Review. *Nutrients* **2017**, *9*, 1274. [CrossRef]
65. Gianini, L.M.; White, M.A.; Masheb, R.M. Eating pathology, emotion regulation, and emotional overeating in obese adults with binge eating disorder. *Eat. Behav.* **2013**, *14*, 309–313. [CrossRef]
66. Wang, S.B.; Lydecker, J.A.; Grilo, C.M. Rumination in Patients with Binge-Eating Disorder and Obesity: Associations with Eating-Disorder Psychopathology and Weight-bias Internalization. *Eur. Eat. Disord. Rev.* **2017**, *25*, 98–103. [CrossRef]
67. Maraldo, T.M.; Zhou, W.; Dowling, J.; Wal, J.S.V. Replication and extension of the dual pathway model of disordered eating: The role of fear of negative evaluation, suggestibility, rumination, and self-compassion. *Eat. Behav.* **2016**, *23*, 187–194. [CrossRef]
68. Smith, E.; Hay, P.; Campbell, L.; Troller, J.N. A review of the association between obesity and cognitive function across the lifespan: Implications for novel approaches to prevention and treatment. *Obes. Rev.* **2011**, *12*, 740–755. [CrossRef]
69. Gordon, E.L.; Ariel-Donges, A.H.; Bauman, V.; Merlo, L.J. What is the evidence for "food addiction"? A systematic review. *Nutrients* **2018**, *10*, 477. [CrossRef] [PubMed]
70. Schulte, E.M.; Smeal, J.K.; Lewis, J.; Gearhardt, A.N. Development of the Highly Processed Food Withdrawal Scale. *Appetite* **2018**, *131*, 148–154. [CrossRef] [PubMed]
71. Schulte, E.M.; Potenza, M.N.; Gearhardt, A.N. A commentary on the "eating addiction" versus "food addiction" perspectives on addictive-like food consumption. *Appetite* **2017**, *115*, 9–15. [CrossRef] [PubMed]
72. Brewerton, T.D. *Posttraumatic Stress Disorder and Disordered Eating: Food Addiction as Self-Medication*; Mary Ann Liebert, Inc.: New Rochelle, NY, USA, 2011.

nutrients

MDPI

Review

Neuroimaging of Sex/Gender Differences in Obesity: A Review of Structure, Function, and Neurotransmission

Danielle S. Kroll [1], Dana E. Feldman [1], Catherine L. Biesecker [1], Katherine L. McPherson [1], Peter Manza [1], Paule Valery Joseph [1,2], Nora D. Volkow [1,3] and Gene-Jack Wang [1,*]

1 Laboratory of Neuroimaging, National Institute of Alcohol Abuse and Alcoholism, Bethesda, MD 20814, USA; danielle.kroll@nih.gov (D.S.K.); dana.feldman@nih.gov (D.E.F.); catherine.biesecker@nih.gov (C.L.B.); katherine.mcpherson@nih.gov (K.L.M.); peter.manza@nih.gov (P.M.); paule.joseph@nih.gov (P.V.J.); nora.volkow@nih.gov (N.D.V.)

2 Sensory Science and Metabolism Unit, National Institute of Nursing Research, Bethesda, MD 20814, USA

3 Office of the Director, National Institute of Drug Abuse, Bethesda, MD 20814, USA

* Correspondence: gene-jack.wang@nih.gov

Received: 5 June 2020; Accepted: 25 June 2020; Published: 30 June 2020

Abstract: While the global prevalence of obesity has risen among both men and women over the past 40 years, obesity has consistently been more prevalent among women relative to men. Neuroimaging studies have highlighted several potential mechanisms underlying an individual's propensity to become obese, including sex/gender differences. Obesity has been associated with structural, functional, and chemical alterations throughout the brain. Whereas changes in somatosensory regions appear to be associated with obesity in men, reward regions appear to have greater involvement in obesity among women than men. Sex/gender differences have also been observed in the neural response to taste among people with obesity. A more thorough understanding of these neural and behavioral differences will allow for more tailored interventions, including diet suggestions, for the prevention and treatment of obesity.

Keywords: obesity; sex; gender; taste; neuroimaging; MRI; PET; opioid; dopamine; serotonin

1. Introduction

The global prevalence of overweight (body mass index (BMI) > 25 kg/m^2) and obesity (BMI > 30 kg/m^2) has risen from 24.6% in 1980 to over one-third of the world's population in 2015 [1]. Although this pattern has been seen in both sexes/genders, obesity is more common in women relative to men, independent of age, geographic region, or socioeconomic status [1]. The deleterious effects of obesity on many of the body's organ systems in both sexes/genders have been well documented. Individuals with obesity are more likely to develop type 2 diabetes mellitus [2], cardiovascular disease [3], certain types of cancer [4], musculoskeletal disorders [5], and psychiatric illness [6,7]. However, obesity can present differently in women than in men. Premenopausal women tend to have a higher subcutaneous to visceral fat ratio due to their high levels of estrogen. This pattern of fat distribution has been shown to protect against some metabolic complications [8]. Nevertheless, given the wide-spread adverse effects of obesity, it is important to understand the disparate prevalence of obesity among men and women.

Neuroimaging can elucidate aspects of brain structure, function, and chemistry that are associated with sex/gender differences in compulsive eating behaviors in obesity. Several groups have linked the neural mechanisms underlying obesity to those for substance use disorders (SUDs) [9–11]. In obesity, palatable food consumption activates the mesolimbic dopaminergic pathway, begetting its rewarding and conditioning effects, as is the case for drug consumption in SUD. Repeated activation of this pathway in response to palatable food intake causes neuroadaptations in downstream inhibitory cortico-striatal circuits, leading to compulsive food intake [12]. While both sexes/genders are believed to share this common neuroadaptation, male and female individuals exhibit unique neural signatures in response to food cues (recently reviewed in Chao et al., 2017 [13]) and taste cues [14,15]. Women demonstrate greater neural responses in striato-limbic and frontal-cortical regions in response to food cues than men, perhaps underlying the elevated prevalence of obesity among women [13].

Sex/gender differences also present behaviorally in populations with obesity. Several behavioral studies have indicated that men and women have different food preferences. Women are more likely to prefer sweet tastes, high-calorie foods, and snack-based comfort foods, such as candy, whereas men tend to prefer savory tastes, low-calorie foods, and meal-based comfort foods, such as pizza [16–18]. Further, men and women show different patterns when engaging in dieting efforts, with women being more likely to report a history of dieting than men [19]. Sex/gender differences in dieting are also linked to societal norms that impact self-perception [20,21]. Women and men perceive their desired weight differently, and women tend to display lower self-esteem and higher levels of dieting than men [20,21]. Diet interventions among women have further been associated with increased activation of cortical areas involved with inhibitory control and self-regulation following food consumption, perhaps normalizing brain function in obesity [22]. However, to our knowledge, no studies have been done on the neural response of diet interventions among men, likely due to the lower frequency of dieting among men [23]. As such, a better understanding of the neural underpinnings of sex/gender behavioral differences in obesity can be used to devise more effective lifestyle interventions.

We recognize that there are many semantic ambiguities in the field of obesity and sex/gender difference research. Here, we digress briefly to clarify some important distinctions between commonly used terms and definitions. For the purposes of this review, as is common in the existing literature, we focus on obesity, which is defined by BMI. However, BMI, and other common metrics used to describe obesity, including total body fat, are not strong predictors of obesity's metabolic comorbidities. The ratio of subcutaneous to visceral fat, while outside of the scope of this review, is a more appropriate indicator of these complications (reviewed in [8]). We further acknowledge that obesity is highly comorbid but not synonymous with binge eating disorder (BED) and food addiction, and that those conditions also exist in lean individuals [9]. In this review, we also reference sex/gender, as we recognize that a difference exists between the concepts of biological "sex" and socio-cultural "gender" [24]. For a more thorough discussion on the use of this term, see Chao et al. (2017), who highlight the practical complexities of disentangling these two distinct, yet related, concepts [13].

In this paper, we review multimodal neuroimaging findings as they relate to sex/gender differences in obesity. We describe the differences in brain structure, resting-state function, and neurotransmission that exist between male and female individuals with obesity. We also summarize the sex/gender differences in the neural response to taste cues. These findings are tabulated in Tables 1–4 and structural and functional changes are further demonstrated visually in Figures 1 and 2. After integrating these results, we discuss the evidence for the development of sex/gender-specific diet interventions for individuals with obesity.

2. Structure

Several studies have examined gray matter volume (GMV) differences in men and women with obesity, as outlined and illustrated in Table 1 and Figure 1, respectively. One study found a positive association between GMV and BMI in the right orbitofrontal cortex (OFC) and nucleus accumbens (NAcc) in both men and women; GMV also positively correlated with central leptin levels, another positive marker of obesity [25]. However, in women, but not men, BMI and central leptin also positively correlated with GMV in the left putamen, and central leptin levels negatively correlated with GMV in the right dorsolateral prefrontal cortex [25]. These findings suggest a link between obesity and greater GMV in reward-associated regions that appears to be stronger among women.

This structural difference perhaps could underlie the greater preference for immediate rewards despite long-term negative consequences among obese versus lean women compared to that between obese and lean men [25]. In another study among a large cohort of lean, overweight, and obese adults, both sexes/genders demonstrated a negative correlation between total body fat and GMV in the globus pallidus. In men, but not women, total body fat was also negatively associated with GMV in subcortical regions, including the thalamus, caudate nucleus, putamen, hippocampus, and NAcc [26] These associations between obesity metrics and GMV in brain reward regions, which may be specific to each sex/gender, might contribute to some of the behavioral manifestations associated with sex/gender differences in obesity. Due to the cross-sectional nature of these studies, the temporal relationship between GMV and obesity remains unknown.

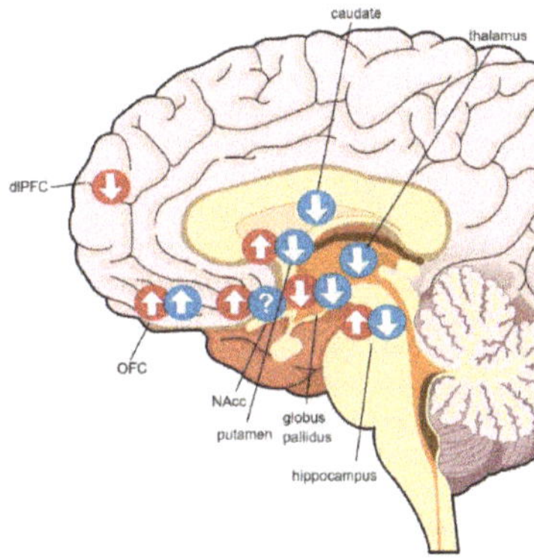

Figure 1. Gray matter volume differences in obesity by sex/gender. Red circles indicate changes among females and blue circles indicate changes among males. Arrows indicate the direction of the correlation between gray matter volume and obesity metrics (i.e., BMI, total body fat, serum leptin). dlPFC: dorsolateral PFC; OFC: orbitofrontal cortex; NAcc: nucleus accumbens.

Obesity also appears to alter white matter (WM) structure and organization in a sex/gender-dependent manner. Many measures derived from diffusion tensor imaging provide insight into this relationship, including axial diffusivity, a measure of axonal integrity; radial diffusivity, a negative measure of myelination; and fractional anisotropy, a measure of WM microstructural integrity [27]. One study of individuals with obesity found that BMI negatively correlated with axial diffusivity in the corpus callosum for both sex/genders [28]. In females only, BMI and serum leptin levels also positively correlated with radial diffusivity and negatively correlated with fractional anisotropy in the corpus callosum [28]. However, Dekkers et al. (2019) found that in women, but not men, total body fat negatively associated with global mean diffusivity, a measure negatively associated with WM integrity [26]. Together, these findings suggest a stronger relationship between obesity and WM integrity in women than in men, though the direction remains unclear. One limitation of these findings is that, to our knowledge, only Dekkers et al. (2019) controlled for total brain volume or intracranial volume; recent studies have shown that when these measures are accounted for, many regional sex/gender differences disappear [29,30].

Age may also interact with sex/gender and obesity in the brain structure. Ronan et al. (2016) demonstrated that participants who were obese or overweight displayed a lower cerebral WM volume, a finding typically observed in normal aging [31]. This effect was independent of sex/gender and was age dependent, with the greatest volume loss, adding an estimated 10 years of 'brain age', occurring at around age 40 [31]. A longitudinal study among women with obesity corroborated this finding by demonstrating that women were more likely to experience atrophy of the temporal lobe as both BMI and age increased [32]. However, not all studies on brain volume in people with obesity have been consistent. In one study, obesity was positively associated with frontal and temporal WM as well as hippocampal volume among women aged 70–89 [33]. Other studies similarly found that among women, but not men, obesity appears to be protective against GMV loss through aging, slowing down both hippocampal volume decline and ventricular enlargement [34]. Finally, sex/gender differences in obesity seem to only arise when participants reach a certain age. Children with obesity with a mean age of 9.0 years (range: 9.0 ± 0.9 years) displayed no significant sex/gender differences in brain structure, although obese children exhibited lower GMV than lean children in the temporal lobe, dorsolateral and medial prefrontal cortices, and the right anterior cingulate cortex [35]. Preclinical studies seem to yield opposite results; at a young age, male mice were more susceptible to the negative effects of a high-fat diet, such as decreased WM integrity and cerebral blood flow, than female mice [36]. Age, therefore, seems to interact with sex/gender and obesity, but this relationship is not fully understood.

Limited evidence suggests that low-calorie diets can contribute to brain structure recovery in a sex/gender-dependent manner. A longitudinal study that did not report sex/gender differences indicated that at baseline, participants with obesity have greater WM volumes in the temporal lobes, brainstem, and cerebellum [37]. After a 6-week low-calorie diet, these participants demonstrated that this expansion of WM volume partially recovered [37]. Some preclinical studies, though, have shown that low-calorie diets can affect these brain changes in a sex-specific manner. For instance, in control mice, levels of proliferating cells and neuroblasts are significantly higher in females than in males. However, when both sexes eat a high-fat diet, this difference disappears, suggesting that a high-fat diet contributes to a reduction in adult hippocampal neurogenesis in females only [38]. These preclinical results have yet to be replicated in humans. Nonetheless, low-calorie diets appear to contribute to structural brain changes in obesity.

Table 1. Brain structure study sample characteristics and findings.

Paper	Age (Years), Mean (SD)	Sample Size	Female, %	Obesity Metrics	Neuro-Imaging Modality	Pertinent Findings
Hortsmann et al. (2011) [25]	Male: 25.46 (4.25) Female: 25.11 (4.43)	122	50	BMI ≥ 30 kg/m^2 Serum leptin	MRI	Men and women show a positive association between GMV and BMI in the right OFC and NAcc. Women's BMI and leptin levels positively correlate with GMV in the left putamen; leptin negatively correlates with GMV in the right dorsolateral prefrontal cortex. Women with obesity were more likely to prefer immediate rewards, despite long-term negative consequences, than lean women.
Dekkers et al. (2019) [26]	62.0 (7.3)	12087	52.8	Overweight: BMI ≥ 25 kg/m^2 Obese: BMI ≥ 30 kg/m^2 Total Body Fat	MRI & DTI	Men and women show a negative correlation between total body fat and GMV in the globus pallidus. In men, total body fat was also negatively associated with subcortical GMV in the thalamus, caudate nucleus, putamen, hippocampus, and NAcc. In women, total body fat was negatively associated with global mean diffusivity.
Mueller et al. (2011) [28]	26.4 (5.0) Men: 25.5 (5.1) Women: 27.1 (5.0)	49	46.9	BMI ≥ 30 kg/m^2 Serum leptin	T1w MRI & DTI	In men and women, BMI negatively correlated with axial diffusivity in the corpus callosum. In females only, BMI and serum leptin levels also positively correlated with radial diffusivity and negatively correlated with fractional anisotropy in the corpus callosum.
Ronan et al. (2016) [31]	Lean: 48(16) Overweight: 57(17) Obese: 61 (16)	Lean: 246 Over-weight: 150 Obese: 77	Lean: 49.6 Over-weight: 44 Obese: 63.6	BMI ≥ 30 kg/m^2	T1w MRI	Greater atrophy of cerebral WM volume in participants who were obese or overweight, independent of sex/gender. This effect was age-dependent, with the greatest atrophy, adding an estimated 10 years of 'brain age', occurring at around age 40.
Gustafson et al. (2004) [32]	NR Range: 70–84 years	290	100	BMI ≥ 25 kg/m^2	CT	Women were more likely to experience atrophy of the temporal lobe as both BMI and age increased.
Driscoll et al. (2016) [33]	NR Range: 65-89 years	1366	100	ND subjects grouped by BMI	T1w MRI	In women aged 70–89, obesity was positively associated with frontal WM, temporal WM, and hippocampal volume.
Armstrong et al. (2019) [34]	71.2 (8.7) Men: 72.2 (8.5) Women: 70.3 (8.7)	617	52.8	BMI ≥ 30 kg/m^2	T1w MRI	In women, obesity protected against GMV loss as age increased, slowing hippocampal volume decline, and ventricular enlargement.
Xu et al. (2019) [35]	Control: 8.3 (0.9) Prader-Willi: 7.2(1.2) Obese: 9.0(0.9)	Control: 18 Prader-Willi: 12 Obese: 18	66.7	BMI percentile > 95%.	T1w MRI & DTI	No sex/gender differences were found in children with obesity, although subjects with obesity had lower GMV in the temporal lobe, dorsolateral, and medial prefrontal cortices, and the right anterior cingulate cortex.
Haltia et al. (2007) [37]	Lean: 37(21) Obese: 37(12)	Lean: 16 Obese: 30	Lean: 50 Obese: 60	BMI > 27 kg/m^2	T1w MRI	Both men and women with obesity had greater WM volumes in temporal lobes, brainstem, and cerebellum, but this expansion could recover after a 6-week low-calorie diet.

NR: Not reported. ND: Not defined.

3. Function

3.1. Resting State

It is well established that sex/gender differences exist in human brain functional connectivity [39,40]. Elevated BMI, the most common marker of obesity (BMI > 30 kg/m^2), has also been associated with brain connectivity [41,42]. However, the interaction between sex/gender and obesity remains unclear. Resting state connectivity in functional magnetic resonance imaging (fMRI) is an important but understudied area as it pertains to sex/gender differences in obesity. Few studies by one group of researchers [43–45] provide a preliminary understanding of the effects of sex/gender and BMI on brain connectivity, though these results, to our knowledge, have yet to be replicated by others. These results are illustrated in Figure 2 and outlined in Table 2.

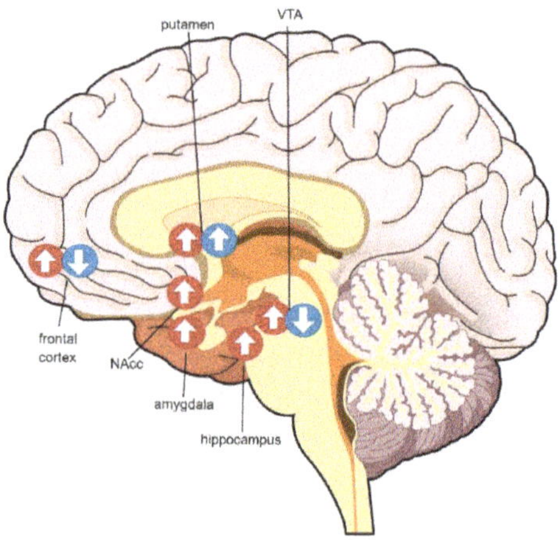

Figure 2. Resting state connectivity differences in obesity by sex/gender. Red circles indicate changes among females and blue circles indicate changes among males. Arrows indicate the direction of the correlation between centrality and obesity metrics (i.e., BMI, total body fat, serum leptin, Yale Food Addiction Scale (YFAS) score). VTA: ventral tegmental area.

There are many methodological approaches to examine resting state fMRI signals, but existing studies have focused on network centrality and slow-wave connectivity. Centrality generally assesses the connectedness of a given brain region to many others [46]. Gupta et al. (2017) found that women with obesity have higher centrality measures in several regions of the reward network, including the left amygdala, right NAcc, and bilateral hippocampus, than men with obesity, while men have higher centrality measures in the bilateral putamen [43]. Measures of food addiction also seem to contribute to the centrality of reward regions among women; scores on the Yale Food Addiction Scale, a standardized and widely used metric to quantify food addiction, were positively associated with the centrality of the ventral tegmental area (VTA) among women whereas they were negatively associated among men [45,47]. While men with obesity have lower centrality in reward regions than their female counterparts, Gupta et al. (2017) demonstrated that men with obesity have greater centrality in the right putamen, hippocampus, and medial orbitofrontal gyrus relative to lean men, and women with obesity have greater centrality in the left amygdala than lean women [43]. Thus, alterations in the centrality of the reward network tend to occur in a sex/gender-dependent manner, with female individuals showing greater centrality than their male counterparts. However, the interaction between sex/gender and obesity on brain network centrality has yet to be formally tested, so it remains unclear.

Table 2. Brain resting-state study sample characteristics and findings.

Paper	Age (years), Mean (SD)	Sample Size	Female, %	Obesity Metrics	Neuro-Imaging Modality	Pertinent Findings
Gupta et al. (2017) [43]	30.96 (11.26)	124	50.8	BMI ≥ 25 kg/m^2	Resting state fMRI	Women with high BMI had higher degrees of centrality in the left amygdala, right NAcc, and bilateral hippocampus than men with high BMI. Men with high BMI have higher centrality measures in the bilateral putamen than women with high BMI. Men with high BMI have greater centrality in the right putamen, hippocampus, and medial orbitofrontal gyrus relative to lean men. Women with high BMI have greater centrality in the left amygdala than lean women.
Osadchiy et al. (2019) [45]	Normal BMI: 28.95 (11.15) High BMI: 33.42 (10.83)	186	54.8	BMI ≥ 25 kg/m^2, YFAS *	Resting state fMRI	The association between the centrality of VTA and YFAS was positive in females but negative in males. The association between the centrality of the ventrolateral prefrontal cortex and YFAS was positive in males but negative in females.
Gupta et al. (2018) [44]	Female: 29.84 (7.45) Male: 31.79 (10.62)	86	50	BMI ≥ 25 kg/m^2	Resting state fMRI	Slow-4 signal in the right globus pallidus and bilateral putamen was associated with BMI in the female cohort, but not in the male cohort. Slow-5 connectivity between the left globus pallidus and substantia nigra with the bilateral posterior mid cingulate cortex and frontal cortical regions was negatively associated with BMI among females. Slow-5 connectivity between the left globus pallidus and substantia nigra and the medial frontal cortex was positively associated with BMI in the male cohort.

* Yale food addiction scale.

Obesity also shows a correlation with slow-wave connectivity throughout the reward network in a sex/gender-dependent manner. Another study by Gupta et al. (2018) examined two classes of slow-wave neural signals, labelled slow-4 and slow-5, in obese and lean men and women [44]. The slow-4 signal (medium frequency) is thought to arise from the basal ganglia, while the slow-5 signal (low frequency) is thought to arise from the cortex [48]. The slow-4 signal in the right globus pallidus and bilateral putamen was associated with BMI in females but not in males [44]. Further, among females, higher BMI was associated with lower slow-5 connectivity between the left globus pallidus and substantia nigra with the bilateral posterior mid cingulate cortex and frontal cortical regions [44]. This finding is reminiscent of the aberrant cortico-striatal signaling characteristic of obesity and substance use disorders [12,44]. Conversely, among males, greater BMI was associated with higher left globus pallidus and substantia nigra slow-5 connectivity with the medial frontal cortex [44]. Centrality measures have corroborated these findings; in women, higher food addiction scores were associated with lower centrality in frontal areas and higher centrality in VTA and this pattern is reversed in men [45]. This further suggests that neuroadaptations in reward regions appear to play a larger role in compulsive eating in women than for men.

3.2. Taste Response

Sex/gender differences in neuroimaging taste perception in people with obesity are an underexplored research area. This is partially due to the difficulty in firmly identifying regions in the human brain associated with taste (recently reviewed in Kure Liu et al., 2019 [49]). It has been established that the anterior insula/frontal operculum is the primary taste cortex [50–54]. However, Avery et al. (2020) reported that taste quality is more accurately analyzed using a combinatorial spatial code, in which taste perception is distributed throughout the sensory network [55]. In this model, taste quality refers to the unique pattern of activation throughout the primary taste cortex and regions involved in processing hedonic and aversive tastes for each individual in a population [55]. This contrasts with a topographic perspective in which activated regions are attributed to each separate taste component [55]. The type of taste (i.e., salty, bitter, sweet, sour, or umami) has also been found to influence brain activation differences between sexes/genders [15]. The diversity of responses to taste prevent researchers from finding a specific and reliable way to pinpoint regions activated by taste using neuroimaging.

Despite these existing limitations, related studies can shed light on neural taste perception in men and women with obesity (Table 3). Specifically, sex/gender differences in anticipation of taste may improve our understanding of underlying neural differences leading to obesity. For example, Cornier et al. (2015) examined taste anticipation in men and women who were identified as obese prone or obese resistant based on a history of diet and weight gain but not current BMI [56]. After undergoing a cue reactivity fMRI task associating sucrose and artificial saliva with visual cues, males in both populations displayed greater neuronal response to the sucrose-associated visual cue in the right caudate nucleus relative to women [56]. The study supports that there are brain activation sex/gender differences in anticipation of receiving sugar but not the receipt of the sugar itself, emphasizing the importance of sex/gender differences in conditioning in obesity. Geliebter et al. (2013) presented a related analysis of sex/gender differences in obese individuals [57]. When presented with high- instead of low-energy dense auditory food cues, male participants with obesity portrayed brain activation in supplementary motor areas (precentral gyrus) in a sated state. Female participants, though, showed activation in cognitive-related regions (parahippocampal gyrus) in a fasted state [57]. The same obese population further demonstrated that otherwise healthy men and women displayed different patterns of functional connectivity in the amygdala and ventral striatum when responding to food cues in both a sated and hungry state [58]. When the subjects were in a sated state, men tended to show greater connectivity in the amygdala than women, while women displayed greater connectivity in the angular gyrus and precentral gyrus than men [58]. However, in the fasted state, the motor/visual processing centers and emotion/reward-related regions (supplementary motor area, precentral gyrus, precuneus,

cuneus) in men were more highly connected, while women had greater connectivity in areas involving response inhibition and cognitive control (i.e., inferior frontal gyrus). Atalayer et al. (2014) suggest their results support the hypothesis that men may process hunger in relation to emotional cues, while women relate it to cognitive processing. The combination of findings from these studies illustrate how satiety and food anticipation may impact brain activation differently in men and women [56–58].

Preliminary evidence suggests that bariatric surgery can impact taste reward anticipation in both sex/genders. In one study, 13 patients with obesity who had undergone a gastric bypass experienced an expected weight loss, in addition to changes in the neural response to the expectation of tastants from before surgery to one-month post-operative [59]. Anticipation of sweet and salty stimuli evoked responses in reward regions, including the NAcc, caudate, VTA, OFC, and prefrontal cortex, as measured by fMRI [59]. While this neural response decreased from baseline to 1-month post-operative in anticipation of sucrose, it increased in anticipation of sodium chloride.

However, lean control participants, who did not have a gastric bypass and were scanned 1 month apart also had a similar decrease in reward response to sucrose but no change in response to sodium chloride, and so it is unclear if this change in sucrose anticipation is due to habituation or the gastric bypass [59]. Some sex/gender differences have also been identified regarding changes in taste response following bariatric surgery. Another group found men (n = 35) exhibited a greater decrease in taste and smell ability than women (n = 120) as assessed using a subjective taste questionnaire five years post-sleeve gastrectomy operation, especially among those with type-2 diabetes [60]. The authors did not report sex/gender differences in starting BMI, nor what type of taste changes the subjects underwent. Further, there were no sex/gender differences following roux-en-Y gastric bypass in taste or taste aversion [60]. Since these are pilot results, their replication as well as clarification of what types of taste changes occur following bariatric surgery and how they differ between sexes/genders are still needed.

Table 3. Brain taste response study sample characteristics and findings.

Paper	Age (years), Mean (SD)	Sample Size	Female%	Obesity Metrics	Neuro-Imaging Modality	Pertinent Findings
Cornier et al. (2015) [56]	Obese resistant: 30.4 (2.6) Obese prone: 30.2 (3.7)	49	49.0	Obesity proneness defined by history of diet and weight-gain	Task fMRI (cue anticipation task)	Obese-prone and -resistant males had greater neuronal response to the sucrose-associated visual cue in the right caudate nucleus relative to women.
Geliebter et al. (2013) [57]	Female: 35(6.9) Male: 35(9.0)	31	45.2	BMI ≥ 30 kg/m^2	Task fMRI (cue reactivity task)	Male participants with obesity portrayed brain activation in response to high energy dense auditory food cues (rel. to low energy dense) in supplementary motor areas (precentral gyrus) in a sated state. Female participants showed activation in response to high energy dense auditory food cues (rel. to low energy dense) in parahippocampal gyrus in a fasted state.
Atalayer et al. (2014) [58]	Female: 35 (6.9) Male: 35 (9.0)	31	45.2	BMI ≥ 30 kg/m^2	Task fMRI (cue reactivity task)	In a sated state, men demonstrated greater connectivity in the amygdala than women, while women displayed greater connectivity in the angular gyrus and precentral gyrus than men. In a fasted state, men displayed greater connectivity in the supplementary motor area, precentral gyrus, precuneus, cuneus, while women had greater connectivity in the inferior frontal gyrus.
Haase et al. (2011) [15]	Female: 21.94 (1.9) Male: 22.25 (2.7)	21	57.1	N/A	Task fMRI (cue reactivity task)	Men had greater brain activation decreases than women in response to all four tastes in the middle frontal gyrus, insula, and cerebellum when changing from a hunger to satiety state. Men had greater activational changes relative to women in reaction to sucrose, citric acid, and caffeine in the inferior frontal gyrus; sucrose and NaCl within the parahippocampal gyrus, entorhinal cortex, perirhinal cortex, and amygdala; and sucrose within the dorsal striatum (caudate, putamen) and posterior cingulate.
Wang et al. (2016) [59]	46.5 (9.3)	13	61.5	1991 NIH guidelines for obesity surgery *	Task fMRI (cue anticipation task)	Participants had decreased response in the reward center (including NAcc, caudate nucleus, VTA, OFC, and prefrontal cortex) in response to sucrose after gastric bypass and increased response in the same region in response to NaCl.

* NIH consensus development program, office of disease prevention. Gastrointestinal surgery for severe obesity. http://consensus.nih.gov/1991/1991gisurgeryobesity084html.htm. Mar 1991.

Research in a lean population analyzing sex/gender differences in taste also contributes valuable insight on how weight and taste interact between sexes/genders. While maintaining insignificant differences in BMI (23.15 kg/m^2 average for men and 22.76 kg/m^2 average for women), one cohort displayed notable sex/gender differences in the neural response to the transition from hunger to satiety to four different taste types: Sour, bitter, sweet, and salty [15]. fMRI results indicated greater brain activation decreases in men compared to women in the middle frontal gyrus, insula, and cerebellum when changing from a hunger to satiety state for all four tastes [15]. The middle frontal gyrus has been identified as an area critical to dual-task performance and decision-making [61,62]. Since women demonstrated consistently high activation through both hunger and satiety while men experienced a decrease in activation following satiety, Haase et al. (2011) speculated that women exhibit greater top-down functioning regarding taste salience [15]. In other words, female brains may process taste input more cognitively than males' brains since they were activated even when sated. In comparison, male brains displayed less activation after feeling full, suggesting their taste processing is built up from a small piece of sensory information, which dissipates after it has been resolved. Compared to women, men also displayed greater activation changes in reaction to sucrose, citric acid, and caffeine in the inferior frontal gyrus; sucrose and NaCl within the parahippocampal gyrus, entorhinal cortex, perirhinal cortex, and amygdala; and sucrose within the dorsal striatum (caudate, putamen) and posterior cingulate [15]. Because these areas are involved in reward and memory processing, these findings imply that women encode and learn about their food differently than men [63–65]. However, how these neural differences play out behaviorally in obese and overweight populations remains unclear.

Another study examined first-year college students and found that weight gain over the first eight months of school was associated with concurrent taste changes: For every 1% body weight increase, male students displayed a decrease in their tasting ability for sweetness (by ≤ 11.0%) and saltiness (by 7.5%) [66]. This finding aligns with similar studies that observed a decrease in sweet and salty perception accompanying weight gain in male more than in female students [67,68]. In comparison, female students did not display this taste decrease, and instead displayed a 6.5% increase in sourness perception ability for every 1% gain in body weight [66]. However, this correlational study cannot conclude a causal relationship between weight gain and changes in taste perception. While the authors suggested that taste differences between sexes/genders are due to hormonal effects, the neural findings from Haase et al. (2011) suggest that functional differences in the limbic system also contribute to differences in taste perception [15,66,69].

4. Neurotransmission

4.1. Serotonin Signaling

Serotonin (5-HT) is involved with regulation of appetite and has shown aberrant signaling in animal models of obesity [70,71]. 5-HT signals through several subtypes of 5-HT receptors, including the serotonin 1A (5-HT1A) and serotonin 2A (5-HT2A) receptors. The 5-HT1A receptor has widespread inhibitory actions, with autoreceptor negative feedback function in some cells and postsynaptic distribution in others (reviewed by Carhart-Harris and Nutt, 2017) [72]. The 5-HT2A receptor has postsynaptic distribution throughout the brain, with generally excitatory action [72]. To date, no clinical positron emission tomography (PET) or single-photon emission computerized tomography (SPECT) studies have assessed 5-HT1A receptors in obesity; however, preclinical studies provide insight into a relationship between 5-HT1A and food intake [73,74]. 5-HT signaling has also been associated with sex hormones in rats [75], but clinical imaging studies have shown conflicting results, with Jovanovic et al. (2008) and Parsey (2002) finding higher radiotracer binding to 5-HT1A in women, and Moses-Kolko et al. (2011) finding higher binding in men [76–78]. Clinical studies of sex/gender differences in obese individuals are needed.

Sex/gender differences in the 5-HT2A receptor have also been investigated (Table 4). Although positive correlations between binding of the 5HT2A receptor [^{18}F]altanserin and estradiol levels have been shown in men [79] and between [^{18}F]deuteroaltanserin binding and estradiol levels in women [80,81], sex/gender differences have not been observed [77,82–86]. To our knowledge, no PET or SPECT studies have compared 5-HT2A receptor binding between obese and normal-weight individuals, though one study reported a positive correlation between BMI and 5-HT2A receptor binding [82]. Future studies considering obesity, sex/gender, and 5-HT2A receptor binding are needed.

Sex/gender differences in serotonin transporter (5-HTT) availability have shown inconsistent results among normal-weight participants [76,87,88]. Additionally, not much is known about the relationship between obesity and 5-HTT. While 5-HTT availability was found to be negatively correlated with BMI [89] in one [^{11}C]3-amino-4-(2-dimethylaminomethyl-phenylsulfanyl)benzonitrile (DASB) PET study, a [^{123}I]2beta-Carbomethoxy-3beta-(4-iodophenyl)nortropane (nor-β-CIT) SPECT study did not demonstrate this [90]. This could be explained by differences between tracer and imaging modalities; both tracers are selective for 5-HTT, but [^{123}I]nor-β-CIT is also selective for the dopamine transporter in the striatum [91], whereas [^{11}C]DASB is selective for 5-HTT in the striatum [92]. Further, PET and SPECT differ in spatial resolution [93]. Female monozygotic twins with higher BMIs had higher [^{123}I]nor-β-CIT binding in the hypothalamus and thalamus than their leaner co-twins. This was not observed in male twinsets, suggesting obesity may have a sex/gender-dependent association with 5-HTT availability [90]. Further, [^{123}I]nor-β-CIT SPECT studies of BED may elucidate how obesity and 5-HTT availability are related: Compared to control women with obesity, women with obesity and BED had lower 5-HTT binding [94]. In an intervention study, women with obesity and BED showed improved midbrain 5-HTT binding during fluoxetine, a selective serotonin reuptake inhibitor used for weight loss in individuals with obesity [95] and therapy-facilitated remission [96]. Control women with obesity who did not receive fluoxetine did not show a change in 5-HTT binding [96]. These studies suggest that BED may be driving 5-HTT binding differences in other studies assessing obesity, assuming that fluoxetine administration would not benefit women without BED.

4.2. Dopamine Signaling

Dopamine (DA) signaling energizes the motivation towards food and preclinical studies have associated DA signaling dysregulation with obesity [97–99]. Clinical brain imaging studies focusing on sex/gender differences in DA D2 and D3 receptor availability have yielded mixed results [100–103]. In obese subjects, brain imaging studies reported that binding of [^{11}C]raclopride, a tracer selective to D2 and D3 receptors, in the striatum was lower in participants with obesity compared to controls [104–106], suggesting downregulation of D2/D3 receptor availability in obesity. Similar findings were observed in overweight and obese participants (BMI > 27 kg/m^2) compared to controls [107]. There are sex/gender differences in both BMI and D2/D3 receptor availability, suggesting there may be a common underlying phenotype driving these associations (Table 4). Women have a higher incidence of obesity, and tend to have lower D2/D3 availability, than men on average [100], and lower D2/D3 availability has been independently associated with high BMI [104–107]. Further, other studies, however, did not demonstrate differences between these groups [108], including one study utilizing the D2 receptor radiotracer N-[^{11}C]-methyl-benperidol [109]. Several of these studies either assessed female-only or mixed-sex/gender samples and lacked the power to investigate sex/gender differences. In one study that did take sex/gender into account, significant sex/gender differences were not observed, although there were only 10 individuals with obesity in that sample [106]. Studies with larger sample sizes are needed to elucidate whether DA signaling in obesity is sex/gender dependent.

In terms of weight-loss interventions, not much is known about their impact on D2/D3 receptor availability. To our knowledge, only three studies to date have assessed gastric bypass surgery. Each study used all-female cohorts and had different findings, with one study showing increased [^{11}C]raclopride binding, another showing decreased binding, and the third showing no significant difference from baseline after the surgery-induced weight loss [110–112]. Due to the inconsistency in results, more research is needed on surgery-induced weight loss, as well as other interventions, to determine whether these treatments impact the DA system. Men should also be included in these investigations to determine whether sex/gender plays a role in weight loss-related changes in DA signaling.

Findings of sex/gender differences in DA release are also mixed [101,103]. In a [^{123}I]iodobenzamide SPECT study, female controls showed significant DA release in response to amphetamine, while women with severe obesity did not show a significant change from baseline [104]. Further, BMI and DA release in response to a caloric glucose stimulus (compared to calorie-free sucralose) were negatively correlated [113]. These findings supported a disruption of DA signaling in individuals with high BMI. In another study, however, DA release after glucose injection (compared to saline) did not differ between participants with BMIs above 27 kg/m^2 and lean controls [107]. It is possible that the conflicting results were due to the difference in the route of administration or that BMI alone may not predict differences in DA release. Another study by Wang et al. (2011) comparing women with obesity to women with obesity and binge eating disorder (BED) showed that those with BED had enhanced DA release to a food stimulus [114]. Thus, binge eating might drive the previous findings discussed. However, van de Giessen et al. (2014) ruled out BED from their group with obesity and still observed differences between women with obesity and normal-weight controls, so more studies are necessary to determine the relationship between obesity and BED with DA release [104]. These opposing findings may be linked to the difference in displacement patterns between [^{123}I] iodobenzamide and [^{11}C]raclopride. However, the two radiotracers have shown similar patterns of displacement in the past (as reviewed by [115]); [^{123}I]iodobenzamide the mono-iodine analog of [^{11}C]raclopride [116]. Further, no studies have yet assessed sex/gender differences in obesity regarding the impact of weight-loss interventions on DA release, which highlights the need for broadened investigations into this area.

Finally, sex/gender differences in the availability and distribution of DA transporter have not been observed [117–119], nor have these studies shown differences between individuals with obesity and controls [118,119] in SPECT studies. However, these studies still need to be replicated, especially because each study cited used a different tracer (Nam et al. (2018) used [^{123}I]FP-CIT, Thomsen et al. (2013) used [^{123}I]PE2I, and Best et al. (2005) used [^{123}I]βCIT) [117–119]. The pharmacokinetics and pharmacodynamics differ between these radiotracers; [^{123}I]PE2I, for example, is faster and has higher affinity for DAT than [^{123}I]FP-CIT and [^{123}I]βCIT [120].

Table 4. Brain neurotransmission study sample characteristics and findings.

Paper	Age (years), Mean (SD)	Sample Size	Female%	Obesity Metrics	Neuro-Imaging Modality	Pertinent Findings
Adams et al. (2004) [82]	Female: 47.4 (19.6) Male: 45.4 (20.1)	52	42.31	N/A	PET [^{18}F]altanserin 5-HT2A	No sex differences in 5-HT2A binding Positive correlation between 5-HT2A binding and BMI
Erritzoe et al. (2010) [89]	35.7(18.2)	60	38.33	Overweight BMI > 25 kg/m^2 Obese BMI ≥ 30 kg/m^2	PET [^{11}C]DASB 5-HTT	Negative correlation between 5-HTT binding and BMI Females > males in midbrain 5-HTT binding No interactions between BMI and gender
Koskela et al. (2008) [90]	25.42 (1.29)	32 (16 mono-zygotic twin pairs)	50	N/A	SPECT [^{123}I]nor-β-CIT 5-HTT	Female, but not male, monozygotic twins with higher BMIs had higher 5-HTT binding in the hypothalamus and thalamus than their leaner co-twins
Wang et al. (2011) [106]	Lean: 37.5 (5.9) Obese: 38.9 (7.3)	20	40	Severely obese BMI > 40 kg/m^2	PET [^{11}C]raclopride D2/D3	Obese < controls in striatal D2/D3 receptor binding; positive correlation between D2/D3 receptor binding and BMI No sex differences
Burghardt et al. (2015) [121]	Lean: 51.43 (11.18) Obese: 52.43 (8.98)	14	0	ND Obese group mean BMI = 37.96 (1.83) kg/m^2	PET [^{11}C]carfentanil MOR	Obese < lean in MOR binding Partial recovery of MOR binding after restricted-calorie intervention in sample with obesity
Joutsa et al. (2018) [122]	Morbidly obese: 41.8 (10.3) Controls for morbidly obese: 44.9 (12.9) BED: 49.4 (5.1) Controls for BED: 43.1 (11.4)	56	100	ND Obese group mean BMI = 40.7 (3.8) kg/m^2 BED group mean BMI = 30.9 (6.6) kg/m^2	PET [^{11}C]carfentanil MOR	Morbid obesity and BED < controls in MOR binding No differences in MOR binding between morbidly obese group and BED group
Karlsson et al. (2015) [108]	Lean: 44.86 (12.88) Obese: 39.08 (10.74)	27	100	ND Obese group mean BMI = 41.89 (3.88) kg/m^2	PET [^{11}C]carfentanil MOR [^{11}C]raclopride D2/D3	Obese < control in MOR binding No differences in D2/D3 receptor binding
Tuominen et al. (2015) [123]	Lean: 42.00 (13.20) Morbidly obese: 41.24 (9.17)	45	100	ND Morbidly obese mean BMI = 41.30 (4.14) kg/m^2 mean fat percentage = 50.34 (3.69)	PET [^{11}C]carfentanil MOR [^{11}C]raclopride D2/D3	Obese < control in MOR binding Positive correlation between MOR and D2/D3 receptor binding in ventral striatum in control, but not morbidly obese group

ND: Not defined.

4.3. Opioid Signaling

Opioid signaling has been implicated in obesity by interacting with other neurotransmitters to regulate feeding and satiety [124]. Sex/gender differences in molecular imaging studies of mu, kappa, and delta opioid signaling are unclear, as very few studies have compared men to women [125–127]. Opioid signaling in obesity may be altered: In two all-female [^{11}C] carfentanil PET studies, obesity was associated with decreased mu-opioid receptor (MOR) availability throughout the brain (see Table 4) [108,122,123]. Further, Tuominen et al. (2015) found a correlation between [^{11}C] carfentanil and [^{11}C]raclopride binding in the ventral striatum and dorsal caudate nucleus in lean control women but not in the ventral striatum of women with obesity [123]. This suggests alterations in the link between MOR and D2/D3 receptors in the ventral striatum of women with obesity. In a study with all men, MOR availability was lower in the temporal pole, amygdala, prefrontal cortex, and thalamus in obese participants compared to controls [121]. After a restricted calorie intervention, MOR availability among this male cohort partially recovered in the left temporal pole, ventral striatum, thalamus, and medial frontal cortex [121]. Another PET study with a cohort of men and women, however, did not observe an association between BMI and binding of the non-selective opioid receptor radioligand [^{18}F] FDPN, though sex/gender differences were not assessed [125]. Given the differences in MOR, this negative finding may indicate a contribution of kappa and delta opioid receptors in obesity. Interestingly, mice without delta opioid receptors were shown to have resistance to diet-induced obesity [128]. More consistent studies of opioid signaling in people with obesity could help clarify these discrepancies.

5. Conclusions

This review aimed to describe the neural underpinnings of sex/gender differences in obesity. In general, obesity is associated with an abnormal structure, function, and chemistry in the brain's reward system [12]. This is characterized by a smaller volume in the NAcc, OFC, and globus pallidus and downregulation of D2/D3 receptors in the striatum [25,26,104–107]. Women appear to be more susceptible to neural adaptations associated with obesity than men [25,26,28,34,43–45,90,123]. Women with obesity also tend to have a greater volume and centrality measures in subcortical reward regions and lower volume and centrality measures in frontal cortical regions [25,26,28,34,43–45]. Men with obesity, however, seem to have more effects in cortical somatosensory regions, the putamen, and thalamus [26,43]. These findings are consistent with activation patterns in response to food cues among men and women with obesity (reviewed by Chao et al., 2017) and suggest distinct neural mechanisms in obesity for each sex/gender [13]. While many of the papers reviewed here interpreted smaller volume as atrophy [31–33], it is important to acknowledge that smaller volume measures in participants with obesity can also be indicative of neuroplasticity or genetic predisposition, particularly in younger study subjects; this is an important limitation to consider when interpreting sex differences in brain structure [129].

Moreover, sex/gender differences in the neural correlates of taste perception, diet, and weight-loss treatment can provide insight for the development of new therapies. Results surrounding brain activation in response to tastants have been mixed [15,56,57]. Men with obesity seem to have a greater neural response to high-energy food cues while sated than their female counterparts [57]. However, lean men have a greater decrease in neural response to tastants from the fasted to sated state than lean women [15]. This discrepancy may be explained by an interaction between sex/gender and obesity in the feeling of satiety, though this has not yet been studied. Further, intervention studies among women, including those for bariatric surgery, therapy, and medication, have yielded mixed results in terms of recovery of brain chemistry and taste response [59,60,96,110–112]. Similar studies have not been conducted in solely male participants to our knowledge, and so the effect that these interventions have on the male brain remains unclear.

While outside the scope of this review, several endocrine pathways contribute to sex/gender differences in obesity and warrant further research. Preclinical, clinical, and epidemiological studies have demonstrated that estrogen is protective against many metabolic complications associated with obesity (recently reviewed by [130]). As such, the estrogen concentration may in part explain sex/gender differences in the manifestation of obesity and how these differences change with age [130]. However, the role of hormones in modulating dimorphic brain responses is less clear. One study in rats found that insulin and leptin impact feeding behavior in a sex-dependent manner [131]. In male, but not female rats, central administration of insulin led to decreased food consumption. Conversely, in female, but not male, rats, central administration of leptin led to decreased food consumption [131]. These results are consistent with the human studies reviewed here, in which serum leptin was associated with GMV and WM integrity in women but not men [25,28]. Still, animal and human studies of hormonal regulation of the brain are not always consistent. For example, rat models demonstrating the role of sex hormones in 5HT signaling yielded different results than human models [75,76,78,81]. Thus, more work should be done in the field of neuropsychoendocrinology as it pertains to sex/gender differences in obesity to clarify the role of these hormonal pathways.

Though sex/gender differences are a burgeoning area of brain research, many studies lack the power to properly describe them [24]. Many studies reviewed include a sample of only women [32,33,94,96,110–112,122,123] or only men [121]. Samples of only women are particularly prominent in studies of eating behavior and weight-loss interventions, which limits our understanding of the generalizability of these treatments across sex/genders [59,60,96,110–112]. Moreover, the contribution of BED to these findings has not been very well established. In the studies reviewed, the effect of BED diagnosis on the brain among obese individuals has only been examined among women [94,96,122], despite there being similar prevalence of BED in individuals with obesity of both sex/genders worldwide [132]. Sex/gender differences in the brain have been well described among healthy adults [39,40,100–103,117–119]. Given the different patterns of men and women behaviorally in weight-loss interventions and food intake [16–19,66], it is imperative to develop a deeper understanding of how these sex/gender differences manifest in the brain in pathological conditions.

Author Contributions: Conceptualization, D.S.K., P.M., P.V.J., N.D.V., G.-J.W.; literature review, D.S.K., D.E.F., C.L.B., K.L.M.; writing—original draft preparation, D.S.K., D.E.F., C.L.B., K.L.M.; writing—review and editing, D.S.K., D.E.F., C.L.B., K.L.M., P.M., P.V.J., N.D.V., G.-J.W.; funding acquisition, N.D.V. All authors have read and agreed to the published version of the manuscript.

Funding: This research and the APC was funded by the National Institute of Alcohol Abuse and Alcoholism, grant number Y1AA-3009.

Acknowledgments: The figures were created using a ProTeams subscription of MindTheGraph.com.

Conflicts of Interest: The authors declare no conflicts of interest.

References

1. Chooi, Y.C.; Ding, C.; Magkos, F. The epidemiology of obesity. *Metabolism* **2019**, *92*, 6–10. [CrossRef]
2. Ozcan, U.; Cao, Q.; Yilmaz, E.; Lee, A.H.; Iwakoshi, N.N.; Ozdelen, E.; Tuncman, G.; Görgün, C.; Glimcher, L.H.; Hotamisligil, G.S. Endoplasmic reticulum stress links obesity, insulin action, and type 2 diabetes. *Science* **2004**, *306*, 457–461. [CrossRef] [PubMed]
3. Mandviwala, T.; Khalid, U.; Deswal, A. Obesity and Cardiovascular Disease: A Risk Factor or a Risk Marker? *Curr. Atheroscler. Rep.* **2016**, *18*, 21. [CrossRef]
4. Renehan, A.G.; Tyson, M.; Egger, M.; Heller, R.F.; Zwahlen, M. Body-mass index and incidence of cancer: A systematic review and meta-analysis of prospective observational studies. *Lancet* **2008**, *371*, 569–578. [CrossRef]
5. Anandacoomarasamy, A.; Caterson, I.; Sambrook, P.; Fransen, M.; March, L. The impact of obesity on the musculoskeletal system. *Int. J. Obes.* **2008**, *32*, 211–222. [CrossRef]
6. Tronieri, J.S.; Wurst, C.M.; Pearl, R.L.; Allison, K.C. Sex differences in obesity and mental health. *Curr. Psychiatry Rep.* **2017**, *19*, 29. [CrossRef] [PubMed]
7. Anstey, K.J.; Cherbuin, N.; Budge, M.; Young, J. Body mass index in midlife and late-life as a risk factor for dementia: A meta-analysis of prospective studies. *Obes. Rev.* **2011**, *12*, e426–e437. [CrossRef] [PubMed]

8. Wajchenberg, B.L. Subcutaneous and Visceral Adipose Tissue: Their Relation to the Metabolic Syndrome. *Endocr. Rev.* **2000**, *21*, 697–738. [CrossRef]
9. Lindgren, E.; Gray, K.; Miller, G.; Tyler, R.; Wiers, C.E.; Volkow, N.D.; Wang, G.J. Food addiction: A common neurobiological mechanism with drug abuse. *Front. Biosci. (Landmark Ed.)* **2018**, *23*, 811–836.
10. Garcia-Garcia, I.; Horstmann, A.; Jurado, M.A.; Garolera, M.; Chaudhry, S.J.; Margulies, D.S.; Villringer, A.; Neumann, J. Reward processing in obesity, substance addiction and non-substance addiction. *Obes. Rev.* **2014**, *15*, 853–869. [CrossRef]
11. Volkow, N.D.; Wang, G.J.; Tomasi, D.; Baler, R.D. Obesity and addiction: Neurobiological overlaps. *Obes. Rev.* **2013**, *14*, 2–18. [CrossRef]
12. Volkow, N.D.; Wang, G.-J.; Fowler, J.S.; Telang, F. Overlapping neuronal circuits in addiction and obesity: Evidence of systems pathology. *Philos. Trans. R. Soc. B Biol. Sci.* **2008**, *363*, 3191–3200. [CrossRef]
13. Chao, A.M.; Loughead, J.; Bakizada, Z.M.; Hopkins, C.; Geliebter, A.; Gur, R.C.; Wadden, T.A. Sex/gender differences in neural correlates of food stimuli: A systematic review of functional neuroimaging studies. *Obes. Rev.* **2017**, *18*, 687–699. [CrossRef]
14. Martin, L.J.; Sollars, S.I. Contributory role of sex differences in the variations of gustatory function. *J. Neurosci. Res.* **2017**, *95*, 594–603. [CrossRef]
15. Haase, L.; Green, E.; Murphy, C. Males and females show differential brain activation to taste when hungry and sated in gustatory and reward areas. *Appetite* **2011**, *57*, 421–434. [CrossRef] [PubMed]
16. Manippa, V.; Padulo, C.; van der Laan, L.N.; Brancucci, A. Gender Differences in Food Choice: Effects of Superior Temporal Sulcus Stimulation. *Front. Hum. Neurosci.* **2017**, *11*, 597. [CrossRef] [PubMed]
17. Herman, C.P.; Polivy, J. Sex and Gender Differences in Eating Behavior. In *Handbook of Gender Research in Psychology*; Springer: New York, NY, USA, 2009; pp. 455–469.
18. Wardle, J.; Haase, A.M.; Steptoe, A.; Nillapun, M.; Jonwutiwes, K.; Bellisie, F. Gender differences in food choice: The contribution of health beliefs and dieting. Ann. *Behav. Med.* **2004**, *27*, 107–116. [CrossRef] [PubMed]
19. Neumark-Sztainer, D.; Shenvood, N.E.; French, S.A.; Jefsery, R.W. Weight control behaviors among adult men and women: Cause for concern? *Obes. Res.* **1999**, *7*, 179–188. [CrossRef] [PubMed]
20. Davison, T.E.; McCabe, M.P. Relationships between Men's and Women's Body Image and Their Psychological, Social, and Sexual Functioning. *Sex. Roles, J. Res.* **2005**, *52*, 463–475. [CrossRef]
21. Tiggemann, M. Gender differences in the interrelationships between weight dissatisfaction, restraint, and self-esteem. *Sex. Roles J. Res.* **1994**, *30*, 319–330. [CrossRef]
22. Del Parigi, A.; Chen, K.; Salbe, A.D.; Hill, J.O.; Wing, R.R.; Reiman, E.M.; Tataranni, P.A. Successful dieters have increased neural activity in cortical areas involved in the control of behavior. *Int. J. Obes.* **2007**, *31*, 440–448. [CrossRef]
23. Long, J.D.; Dodd, S.L.; Doumit, R.; Boswell, C.; O'Boyle, M.W.; Rogers, T. Integrative Review of Dietary Choice Revealed by fMRI: Considerations for Obesity Prevention and Weight-Loss Education. *Worldviews Evid. Based Nurs.* **2020**, *17*, 151–157. [CrossRef] [PubMed]
24. Clayton, J.A.; Tannenbaum, C. Reporting Sex, Gender, or Both in Clinical Research? *JAMA* **2016**, *316*, 1863–1864. [CrossRef] [PubMed]
25. Horstmann, A.; Busse, F.P.; Mathar, D.; Muller, K.; Lepsien, J.; Schlogl, H. Obesity-Related Differences between Women and Men in Brain Structure and Goal-Directed Behavior. *Front. Hum. Neurosci.* **2011**, *5*, 58. [CrossRef] [PubMed]
26. Dekkers, I.A.; Jansen, P.R.; Lamb, H.J. Obesity, Brain Volume, and White Matter Microstructure at MRI: A Cross-sectional UK Biobank Study. *Radiology* **2019**, *291*, 763–771. [CrossRef]
27. Basser, P.J. Inferring microstructural features and the physiological state of tissues from diffusion-weighted images. *NMR Biomed.* **1995**, *8*, 333–344. [CrossRef]
28. Mueller, K.; Anwander, A.; Möller, H.E.; Horstmann, A.; Lepsien, J.; Busse, F.; Mohammadi, S.; Schroeter, M.L.; Stumvoll, M.; Villringer, A.; et al. Sex-dependent influences of obesity on cerebral white matter investigated by diffusion-tensor imaging. *PLoS ONE* **2011**, *6*, e18544. [CrossRef]
29. Kijonka, M.; Borys, D.; Psiuk-Maksymowicz, K.; Gorczewski, K.; Wojcieszek, P.; Kossowski, B.; Marchewka, A.; Swierniak, A.; Sokol, M.; Bobek-Billewicz, B. Whole Brain and Cranial Size Adjustments in Volumetric Brain Analyses of Sex- and Age-Related Trends. *Front. Neurosci* **2020**, *14*, 278. [CrossRef]

30. Ritchie, S.J.; Cox, S.R.; Shen, X.; Lombardo, M.V.; Reus, L.M.; Alloza, C.; Harris, M.A.; Alderson, H.L.; Hunter, S.; Neilson, E.; et al. Sex Differences in the Adult Human Brain: Evidence from 5216 UK Biobank Participants. *Cereb. Cortex* **2018**, *28*, 2959–2975. [CrossRef]
31. Ronan, L.; Alexander-Bloch, A.F.; Wagstyl, K.; Farooqi, S.; Brayne, C.; Tyler, L.K.; Fletcher, P.C. Obesity associated with increased brain age from midlife. *Neurobiol. Aging* **2016**, *47*, 63–70. [CrossRef] [PubMed]
32. Gustafson, D.; Lissner, L.; Bengtsson, C.; Björkelund, C.; Skoog, I. A 24-year follow-up of body mass index and cerebral atrophy. *Neurology* **2004**, *63*, 1876–1881. [CrossRef]
33. Driscoll, I.; Gaussoin, S.A.; Wassertheil-Smoller, S.; Limacher, M.; Casanova, R.; Yaffe, K.; Resnick, S.M.; Espeland, M.A. Obesity and Structural Brain Integrity in Older Women: The Women's Health Initiative Magnetic Resonance Imaging Study. *J. Gerontol. Ser. A Biol. Sci. Med. Sci.* **2016**, *71*, 1216–1222. [CrossRef]
34. Armstrong, N.M.; An, Y.; Beason-Held, L.; Doshi, J.; Erus, G.; Ferrucci, L.; Davatzikos, C.; Resnick, S.M. Sex differences in brain aging and predictors of neurodegeneration in cognitively healthy older adults. *Neurobiol. Aging* **2019**, *81*, 146–156. [CrossRef]
35. Xu, M.; Zhang, Y.; Von Deneen, K.M.; Zhu, H.Q.; Gao, J.H. Brain structural alterations in obese children with and without Prader-Willi Syndrome. *Hum. Brain Mapp.* **2017**, *38*, 4228–4238. [CrossRef]
36. Jacobs, S.A.; Gart, E.; Vreeken, D.; Franx, B.A.; Wekking, L.; Verweij, V.G.; Worms, N.; Schoemaker, M.H.; Gross, G.; Morrison, M.C.; et al. Sex-Specific Differences in Fat Storage, Development of Non-Alcoholic Fatty Liver Disease and Brain Structure in Juvenile HFD-Induced Obese Ldlr-/-.Leiden Mice. *Nutrients* **2019**, *11*, 1861. [CrossRef]
37. Haltia, L.T.; Viljanen, A.; Parkkola, R.; Kemppainen, N.; Rinne, J.O.; Nuutila, P.; Kaasinen, V. Brain white matter expansion in human obesity and the recovering effect of dieting. *J. Clin. Endocrinol. Metab.* **2007**, *92*, 3278–3284. [CrossRef]
38. Robison, L.S.; Albert, N.M.; Camargo, L.A.; Anderson, B.M.; Salinero, A.E.; Riccio, D.A.; Abi-Ghanem, C.; Gannon, O.J.; Zuloaga, K.L. High-Fat Diet-Induced Obesity Causes Sex-Specific Deficits in Adult Hippocampal Neurogenesis in Mice. *eNeuro* **2020**, *7*. [CrossRef] [PubMed]
39. Tomasi, D.; Volkow, N.D. Gender differences in brain functional connectivity density. *Hum. Brain Mapp.* **2012**, *33*, 849–860. [CrossRef]
40. Gong, G.; He, Y.; Evans, A.C. Brain connectivity: Gender makes a difference. *Neuroscientist* **2011**, *17*, 575–591. [CrossRef] [PubMed]
41. García-García, I.; Jurado, M.A.; Garolera, M.; Marques-Iturria, I.; Horstmann, A.; Segura, B.; Pueyo, R.; Sender-Palacios, M.J.; Vernet-Vernet, M.; Villringer, A.; et al. Functional network centrality in obesity: A resting-state and task fMRI study. *Psychiatry Res.* **2015**, *233*, 331–338. [CrossRef]
42. Doucet, G.E.; Rasgon, N.; McEwen, B.S.; Micali, N. Elevated Body Mass Index is Associated with Increased Integration and Reduced Cohesion of Sensory-Driven and Internally Guided Resting-State Functional Brain Networks. *Cereb. Cortex* **2018**, *28*, 988–997. [CrossRef]
43. Gupta, A.; Mayer, E.A.; Hamadani, K.; Bhatt, R.; Fling, C.; Alaverdyan, M.; Torgerson, C.; Ashe-McNalley, C.; Van Horn, J.D.; Naliboff, B.; et al. Sex differences in the influence of body mass index on anatomical architecture of brain networks. *Int. J. Obesity.* **2017**, *41*, 1185–1195. [CrossRef]
44. Gupta, A.; Mayer, E.A.; Labus, J.S.; Bhatt, R.R.; Ju, T.; Love, A.; Bal, A.; Tillisch, K.; Naliboff, B.; SanMiguel, C.P.; et al. Sex commonalities and differences in obesity-related alterations in intrinsic brain activity and connectivity. *Obesity* **2017**, *26*, 340–350. [CrossRef] [PubMed]
45. Osadchiy, V.; Mayer, E.A.; Bhatt, R.; Labus, J.S.; Gao, L.; Kilpatrick, L.A.; Liu, C.; Tillisch, K.; Naliboff, B.; Chang, L.; et al. History of early life adversity is associated with increased food addiction and sex-specific alterations in reward network connectivity in obesity. *Obes. Sci. Pract.* **2019**, *5*, 416–436.
46. Rubinov, M.; Sporns, O. Complex network measures of brain connectivity: Uses and interpretations. *Neuroimage* **2010**, *52*, 1059–1069. [CrossRef] [PubMed]
47. Gearhardt, A.N.; Corbin, W.R.; Brownell, K.D. Preliminary validation of the Yale Food Addiction Scale. *Appetite* **2009**, *52*, 430–436. [CrossRef] [PubMed]
48. Zuo, X.N.; Di Martino, A.; Kelly, C.; Shehzad, Z.E.; Gee, D.G.; Klein, D.F.; Castellanos, F.X.; Biswal, B.B.; Milham, M.P. The oscillating brain: Complex and reliable. *Neuroimage.* **2010**, *49*, 1432–1445. [CrossRef] [PubMed]
49. Kure Liu, C.; Joseph, P.V.; Feldman, D.E.; Kroll, D.S.; Burns, J.A.; Manza, P.; Volkow, N.D.; Wang, G.J. Brain Imaging of Taste Perception in Obesity: A Review. *Curr. Nutr. Rep.* **2019**, *8*, 108–119. [CrossRef] [PubMed]

50. Small, D.M.; Zald, D.H.; Jones-Gotman, M.; Zatorre, R.J.; Pardo, J.V.; Frey, S.; Petrides, M. Human cortical gustatory areas: A review of functional neuroimaging data. *NeuroReport* **1999**, *10*, 7–14. [CrossRef]
51. O'Doherty, J.; Rolls, E.; Francis, S.; Bowtell, R.; McGlone, F. Representation of Pleasant and Aversive Taste in the Human Brain. *J. Neurophysiol.* **2001**, *85*, 1315–1321. [CrossRef]
52. De Araujo, I.E.T.; Rolls, E.T.; Kringelbach, M.L.; McGlone, F.; Phillips, N. Taste-olfactory convergence, and the representation of the pleasantness of flavour, in the human brain. *Eur. J. Neurosci.* **2003**, *18*, 2059–2068. [CrossRef]
53. Grabenhorst, F.; Rolls, E.T. Selective attention to affective value alters how the brain processes taste stimuli. *Eur. J. Neurosci.* **2008**, *27*, 723–729. [CrossRef] [PubMed]
54. Small, D.M. Taste representation in the human insula. *Brain Struct. Funct.* **2010**, *214*, 551–561. [CrossRef] [PubMed]
55. Avery, J.A.; Liu, A.G.; Ingeholm, J.E.; Riddell, C.D.; Gotts, S.J.; Martin, A. Taste Quality Representation in the Human Brain. *J. Neurosci.* **2020**, *40*, 1042–1052. [CrossRef] [PubMed]
56. Cornier, M.A.; Shott, M.E.; Thomas, E.A.; Bechtell, J.L.; Bessesen, D.H.; Tregellas, J.R.; Frank, G.K. The effects of energy balance, obesity-proneness and sex on the neuronal response to sweet taste. *Behav. Brain Res.* **2015**, *278*, 446–452. [CrossRef]
57. Geliebter, A.; Pantazatos, S.P.; McOuatt, H.; Puma, L.; Gibson, C.D.; Atalayer, D. Sex-based fMRI differences in obese humans in response to high vs. low energy food cues. *Behav. Brain Res.* **2013**, *243*, 91–96. [CrossRef]
58. Atalayer, D.; Pantazatos, S.P.; Gibson, C.D.; McOuatt, H.; Puma, L.; Astbury, N.M.; Geliebter, A. Sexually dimorphic functional connectivity in response to high vs. low energy-dense food cues in obese humans: An fMRI study. *Neuroimage* **2014**, *100*, 405–413. [CrossRef]
59. Wang, J.L.; Yang, Q.; Hajnal, A.; Rogers, A.M. A pilot functional MRI study in Roux-en-Y gastric bypass patients to study alteration in taste functions after surgery. *Surg. Endosc.* **2016**, *30*, 892–898. [CrossRef]
60. Makaronidis, J.M.; Batterham, R.L. Potential Mechanisms Mediating Sustained Weight Loss Following Roux-en-Y Gastric Bypass and Sleeve Gastrectomy. *Endocrinol. Metab. Clin. N. Am.* **2016**, *45*, 539–552. [CrossRef]
61. Collette, F.; Olivier, L.; Van der Linden, M.; Laureys, S.; Delfiore, G.; Luxen, A.; Salmon, E. Involvement of both prefrontal and inferior parietal cortex in dual-task performance. *Brain Res. Cogn. Brain Res.* **2005**, *24*, 237–251. [CrossRef]
62. Rolls, E.T.; Grabenhorst, F. The orbitofrontal cortex and beyond: From affect to decision-making. *Prog. Neurobiol.* **2008**, *86*, 216–244. [CrossRef] [PubMed]
63. Elliott, R.; Dolan, R.J.; Frith, C.D. Dissociable functions in the medial and lateral orbitofrontal cortex: Evidence from human neuroimaging studies. *Cereb. Cortex* **2000**, *10*, 308–317. [CrossRef]
64. O'Doherty, J.P. Reward representations and reward-related learning in the human brain: Insights from neuroimaging. *Curr. Opin. Neurobiol.* **2004**, *14*, 769–776. [CrossRef] [PubMed]
65. Volkow, N.D.; Fowler, J.S.; Wang, G.J. Role of dopamine in drug reinforcement and addiction in humans: Results from imaging studies. *Behav. Pharmacol.* **2002**, *13*, 355–366. [CrossRef]
66. Noel, C.A.; Cassano, P.A.; Dando, R. College-Aged Males Experience Attenuated Sweet and Salty Taste with Modest Weight Gain. *J. Nutr.* **2017**, *147*, 1885–1891. [CrossRef] [PubMed]
67. Sartor, F.; Donaldson, L.F.; Markland, D.A.; Loveday, H.; Jackson, M.J.; Kubis, H.-P. Taste perception and implicit attitude toward sweet related to body mass index and soft drink supplementation. *Appetite* **2011**, *57*, 237–246. [CrossRef]
68. Simchen, U.; Koebnick, C.; Hoyer, S.; Issanchou, S.; Zunft, H.J. Odour and taste sensitivity is associated with body weight and extent of misreporting of body weight. *Eur. J. Clin. Nutr.* **2006**, *60*, 698–705. [CrossRef]
69. Loper, H.B.; La Sala, M.; Dotson, C.; Steinle, N. Taste perception, associated hormonal modulation, and nutrient intake. *Nutr. Rev.* **2015**, *73*, 83–91. [CrossRef]
70. Halford, J.C.; Blundell, J.E. Separate systems for serotonin and leptin in appetite control. *Ann. Med.* **2000**, *32*, 222–232. [CrossRef]
71. Hassanain, M.; Levin, B.E. Dysregulation of hypothalamic serotonin turnover in diet-induced obese rats. *Brain Res.* **2002**, *929*, 175–180. [CrossRef]
72. Carhart-Harris, R.L.; Nutt, D.J. Serotonin and brain function: A tale of two receptors. *J. Psychopharmacol.* **2017**, *31*, 1091–1120. [CrossRef]

73. De Vry, J.; Schreiber, R. Effects ofselected serotonin 5-HT1 and 5-HT2 receptor agonists on feeding behavior: Possible mechanisms of action. *Neurosci. Biobehav. Rev.* **2000**, *24*, 341–353. [CrossRef]
74. Dill, M.J.; Shaw, J.; Cramer, J.; Sindelar, D.K. 5-HT1A receptor antagonists reduce food intake and body weight by reducing total meals with no conditioned taste aversion. *Pharmacol. Biochem. Behav.* **2013**, *112*, 1–8. [CrossRef] [PubMed]
75. Zhang, L.; Ma, W.; Barker, J.L.; Rubinow, D.R. Sex differences in expression of serotonin receptors (subtypes 1A and 2A) in rat brain: A possible role of testosterone. *Neuroscience* **1999**, *94*, 251–254. [CrossRef]
76. Jovanovic, H.; Lundberg, J.; Karlsson, P.; Cerin, Å.; Saijo, T.; Varrone, A.; Halldin, C.; Nordström, A.L. Sex differences in the serotonin 1A receptor and serotonin transporter binding in the human brain measured by PET. *Neuroimage* **2008**, *39*, 1408–1419. [CrossRef]
77. Moses-Kolko, E.L.; Price, J.C.; Shah, N.; Berga, S.; Sereika, S.M.; Fisher, P.M.; Coleman, R.; Becker, C.; Mason, N.S.; Loucks, T.; et al. Age, sex, and reproductive hormone effects on brain serotonin-1A and serotonin-2A receptor binding in a healthy population. *Neuropsychopharmacology* **2011**, *36*, 2729–2740. [CrossRef]
78. Parsey, R.V.; Oquendo, M.A.; Simpson, N.R.; Ogden, R.T.; Van Heertum, R.; Arango, V.; Mann, J.J. Effects of sex, age, and aggressive traits in man on brain serotonin 5-HT1A receptor binding potential measured by PET using [C-11]WAY-100635. *Brain Res.* **2002**, *954*, 173–182. [CrossRef]
79. Frokjaer, V.G.; Erritzoe, D.; Juul, A.; Nielsen, F.Å.; Holst, K.; Svarer, C.; Madsen, J.; Paulson, O.B.; Knudsen, G.M. Endogenous plasma estradiol in healthy men is positively correlated with cerebral cortical serotonin 2A receptor binding. *Psychoneuroendocrinology* **2010**, *35*, 1311–1320. [CrossRef]
80. Kugaya, A.; Epperson, C.N.; Zoghbi, S.; van Dyck, C.H.; Hou, Y.; Fujita, M.; Staley, J.K.; Garg, P.K.; Seibyl, J.P.; Innis, R.B. Increase in Prefrontal Cortex Serotonin2A Receptors Following Estrogen Treatment in Postmenopausal Women. *Am. J. Psychiatry* **2003**, *160*, 1522–1524. [CrossRef]
81. Moses, E.L.; Drevets, W.C.; Smith, G.; Mathis, C.A.; Kalro, B.N.; Butters, M.A.; Leondires, M.P.; Greer, P.J.; Lopresti, B.; Loucks, T.L.; et al. Effects of Estradiol and Progesterone Administration on Human Serotonin 2A Receptor Binding: A PET Study. *Biol. Psychiatry* **2000**, *48*, 854–860. [CrossRef]
82. Adams, K.H.; Pinborg, L.H.; Svarer, C.; Hasselbalch, S.G.; Holm, S.; Haugbøl, S.; Madsen, K.; Frøkjær, V.; Martiny, L.; Paulson, O.B.; et al. A database of [(18)F]-altanserin binding to 5-HT(2A) receptors in normal volunteers: Normative data and relationship to physiological and demographic variables. *Neuroimage* **2004**, *21*, 1105–1113. [CrossRef]
83. Biver, F.; Lotstra, F.; Monclus, M.; Wikler, D.; Damhaut, P.; Mendlewicz, J.; Goldman, S. Sex difference in 5HT2 receptor in the living human brain. *Neurosci. Lett.* **1996**, *204*, 25–28. [CrossRef]
84. Frokjaer, V.G.; Erritzoe, D.; Madsen, J.; Paulson, O.B.; Knudsen, G.M. Gender and the use of hormonal contraception in women are not associated with cerebral cortical 5-HT 2A receptor binding. *Neuroscience* **2009**, *163*, 640–645. [CrossRef]
85. Meyer, J.H.; Kapur, S.; Houle, S.; DaSilva, J.; Owczarek, B.; Brown, G.M.; Wilson, A.A.; Kennedy, S.H. Prefrontal Cortex 5-HT2 Receptors in Depression: An [18F]Setoperone PET Imaging Study. *Am. J. Psychiatry* **1999**, *156*, 1029–1034. [PubMed]
86. Rosier, A.; Dupont, P.; Peuskens, J.; Bormans, G.; Vandenberghe, R.; Maes, M.; de Groot, T.; Schiepers, C.; Verbruggen, A.; Mortelmans, L. Visualisation of loss of 5-HT2A receptors with age in healthy volunteers using [18F]altanserin and positron emission tomographic imaging. *Psychiatry Res. Neuroimaging* **1996**, *68*, 11–22. [CrossRef]
87. Staley, J.K.; Zoghbi, S.; Tamagnan, G.; Fujita, M.; Seibyl, J.P.; Maciejewski, P.K.; O'Malley, S.; Innis, R.B. Sex differences in [123I]β-CIT SPECT measures of dopamine and serotonin transporter availability in healthy smokers and nonsmokers. *Synapse* **2001**, *41*, 275–284. [CrossRef] [PubMed]
88. Van Dyck, C.H.; Malison, R.T.; Seibyl, J.P.; Laruelle, M.; Klumpp, H.; Zoghbi, S.S.; Baldwin, R.M.; Innis, R.B. Age-related decline in central serotonin transporter availability with [123I]β-CIT SPECT. *Neurobiol. Aging* **2000**, *21*, 497–501. [CrossRef]
89. Erritzoe, D.; Frokjaer, V.G.; Haahr, M.T.; Kalbitzer, J.; Svarer, C.; Holst, K.K.; Hansen, D.L.; Jernigan, T.L.; Lehel, S.; Knudsen, G.M. Cerebral serotonin transporter binding is inversely related to body mass index. *Neuroimage* **2010**, *52*, 284–289. [CrossRef]

90. Koskela, A.K.; Kaurijoki, S.; Pietiläinen, K.H.; Karhunen, L.; Pesonen, U.; Kuikka, J.T.; Kaprio, J.; Rissanen, A. Serotonin transporter binding and acquired obesity—An imaging study of monozygotic twin pairs. *Physiol. Behav.* **2008**, *93*, 724–732. [CrossRef]
91. Laruelle, M.; Baldwin, R.M.; Malison, R.T.; Zea-Ponce, Y.; Zoghbi, S.S.; Al-Tikeriti, M.S.; Sybirska, E.H.; Zimmermann, R.C.; Wisniewski, G.; Neumeyer, J.L. SPECT imaging of dopamine and serotonin transporters with [123I]β-CIT: Pharmacological characterization of brain uptake in nonhuman primates. *Synapse* **1993**, *13*, 295–309. [CrossRef]
92. Kim, J.S.; Ichise, M.; Sangare, J.; Innis, R.B. PET Imaging of Serotonin Transporters with [^{11}C]DASB: Test–Retest Reproducibility Using a Multilinear Reference Tissue Parametric Imaging Method. *J. Nucl. Med.* **2006**, *47*, 208–214.
93. Rahmim, A.; Zaidi, H. PET versus SPECT: Strengths, limitations and challenges. *Nucl. Med. Commun.* **2008**, *29*, 193–207. [CrossRef]
94. Kuikka, J.T.; Tammela, L.; Karhunen, L.; Rissanen, A.; Bergström, K.A.; Naukkarinen, H.; Vanninen, E.; Karhu, J.; Lappalainen, R.; Repo-Tiihonen, E.; et al. Reduced serotonin transporter binding in binge eating women. *Psychopharmacology* **2001**, *155*, 310–314. [CrossRef]
95. Wise, S.D. Clinical studies with fluoxetine in obesity. *Am. J. Clin. Nutr.* **1992**, *55*, 181S–184S. [CrossRef] [PubMed]
96. Tammela, L.I.; Rissanen, A.; Kuikka, J.T.; Karhunen, L.J.; Bergström, K.A.; Repo-Tiihonen, E.; Naukkarinen, H.; Vanninen, E.; Tiihonen, J.; Uusitupa, M. Treatment improves serotonin transporter binding and reduces binge eating. *Psychopharmacology* **2003**, *170*, 89–93. [CrossRef] [PubMed]
97. Figlewicz, D.P.; Patterson, T.A.; Johnson, L.B.; Zavosh, A.; Israel, P.A.; Szot, P. Dopamine transporter mRNA is increased in the CNS of Zucker fatty (fa/fa) rats. *Brain Res. Bull.* **1998**, *46*, 199–202. [CrossRef]
98. Johnson, P.M.; Kenny, P.J. Dopamine D2 receptors in addiction-like reward dysfunction and compulsive eating in obese rats. *Nat. Neurosci.* **2010**, *13*, 635–641. [CrossRef] [PubMed]
99. Wang, G.J.; Volkow, N.D.; Fowler, J.S. The role of dopamine in motivation for food in humans: Implications for obesity. *Expert Opin. Ther. Targets* **2002**, *6*, 601–609. [CrossRef]
100. Laakso, A.; Vilkman, H.; Bergman, J.; Haaparanta, M.; Solin, O.; Syvälahti, E.; Salokangas, R.K.; Hietala, J. Sex differences in striatal presynaptic dopamine synthesis capacity in healthy subjects. *Biol. Psychiatry* **2002**, *52*, 759–763. [CrossRef]
101. Munro, C.A.; McCaul, M.E.; Wong, D.F.; Oswald, L.M.; Zhou, Y.; Brasic, J.; Kuwabara, H.; Kumar, A.; Alexander, M.; Ye, W.; et al. Sex differences in striatal dopamine release in healthy adults. *Biol. Psychiatry* **2006**, *59*, 966–974. [CrossRef]
102. Pohjalainen, T.; Rinne, J.O.; Någren, K.; Syvälahti, E.; Hietala, J. Sex differences in the striatal dopamine d2 receptor binding characteristics in vivo. *Am. J. Psychiatry* **1998**, *155*, 768–773.
103. Riccardi, P.; Zald, D.; Li, R.; Park, S.; Ansari, M.S.; Dawant, B.; Anderson, S.; Woodward, N.; Schmidt, D.; Baldwin, R.; et al. Sex differences in amphetamine-induced displacement of [18 F] fallypride in striatal and extrastriatal regions: A PET study. *Am. J. Psychiatry* **2006**, *163*, 1639–1641. [CrossRef]
104. Van de Giessen, E.; Celik, F.; Schweitzer, D.H.; van den Brink, W.; Booij, J. Dopamine D2/D3 receptor availability and amphetamine-induced dopamine release in obesity. *J. Psychopharmacol.* **2014**, *28*, 866–873. [CrossRef] [PubMed]
105. Volkow, N.D.; Wang, G.J.; Telang, F.; Fowler, J.S.; Thanos, P.K.; Logan, J.; Alexoff, D.; Ding, Y.S.; Wong, C.; Ma, Y.; et al. Low dopamine striatal D2 receptors are associated with prefrontal metabolism in obese subjects: Possible contributing factors. *Neuroimage* **2008**, *42*, 1537–1543. [CrossRef]
106. Wang, G.J.; Volkow, N.D.; Logan, J.; Pappas, N.R.; Wong, C.T.; Zhu, W.; Netusil, N.; Fowler, J.S. Brain dopamine and obesity. *Lancet* **2001**, *357*, 354–357. [CrossRef]
107. Haltia, L.T.; Rinne, J.O.; Merisaari, H.; Maguire, R.P.; Savontaus, E.; Helin, S.; Nagren, K.; Kaasinen, V. Effects of intravenous glucose on dopaminergic function in the human brain in vivo. *Synapse* **2007**, *61*, 748–756. [CrossRef]
108. Karlsson, H.K.; Tuominen, L.; Tuulari, J.J.; Hirvonen, J.; Parkkola, R.; Helin, S.; Salminen, P.; Nuutila, P.; Nummenmaa, L. Obesity is associated with decreased mu-opioid but unaltered dopamine D_2 receptor availability in the brain. *J. Neurosci.* **2015**, *35*, 3959–3965. [CrossRef] [PubMed]

109. Eisenstein, S.A.; Antenor-Dorsey, J.A.; Gredysa, D.M.; Koller, J.M.; Bihun, E.C.; Ranck, S.A.; ArbelÁez, A.M.; Klein, S.; Perlmutter, J.S.; Moerlein, S.M.; et al. A comparison of D2 receptor specific binding in obese and normal-weight individuals using PET with (N-[(11)C]methyl)benperidol. *Synapse* **2013**, *67*, 748–756. [CrossRef] [PubMed]
110. De Weijer, B.A.; van de Giessen, E.; Janssen, I.; Berends, F.J.; van de Laar, A.; Ackermans, M.T.; Fliers, E.; La Fleur, S.E.; Booij, J.; Serlie, M.J. Striatal dopamine receptor binding in morbidly obese women before and after gastric bypass surgery and its relationship with insulin sensitivity. *Diabetologia* **2014**, *57*, 1078–1080. [CrossRef]
111. Dunn, J.P.; Cowan, R.L.; Volkow, N.D.; Feurer, I.D.; Li, R.; Williams, D.B.; Kessler, R.M.; Abumrad, N.N. Decreased dopamine type 2 receptor availability after bariatric surgery: Preliminary findings. *Brain Res.* **2010**, *1350*, 123–130. [CrossRef]
112. Steele, K.E.; Prokopowicz, G.P.; Schweitzer, M.A.; Magunsuon, T.H.; Lidor, A.O.; Kuwabawa, H.; Kuma, A.; Brasic, J.; Wong, D.F. Alterations of central dopamine receptors before and after gastric bypass surgery. *Obes. Surg.* **2010**, *20*, 369–374. [CrossRef] [PubMed]
113. Wang, G.J.; Tomasi, D.; Convit, A.; Logan, J.; Wong, C.T.; Shumay, E.; Fowler, J.S.; Volkow, N.D. BMI modulates calorie-dependent dopamine changes in accumbens from glucose intake. *PLoS ONE* **2014**, *9*, e101585. [CrossRef]
114. Wang, G.J.; Geliebter, A.; Volkow, N.D.; Telang, F.W.; Logan, J.; Jayne, M.C.; Galanti, K.; Selig, P.A.; Han, H.; Zhu, W.; et al. Enhanced striatal dopamine release during food stimulation in binge eating disorder. *Obesity* **2011**, *19*, 1601–1608. [CrossRef] [PubMed]
115. Ginovart, N. Imaging the Dopamine System with In Vivo [11C]raclopride Displacement Studies: Understanding the True Mechanism. *Mol. Imaging Biol.* **2005**, *7*, 45–52. [CrossRef]
116. Kung, H.F.; Alavi, A.; Chang, W.; Kung, M.P.; Keyes, J.; Velchik, M.G.; Billings, J.; Pan, S.; Noto, R.; Rausch, A. In vivo SPECT imaging of CNS D-2 dopamine receptors: Initial studies with iodine-123-IBZM in humans. *J. Nucl. Med.* **1990**, *31*, 573–579. [PubMed]
117. Best, S.E.; Sarrel, P.M.; Malison, R.T.; Laruelle, M.; Zoghbi, S.S.; Baldwin, R.M.; Seibyl, J.P.; Innis, R.B.; Van Dyck, C.H. Striatal dopamine transporter availability with [123I]beta-CIT SPECT is unrelated to gender or menstrual cycle. *Psychopharmacology* **2005**, *183*, 181–189. [CrossRef]
118. Nam, S.B.; Kim, K.; Kim, B.S.; Im, H.J.; Lee, S.H.; Kim, S.J.; Kim, I.J.; Pak, K. The Effect of Obesity on the Availabilities of Dopamine and Serotonin Transporters. *Sci. Rep.* **2018**, *8*, 4924. [CrossRef] [PubMed]
119. Thomsen, G.; Ziebell, M.; Jensen, P.S.; da Cuhna-Bang, S.; Knudsen, G.M.; Pinborg, L.H. No correlation between body mass index and striatal dopamine transporter availability in healthy volunteers using SPECT and [123I]PE2I. *Obesity* **2013**, *21*, 1803–1806. [CrossRef] [PubMed]
120. Ziebell, M. Evaluation of the superselective radioligand [123I]PE2I for imaging of the dopamine transporter in SPECT. *Dan. Med. Bull.* **2011**, *58*, B4279. [PubMed]
121. Burghardt, P.R.; Rothberg, A.E.; Dykhuis, K.E.; Burant, C.F.; Zubieta, J.K. Endogenous Opioid Mechanisms Are Implicated in Obesity and Weight Loss in Humans. *J. Clin. Endocrinol. Metab.* **2015**, *100*, 3193–3201. [CrossRef]
122. Joutsa, J.; Karlsson, H.K.; Majuri, J.; Nuutila, P.; Helin, S.; Kaasinen, V.; Nummenmaa, L. Binge eating disorder and morbid obesity are associated with lowered mu-opioid receptor availability in the brain. *Psychiatry Res. Neuroimaging* **2018**, *276*, 41–45. [CrossRef] [PubMed]
123. Tuominen, L.; Tuulari, J.; Karlsson, H.; Hirvonen, J.; Helin, S.; Salminen, P.; Parkkola, R.; Hietala, J.; Nuutila, P.; Nummenmaa, L. Aberrant mesolimbic dopamine–opiate interaction in obesity. *NeuroImage* **2015**, *122*, 80–86. [CrossRef] [PubMed]
124. Gosnell, B.A.; Levine, A.S. Reward systems and food intake: Role of opioids. *Int. J. Obes.* **2009**, *33*, S54–S58. [CrossRef]
125. Henriksen, G.; Spilker, M.E.; Sprenger, T.; Hauser, A.; Platzer, S.; Boecker, H.; Toelle, T.R.; Schwaiger, M.; Wester, H.J. Gender dependent rate of metabolism of the opioid receptor-PET ligand [^{18}F]fluoroethyldiprenorphine. *Nuklearmedizin* **2006**, *45*, 197–200. [PubMed]
126. Vijay, A.; Wang, S.; Worhunsky, P.; Zheng, M.-Q.; Nabulsi, N.; Ropchan, J.; Krishnan-Sarin, S.; Huang, Y.; Morris, E.D. PET imaging reveals sex differences in kappa opioid receptor availability in humans, in vivo. *Am. J. Nucl. Med. Mol. Imaging* **2016**, *6*, 205–214. [PubMed]

127. Zubieta, J.K.; Dannals, R.F.; Frost, J.J. Gender and Age Influences on Human Brain Mu-Opioid Receptor Binding Measured by PET. *Am. J. Psychiatry* **1999**, *156*, 842–848. [CrossRef]
128. Czyzyk, T.A.; Romero-Picó, A.; Pintar, J.; McKinzie, J.H.; Tschöp, M.H.; Statnick, M.A.; Nogueiras, R. Mice lacking delta-opioid receptors resist the development of diet-induced obesity. *FASEB J.* **2012**, *26*, 3483–3492. [CrossRef]
129. Paus, T.; Keshavan, M.; Giedd, J.N. Why do many psychiatric disorders emerge during adolescence? *Nat. Rev. Neurosci.* **2008**, *9*, 947–957. [CrossRef]
130. Tramunt, B.; Smati, S.; Grandgeorge, N.; Lenfant, F.; Arnal, J.F.; Montagner, A.; Gourdy, P. Sex differences in metabolic regulation and diabetes susceptibility. *Diabetologia* **2020**, *63*, 453–461. [CrossRef]
131. Clegg, D.J.; Riedy, C.A.; Smith, K.A.; Benoit, S.C.; Woods, S.C. Differential sensitivity to central leptin and insulin in male and female rats. *Diabetes* **2003**, *52*, 682–687. [CrossRef]
132. Kessler, R.C.; Berglund, P.A.; Chiu, W.T.; Deitz, A.C.; Hudson, J.I.; Shahly, V.; Aguilar-Gaxiola, S.; Alonso, J.; Angermeyer, M.C.; Benjet, C.; et al. The prevalence and correlates of binge eating disorder in the World Health Organization World Mental Health Surveys. *Biol. Psychiatry* **2013**, *73*, 904–914. [CrossRef] [PubMed]

nutrients

MDPI

Article

Prisoners of Addictive Cues: Biobehavioral Markers of Overweight and Obese Adults with Food Addiction

Roni Aviram-Friedman *, Lior Kafri, Guy Baz, Uri Alyagon and Abraham Zangen *

Department of Life Sciences and the Zlotowski Center for Neuroscience, Ben-Gurion University, Be'er Sheva 8410501, Israel; lior363@gmail.com (L.K.); guy.s.baz@gmail.com (G.B.); uri.alyagon@gmail.com (U.A.)

* Correspondence: aviram.f.roni@gmail.com (R.A.-F.); azangen@bgu.ac.il (A.Z.); Tel.: +972-52-6066876 or +972-8-6472646 (R.A.-F. & A.Z.)

Received: 30 September 2020; Accepted: 17 November 2020; Published: 20 November 2020

Abstract: Obesity is associated with food and eating addiction (FA), but the biobehavioral markers of this condition are poorly understood. To characterize FA, we recruited 18 healthy controls and overweight/obese adults with (n = 31) and without (n = 17) FA (H-C, FAOB, NFAOB, respectively) to assess alpha brain asymmetry at rest using electroencephalogram; event-related potentials following exposure to high-calorie food (HCF), low-calorie food (LCF), and nonfood (NF) images in a Stroop paradigm; reaction time reflective of the Stroop bias; and symptoms of depression and disordered eating behavior. The FAOB group had the greatest emotional and uncontrollable eating, depressive, and binge-eating symptoms. The FAOB group displayed lower resting left alpha brain asymmetry than that of the NFAOB group. Differently from the other groups, the FAOB group presented attenuated Stroop bias following exposure to HCF relative to NF images, as well as a lower late positive potential component (LPPb; 450–495 ms) in both frontal and occipital regions. In the total cohort, a correlation was found between the Stroop bias and the LPPb amplitude. These results point to biobehavioral hypervigilance in response to addictive food triggers in overweight/obese adults with FA. This resembles other addictive disorders but is absent in overweight/obesity without FA.

Keywords: food addiction; obesity; brain asymmetry; event-related potentials; food stroop; attention bias; cue responsivity

---

1. Introduction

Obesity is a growing public health condition associated with significant co-morbidities, including diabetes, heart disease, cancer, and depression [1,2]. Consequently, it carries an enormous economic and public health burden [3]. Traditional remedies to reduce the prevalence of overweight and obesity, such as diet and physical activity, are often successful, but most post-obese individuals fail to maintain a healthy body weight in the long run [4]. In the past decade, the clinical construct of addiction to food and eating (i.e., food addiction; FA) has been suggested to explain why some obese individuals are resistant to conventional weight management regimens. The concept of addiction to food is not under a consensus though, and it is not yet recognized in the Diagnostic and Statistical Manual of Mental Disorders (DSM). Proponents of FA posit that there are psychophysiological commonalities between obesity and substance or behavioral addiction [5], while opponents claim that this construct is highly related to binge-eating disorder (BED) [6,7], and that the term "addiction" is inappropriate since food is a legitimate necessity, not a substance individuals can abstain from [8,9]. Nevertheless, eating behavior commonly seen in individuals with overweight and obesity may resemble symptoms of substance or behavioral addiction, as specified in the DSM. This has led researchers to propose FA, by extrapolating criteria for addiction diagnosis in the DSM and translating them to the food and eating behavior domain [10]. These symptoms include frequent and excessive cravings for rewarding food [11], associated with an urgency to relieve stress and negative affect; hypersensitivity to external cues signaling the rewarding food [12]; impulsivity [13] and disinhibition of eating restraint in response to the cues [14]; recurrent overeating past the point of satiety [15], and reduced inhibitory control over eating [16], despite being conscious to the adverse consequences of the behavior [17]. Individuals with these symptoms may experience binge-eating episodes, but many do not satisfy the diagnostic criteria for BED, such as eating binges that occur in a 2-h time window [18]. In fact, in a sample of obese treatment seeking BED patients, only a subset of them met FA criteria [19,20]. Therefore, addictive-like eating symptoms may accelerate chronic obesity rates, above and beyond the incidence of BED.

The psychobiological mechanisms of recurrent overeating in overweight and obesity are partially characterized. A decade of research in this area has been channeled into several theories to explain excessive overeating for reward [21]. The incentive sensitization theory, for example, posits that by repeatedly overconsuming rewarding food, regardless of metabolic hunger, a conditioned food-reward response is formed. The rewarding food is commonly hyperpalatable, high in calories, ultra-processed fat, sugar, and/or salt, and it is considered to have an addictive potential [22,23]. With excessive repetition of this behavior, the conditioned response transfers from a consummatory to an anticipatory food reward (i.e., the reward associated with anticipating the consumption of this food) [24]. In this way, a food cue, such as its olfactory or visual properties, is paired with an anticipatory reward in a conditioned learning mechanism, which stems from neurobiological changes in the mesolimbic dopaminergic system [12]. The anticipatory reward from consuming hyperpalatable food can elicit intense cravings in response to the relevant cue [25]. These food cravings are associated attention bias (AB), or the allocation of attentional resources to the relevant cue. Indeed, multiple studies have confirmed the heightened allocation of attentional systems to food cues in obese, compared with lean individuals [26–28].

Another theory that may explain recurrent overeating in obesity is the inhibitory control deficit theory. This theory emphasizes that individuals keep overeating due to dysfunctional activity in brain regions involved in higher-order control functions, which are under the command of inhibitory control regions in the prefrontal cortex (PFC). According to this theory, a dysfunctional PFC activity may accelerate repeated overconsumption of highly rewarding food, even in the absence of metabolic hunger, due to reduced inhibitory control over eating [15].

The two neurocognitive theories described herein are in line with the approach-avoidance motivational direction theory, according to which individuals with left-sided brain prefrontal asymmetry (i.e., greater left, compared with right, PFC activity) are more likely to be responsive to reward, and/or seek out experiences generating a reward [29]. Left-brain asymmetry is also associated with impaired ability to avoid (or inhibit) behaviors generating undesirable consequences. Differently, individuals with right-sided brain prefrontal asymmetry (i.e., greater right, compared with left, PFC activity) are more likely to avoid experiences generating punishment or disgust, and/or keep away from those generating a reward.

In line with the asymmetry hypothesis of obesity [30], a growing body of literature suggests that left-brain asymmetry may be a neurocognitive characteristic of uncontrollable hedonic overeating (i.e., the overconsumption of highly rewarding food in the absence of hunger), and impaired inhibitory control over food consumption [31,32]. There is evidence in females ranging from lean to obese for the presence of left PFC asymmetry as a possible mediator of the association between AB to highly rewarding food and a high BMI [33]. Moreover, participants with obesity and BED, compared with obesity only, display greater left-hemispheric regional cerebral blood flow in response to rewarding food [34]. These studies support the hypothesis that left-brain asymmetry is a neuronal mechanism that preserves aberrant eating behavior in obesity and BED, but no study to date examined this neurobiological feature directly in FA.

The neurocognitive features of heightened attentional sensitivity to an addictive rewarding cue, coupled with inhibitory control deficits over the addictive behavior, and left-brain asymmetry traits, are commonly seen in Substance Use Disorder (SUD) and behavioral addiction [35–37]. Therefore, there is some support for similarities between FA and other types of addiction. In an fMRI study of participants ranging from lean to obese, a correlation between FA scores and increased reward circuitry reactivity was found in the Anterior Cingulate Cortex (ACC), medial orbitofrontal cortex (OFC), and amygdala during anticipation of chocolate milkshake consumption [38]. This was observed jointly with reduced inhibitory control activation in the dorsolateral prefrontal cortex (DLPFC) and the caudate in response to the actual consumption of the food, when comparing participants with high vs. low FA scores. Similar neuronal alterations have been identified in classical addiction, in anticipation and the actual intake of the substance [39–42].

Previous studies have identified reward responsiveness and AB to highly rewarding food cues in FA. For example, reward-responsive eating mediated the relationship between elevated dopamine signaling and FA scores [43]. Moreover, in female overweight adults with FA, AB to rewarding food was identified in an eye-tracking paradigm following a negative mood induction, but the neurobiological mechanisms have not been studied [44]. Therefore, no study thus far examined brain asymmetry and neuronal correlates of AB to highly rewarding food in FA, to empirically test these aspects of neurocognition. Moreover, no study to date examined the specific psychobiological distinction between overweight/obesity with FA relative to overweight/obesity without FA.

To address the knowledge gap in the FA literature, we measured brain asymmetry at rest, cue-reactivity to images of rewarding food in a Stroop Task while measuring EEG, and several psycho-behavioral parameters, including eating behavior (emotional, uncontrollable, restraint, and binge-eating behavior), and depressive symptoms. We compared these parameters between overweight/obese adults with and without FA (FAOB and NFAOB, respectively), as well as lean controls. We hypothesized that the FAOB group, compared with the other two groups, shows greater left-brain asymmetry at rest and AB to food cues, along with heightened evoked electrophysiological responses, in the Stroop task. We also hypothesized greater depressive, uncontrollable and emotional eating, and binge-eating symptoms in the FAOB group, compared with the other two groups.

2. Materials and Methods

2.1. Participants

As part of a larger clinical trial aiming to test the effect of repetitive transcranial magnetic stimulation)TMS(on FA and obesity, we recruited 66 adults (ages 18–65), 48 of whom were overweight and obese and 18 were healthy controls (H-C) (body mass index (BMI) >= 28 and 19 ≤ BMI ≤ 25, respectively; see Table 1 for age and gender distribution). Participants were recruited by ads and assessed for several demographic parameters, including BMI and age, using a short screening questionnaire. Before recruitment, participants were also assessed for FA with the Yale Food Addiction Scale (YFAS) [45], of which details can be found in Section 2.3. Psycho-Behavioral Questionnaires below. To be recruited for the FAOB group, participants had to meet three or more YFAS symptoms and score positive on the YFAS clinical distress questions. All obese or overweight participants who did not get an FA diagnosis on the YFAS (i.e., they had either lower than three symptoms, or scored negative on the clinical distress questions) were recruited to the non-addicted overweight and obese group (NFAOB). The H-C comparison group had to have a healthy BMI and show no YFAS FA diagnosis. Therefore, the study included three groups of participants: overweight and obese with FA (FAOB, $n = 31$), overweight and obese without FA (NFAOB, $n = 17$), and healthy controls (H-C, $n = 18$).

Table 1. Demographic and clinical data of the participants in the study.

Variable	FAOB (M ± SE)	NFAOB (M ± SE)	H-C (M ± SE)	*p* Value
Age	39.05 (13.01)	37.56 (12.9)	34 (6.38)	0.34
Gender (M; F)	7; 23	6; 10	9; 9	0.24
Education	14.3 (0.33)	15.13 (0.52)	15.07 (0.44)	0.27
BMI	34.54 (0.76)	34.29 (1.17)	22.98 (0.38)	<0.0001 ^#§
VAS (hunger)	4.81 (0.2)	5.5 (0.19)	5.06 (0.28)	0.11
YFAS-S	5.71 (0.26)	3.28 (0.54)	1.38 (0.21)	<0.0001 ^#*§
TFEQ-EE	9.42 (0.55)	7.75 (0.6)	5.56 (0.45)	<0.0001 ^#*
TFEQ-UE	27.96 (0.81)	24.94 (1.18)	16.78 (1.06)	<0.0001 ^#*
TFEQ-CR	13.27 (0.62)	13.72 (0.97)	15.31 (1.1)	0.22
BDI	10.46 (1.34)	5.07 (1.19)	2.88 (0.9)	<0.0001 #*
BE	6.73 (1.19)	1.09 (0.45)	0.07 (0.05)	<0.0001 #*
PANAS	17.87 (7.93)	21.73 (8.35)	17.12 (7.25)	0.21

^ NFAOB vs. H-C ($p < 0.05$); # FAOB vs. H-C ($p < 0.05$); * FAOB vs. NFAOB ($p < 0.05$); § Controlling for BE. FAOB: overweight and obese with food addiction. NFAOB: overweight and obese without food addiction. H-C: healthy controls.

Potential participants were required to have no conventional weight loss attempt currently or in the past three months. As part of the screening, all participants filled out a short medical assessment questionnaire to ensure adherence to the inclusion/exclusion criteria of the TMS trial. Exclusion criteria included a cognitive or functional disability diagnosed within the past year; starting or changing a psychotropic prescription in the past three months; substance abuse, current or in the past 12 months; known or suspected pregnancy or lactation; and practicing veganism. Participants signed an informed consent approved by the Soroka Medical Center in Beer-Sheva, Israel.

2.2. General Procedures

The procedures' timeline is shown in Figure A1. In the current study, we aimed to examine AB to rewarding food cues in the absence of physiological hunger. This is based on evidence pointing to the interaction between physiological hunger and food reward, cravings, and attention bias to food cues [28,46–49]. We therefore asked the participants to follow preparatory guidelines before arriving at the lab. The guidelines included a dietary menu to follow 24-h before they arrive at the lab. Participants arrived between 9 a.m.–12 p.m. on the day of the study, following an overnight fast (starting at 8 p.m. the night before), except for water or unsweetened hot beverages [50]. Adherence to these guidelines was assessed with a nutritionist upon arrival to the study. At the lab, participants' weight was measured with a Charder scale (MS4900), and they filled out several questionnaires about their eating behavior (see Section 2.3. Psycho-Behavioral Questionnaires below). Thereafter, they were provided with a standardized breakfast composed of bland food, totaling 640 calories, 43 g of protein, and 27 g of fat. The purpose of the dietary preparation was to standardize hunger and control for metabolic variations between the participants, and by providing the bland breakfast we aimed to intensify their cravings for highly rewarding food cues [47,49,51]. Following breakfast, participants were escorted to a room where EEG recordings and a Food Stroop task were conducted (see below).

2.3. Psycho-Behavioral Questionnaires

2.3.1. YFAS

This is a 25-item instrument asking about eating highly rewarding food, and it provides two types of score: a symptom count (between 1–7) and a clinical significance score (either 0 or 1) that pertains to clinical distress associated with the symptoms experienced. The symptom count is based on the seven substance dependence criteria in the diagnostic and statistical manual of mental disorders, Fifth Edition (DSV-V), where each criterion is measured with item questions containing frequency (i.e., ranging from 'never' to 'four or more times a week, or daily') or a dichotomous (i.e., 'yes' or 'no') scoring. For each substance dependence criterion, item scores are added up and transformed to a dichotomous score (i.e., '0' or '1'), to reflect if the criterion has been met. All criteria scores are then added up to provide an overall symptom score between 0–7 (see Table A1 for example of items and scoring of the YFAS).

The clinical significance score pertains to impairment or distress associated with the problematic eating symptoms. This score is calculated based on two questions with five response criteria, ranging from 'never' to '4 or more times a week, or daily'. Here too, item scores are added up and transformed to a dichotomous score of either '0' or '1', reflecting the final clinical significance score. Food addiction can be 'diagnosed' when at least three symptoms, plus the criterion of a clinically significant impairment or distress, are met. This instrument is the most frequently used tool to assess FA in research. It has good psychometric properties (Cronbach's alpha: 0.84) [52], and it can reliably distinguish between individuals showing symptoms of addictive eating and those who do not [10].

2.3.2. Three-Factor Eating Questionnaire (TFEQ)

Originally a 51-item self-report inventory designed to assess three aspects of eating behavior in obesity [53], the TFEQ has recently been revised to a more concise 18-item instrument with improved psychometric properties [54]. The TFEQ has three subscales: 1. CR (the conscious restriction of food intake to control body weight or to promote weight loss); 2. UE (uncontrollable eating; the tendency to eat more than usual due to loss of control over food intake); and 3. EE (emotional eating; overeating during dysphoric mood states). The Cronbach's alpha of the TFEQ is 0.79, 0.85, and 0.87 for the three subscales, respectively [54].

2.3.3. Binge Eating (BE) Symptoms

To assess for BE symptoms, we used the Eating Disorders Examination Questionnaire with Instructions (EDE-Q-I) [55], which is a short questionnaire that can accurately assess the frequency of BE symptoms [56]. This instrument has shown comparable to the gold standard, the EDE interview, in assessing BE symptom frequency [56] (see Appendix B for the specific questions used in the EDE-Q-I to assess BE symptoms).

2.3.4. Beck Depression Inventory (BDI)

The BDI is a well-established and highly reliable (Cronbach's alpha: 0.87) self-assessment questionnaire of depressive symptoms [57]. It contains twenty-one questions, each having a 4-point Likert scale, ranging from 0 (no symptoms) to 3 (severe symptoms). Scores on the BDI indicate mild, moderate, severe, or no depression (represented by scores of 14–19, 20–28, 20–63, and 0–13, respectively) [58].

2.3.5. Positive Affectivity Negative Affectivity Schedule (PANAS)

The PANAS is a self-rating measure of positive affect (PA; 10 items) and negative affect (NA; 10 items), that reflects transitory mood states [59]. This scale has good internal consistency reliabilities (Cronbach's alpha of 0.88 for the PA and 0.87 for the NA) [60]. Past research has shown the potential influence of affective states on cognitive function in obesity [61]. Therefore, we administered the PANAS following breakfast consumption, to control for the potential effect of affective variability on AB in the Stroop task (see Statistics below).

2.3.6. Visual Analogue Scale (VAS)-hunger

We administered a Visual Analogue Scale of hunger and fullness (VAS-hunger) before the Food Stroop task, to control for a possible effect of variations in hunger and/or cravings for highly rewarding food on attention bias in the Stroop task [62] (see Statistics below). The VAS-hunger is a 7-point Likert scale ranging from 1 ("extremely hungry") to 7 ("extremely full"), and participants are asked to circle what best describes the way they feel at the moment of measurement.

2.4. The Food Stroop Task

The Stroop task is a reliable and widely used neurocognitive tool to assess AB, including in addiction research [63]. We administered a Food Stroop task using the E-Prime software (Psychology Software Tools, Inc., Sharpsburg, PA, USA) on a 17" computer screen, adjacent to a keyboard with four keys denoted using stickers with different colors (red, green, yellow, and blue; color-key mapping was counterbalanced between the participants). The Food Stroop is a version of the combi-Stroop test [64,65], in which food images precede the Stroop stimuli, to test their influence on participants' emotional attention. Food images have been commonly used in appetite [66], obesity [67,68], and addiction [63] research. The food images are considered more potent than words [69,70], since they can elicit cravings for food in a similar manner to real food exposure, capturing attention, saliency, and reward [69].

The Food Stroop paradigm is illustrated in Figure 1. Each trial started with a fixation cross presented for 800 ms, followed by a food or nonfood image presented for 500 ms. Thereafter, the Stroop word was presented for a maximal duration of 2500 ms, or until the participant responded. The inter-trial interval was randomly set between 1100–1900 ms.

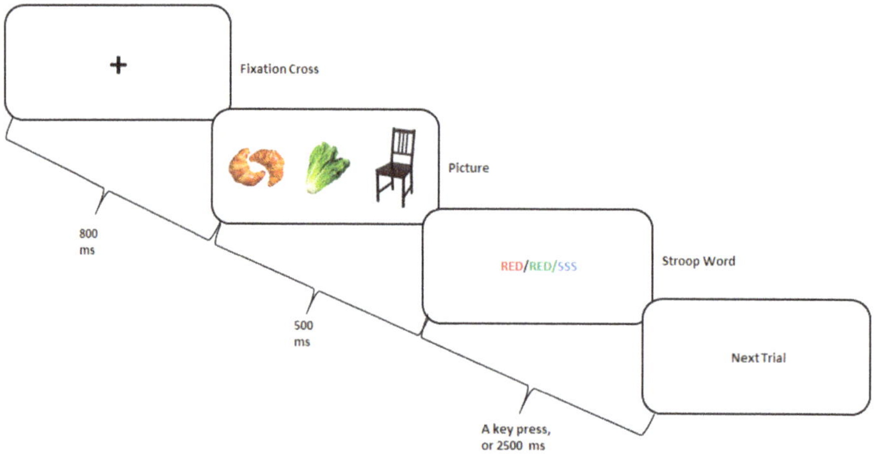

Figure 1. The Food Stroop task experimental paradigm. A fixation cross was presented for 800 ms, followed by a high-calorie food, low-calorie food, or nonfood picture presented for 500 ms, and by a congruent, incongruent, or neutral Stroop word presented for 2500 ms, or until the participant responded. The next trial randomly began 1100–1900 ms thereafter.

There were three picture categories: high-calorie food (HCF), low-calorie food (LCF), and nonfood (NF) items, such as furniture. The pictures were retrieved from a freely accessible, previously validated, database [71] and controlled for size, color, and shade. For vegetarian participants, we omitted pictures containing meat and replaced them with vegetarian dishes.

The pictures were followed by either one of the three Stroop conditions, randomly presented. In the three conditions, the meaning of the Stroop word (i.e., "RED", "GREEN", "YELLOW", or "BLUE") was either congruent, incongruent, or neutral with respect to its color. For example, for the congruent condition, the word "RED" was painted in red; for the incongruent condition, the words "RED" was painted in green, yellow, or blue; and for the neutral condition, a neutral word was presented in any one of the four colors available. The task included 24 practice trials with no pictures presented, followed by 3 blocks of 180 trials each, with a one-minute break in-between. This sums to a total of 540 trials, sixty for each combination of picture typ, and Stroop condition [i.e., (3 picture types × 3 Stroop conditions) × 60]. Overall, 180 pictures were presented, each shown once with each Stroop condition. Participants were instructed to press as quickly and accurately as possible, on the key associated with the color of the word, while ignoring the word's meaning.

Mean color-naming latencies (i.e., reaction time, RT) were calculated, excluding error-response trials or ones with RT > 1200 ms. Prior to the beginning of the Stroop task, the VAS-hunger was administered.

2.5. Electrophysiology (EEG) Procedures

EEG was recorded using a 64 electrode Wave guard cap (ANT neuro, Enschede, The Netherlands) and a TMS compatible EEG amplifier (TMSi, Oldenzaal, The Netherlands) referenced to the Cz electrode. Data were acquired using ASA™ version 4.7.3. Impedance was kept below 5 kOhm, and POz was determined as the ground electrode. The recording frequency was set to 2048 Hz and digitized with a 24-bit AD converter. Off-line data processing was conducted using EEGlab 14.0.0 [72] and the FieldTrip toolbox in Matlab (version R2018a; MathWorks Inc., Natick, MA, USA) [73,74], by a trained researcher.

EEG was recorded over a 5-min resting period with lights dimmed, during which the participants were instructed to keep their eyes closed while remaining alert [73,75]. The first 30 and the last 10 s of the resting state EEG data were removed to prevent state transitional influence. The data were filtered using a 1 Hz high-pass and 45–55 Hz notch FIR filters and then epoched into segments of 2 s each. Noisy channels and epochs were detected using an automatic procedure, which was followed by both manual review and rejection, performed by a trained researcher (mean ± SD of epochs and channels rejected in accordance: 14.65 ± 6.02 and 1 ± 0). The criteria for rejection were an absolute amplitude threshold of ±150 µV and an improbability threshold of 3 standard deviations above the mean of the entropy value. An infomax Independent Components Analysis (ICA) was conducted; eye blinks and movements were manually removed (2.3 ± 0.94 components), and a second automatic and manual scan assured no residual noisy segments were left (6.33 ± 7.25 epochs removed). Next, rejected channels were interpolated using spline interpolation; data were re-referenced to the average and transformed to the frequency domain using a fast Fourier transform (1–100 Hz; frequency resolution of 0.125 Hz). Finally, alpha power (8–12 HZ) was extracted for each electrode, and brain asymmetry scores were calculated as the decibel-transformed ratio in alpha power between each left electrode and its homologues right electrode (midline electrodes excluded), using the formula below, previously described [75]:

$$\begin{aligned}\text{Left alpha brain asymmetry} = 10 \times \log_{10}(\textit{left electrode's alpha power}) \\ - 10\log_{10}(\textit{contralateral homologous electrode's alpha power})\end{aligned} \quad (1)$$

EEG data collected during the food Stroop task were segmented around the appearance of the food/nonfood image (regardless of Stroop condition), starting at 500 ms before, and ending 1000 ms after image appearance, and then baseline corrected. The data were preprocessed similarly to the resting state EEG analysis, with a manual rejection of noisy epochs (8.19 ± 9.36). Then, an ICA was performed on all recordings (5.05 ± 2.73 components rejected), with additional detection of noisy epochs manually (17.13 ± 17.9) and automatically (31.88 ± 11.73), leaving an average of 161.15 ± 11.31 epochs per food condition. The data were then analyzed in the time domain, as Event-Related Potentials (ERP), in response to a food-specific stimulus.

Missing behavioral data were removed from the analysis. Table A2 details the number of participants included in the analysis of each variable studied.

2.6. Statistics

All statistical inference tests were performed using two-tailed tests requiring an a-priory alpha level of 5%. Participants' demographic and clinical data were analyzed using descriptive statistics and a 1-way ANOVA, with the group as a between-subject factor. In the food Stroop task, to capture the maximal emotional and motivational attention elicited by the food cues [63], we focused on the differential effect of HCF and NF images on participants' performance in the task. Therefore, the Stroop bias score (incongruent-congruent) was computed and analyzed using a 2-way mixed model ANOVA (STATISTICA 13; TIBCO Soft Inc., Seattle, WA, USA) design, with image type (HCF and NF) as a within-subjects factor and the group (FAOB, NFAOB, and H-C) as a between-subjects factor. Post-hoc significance tests were Bonferroni corrected. When analyzing the Stroop bias, we controlled for PANAS as well as for VAS-hunger, based on previous addiction Stroop studies [28,40,44], demonstrating the necessity of controlling for variations between participants in hunger and transitory mood states.

All EEG data were analyzed using nonparametric permutation tests (Monte Carlo method), implemented via the FieldTrip toolbox. This method samples the data repeatedly and randomly (10,000 iterations) to evaluate the characteristics of the sample's distribution under the null hypothesis, obviating the need for prior assumptions concerning its normality.

Brain asymmetry power scores were first averaged for three regions of interest (ROI), including frontal (F5, F3, FC5, and FC3), parietal (CP5 and CP3), and occipital (PO5, PO3, and PO7) electrodes. This method was determined based on previous brain asymmetry research in healthy individuals [76] and in the eating behavior domain [77], depicting these ROI and pointing to their validity in studying brain asymmetry. Next, group differences were tested using permutation tests following the ANOVA logic, i.e., pair-wise post-hoc contrasts were tested only if differences between the three groups were first found significant in one of the ROI [73].

ERP components related to the processing of the food images during the food Stroop task were defined prior to the statistical analysis and regardless of group affiliation, using Global Mean Field Power analysis (see Figure A2) and in line with earlier literature [65,78]. We focused our investigation on the Late Positive Potential (LPP) component for two reasons: 1. It reflects the emotional processing of arousing affective pictures [79,80]. 2. It was the only component to show an association with the attention bias induced by the food cues (Figure 5). Note that if not interrupted by another stimulus (i.e., the next trial), the LPP may last a few seconds [79,80]. In contrast, in the current paradigm, this ERP is decreased in preparation for the Stroop word; thus, we subdivided this component into LPPa (300–450 ms) and LPPb (450–495 ms), to better align with differential motivational processes. Three regions of interest (ROI) were defined to capture the anterior and posterior parts of the LPP: frontal (electrodes: F1, Fz, F2, FC1, FCz, FC2), right posterior (P2, P4, P6, P8, PO4, PO6, PO8, O2), and left posterior (P1, P3, P5, P7, PO3, PO5, PO7, O1). Next, mean amplitude differences between HCF and NF were computed for the LPPa and LPPb and subjected to permutation tests as described above.

3. Results

3.1. Participants' Demographics and Clinical Data

Detailed demographic and clinical data of the three groups are shown in Table 1. The groups did not differ in age, gender, or education level. As expected, the three groups differed in FA symptoms; this difference stayed significant when controlling for BE symptoms. As expected, the H-C differed in BMI from both FAOB and NFAOB groups. The three groups differed in their emotional and uncontrollable eating symptoms (TFEQ-EE and TFEQ-UE, respectively), while the FAOB group showed the greatest degree of symptoms. The NFAOB and H-C groups did not differ in depressive or binge-eating symptoms (BDI and BE, respectively), but both groups differed from the FAOB group, who showed the highest level of symptoms. Lastly, the three groups did not differ in cognitive restraint related to food (TFEQ-CR), nor in their PANAS or VAS-hunger scores, but the FAOB group compared with the other two groups, showed a trend for greater hunger (i.e., lower scores on the scale) on the morning of the study.

3.2. Hemispheric Brain Asymmetry

No differences in the raw EEG power were found between the groups (see Figure A3). Non-parametrical permutation tests based on F statistics indicated that brain alpha asymmetry scores differed between the three groups in the frontal (F = 3.45 $_{(n = 49)}$, $p = 0.03$) but not in the parietal (F = 2.33 $_{(n = 49)}$, $p = 0.09$) or the occipital ROI (F = 2.52 $_{(n = 49)}$, $p = 0.08$; (Figure 2)). Post-hoc contrasts indicated that the FAOB group showed lower left alpha brain asymmetry scores (i.e., lower left alpha compared with the right alpha), differently from the NFAOB group, in the frontal and occipital ROI (frontal ($t = 2.36$ $_{(n = 33)}$, $p = 0.02$), parietal (t = 1.68 $_{(n = 33)}$, $p = 0.11$), and occipital (F = 2.54 $_{(n = 33)}$, $p = 0.01$)), but neither of the overweight and obese groups statistically differed from the H-C group. Since the alpha power is a brain frequency inversely related to local brain activity [81], the lower alpha brain asymmetry means greater left, compared with the right, hemispheric dominance in the FAOB group compared with the NFAOB group. Note that nonparametrical tests do not use F or *t* tables; hence, the sample size is detailed rather than the degrees of freedom. Considering previous research [82], we tested for all the participants together the correlation between left alpha brain asymmetry and gender, which was not significant ($r = 0.19$ $p = 0.15$). Therefore, we did not use gender as a covariate in our electrophysiological analyses.

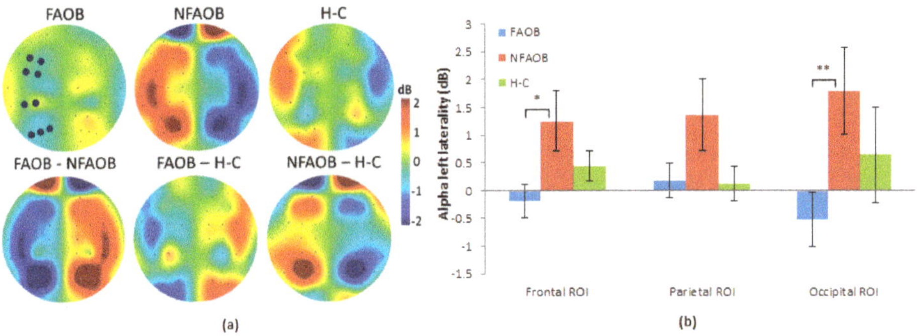

Figure 2. Group differences in brain alpha asymmetry. (**a**) Topographic plots of the mean group left alpha asymmetry power (upper row) and the group contrasts (lower row), for the FAOB, NFAOB, and H-C groups. Note that the maps are based on subtraction of alpha power between symmetric electrode pairs and are thus left/right mirrored. (**b**) Mean and SEM of left alpha asymmetry power for each group in the frontal, parietal and occipital ROI. Electrodes included in each ROI are marked in dark blue in 2a. *$p \leq 0.05$, **$p \leq 0.01$. FAOB: overweight and obese with food addiction. NFAOB: overweight and obese without food addiction. H-C: healthy controls. SEM: standard error of mean. ROI: region of interest.

3.3. The Effect of Food Cues on Attention Bias and Brain Potentials

ANOVA of the Stroop attention bias in response to food cues in the Food Stroop task revealed a significant 2-way group * image type interaction ($F_{(2,56)} = 3.46$, $p = 0.04$) (Figure 3). This interaction was statistically significant when controlling for VAS-hunger and PANAS scores (mean scores: 6.62 and 20.1, respectively). Post-hoc tests indicated a significant reduction of the Stroop bias following HCF compared to the NF image, which was observed in the FAOB but not in the other two groups ($p = 0.05$; Figure 3).

Nonparametrical permutation tests indicated differences between the three groups in brain potentials in response to food cues. There was an amplitude change induced by the food cues (HCF vs. NF), in the LPPb but not in the LPPa, in the frontal (F = 4.09 $_{(n = 52)}$, $p = 0.02$) and right occipital (F = 3.78 $_{(n = 52)}$, $p = 0.02$) but not in the left occipital ROI (F = 1.40 $_{(n = 52)}$, $p = 0.26$) (Figure 4). Post-hoc contrasts revealed that the FAOB group had a lower food cue-induced amplitude change in the LPPb compared with the NFAOB group, in the frontal ($t = 2.84$ $_{(n = 37)}$, $p = 0.0058$) and right posterior ($t = 2.62$ $_{(n = 37)}$, $p = 0.0013$) ROI. Neither the FAOB nor the NFAOB groups statistically differed from the H-C group.

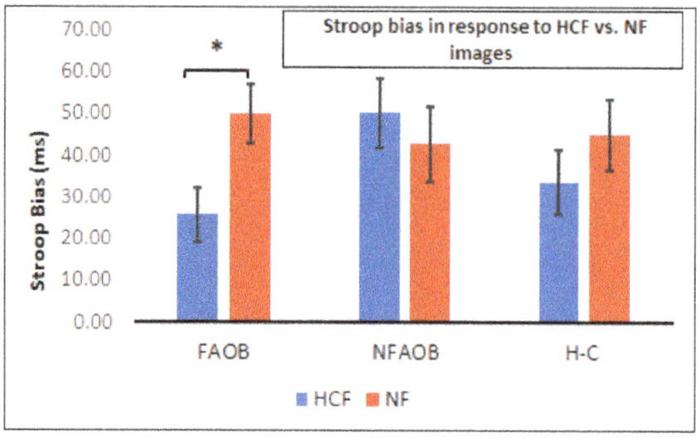

Figure 3. Attention bias following food and nonfood cues. A graphical depiction of mean and SEM of the Stroop bias following HCF and NF images in the FAOB, NFAOB, and H-C participants. * $p < 0.05$. SEM: standard error of mean. HCF: high-calorie food. NF: nonfood. FAOB: overweight and obese with food addiction. NFAOB: overweight and obese without food addiction. H-C: healthy controls.

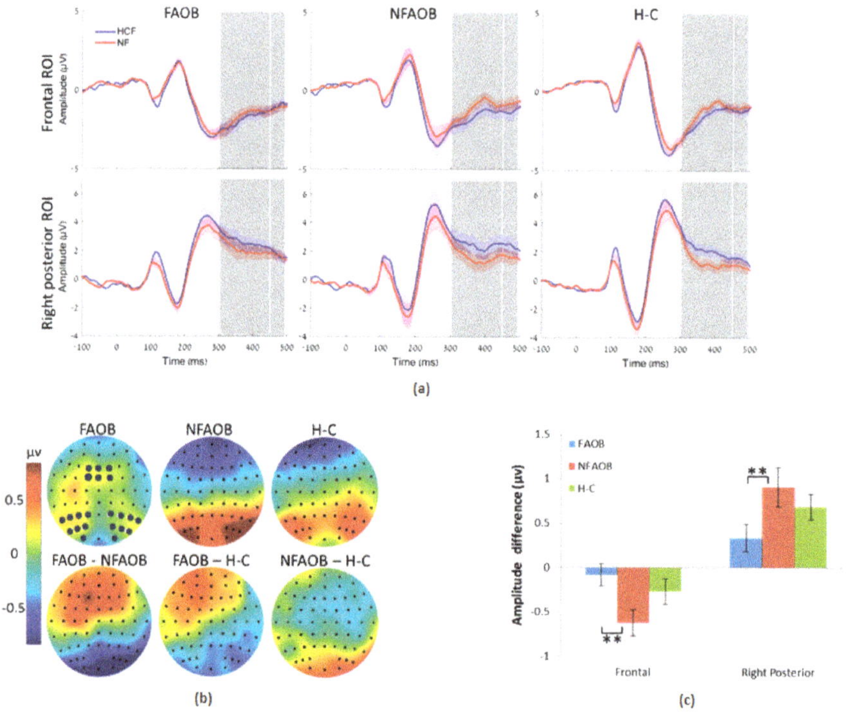

Figure 4. Brain potentials in response to food cues. (**a**) Event related potential curves in the FAOB, NFAOB and H-C groups, following HCF compared with NF images, in the frontal and right posterior ROI (no group differences in the left posterior ROI, hence not shown). Curve's shades mark SEM; TOIs (LPPa, LPPb) are marked in gray. (**b**) Topographic plots of mean group amplitudes (upper row) and group contrasts (lower row) during the LPPb. Electrodes of the different ROI are marked in dark blue (**c**) Mean and SEM of the same in the frontal and right posterior ROI. ** $p \leq 0.01$. FAOB: overweight and obese with food addiction. NFAOB: overweight and obese without food addiction. H-C: healthy controls. HCF: high-calorie food. NF: nonfood. ROI: region of interest. SEM: standard error of mean. TOI: times of interest. LPP: late positive potential.

Correlational analysis between the food cue-induced differences in both the Stroop bias and the LPPb amplitude revealed a negative correlation in the frontal ROI ($r_{(n=51)} = -0.33$, $p = 0.02$) and a positive correlation in the right posterior ROI ($r_{(n=51)} = 0.30$, $p = 0.03$; Figure 5). Therefore, the food cue-induced amplitude change was linearly associated with the Stroop bias change following HCF compared to NF images.

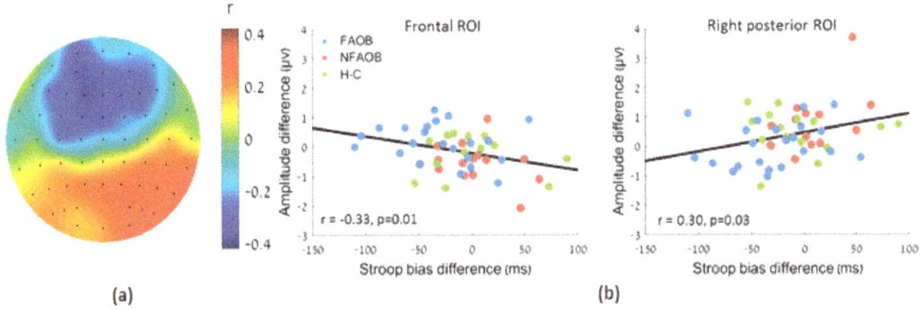

Figure 5. Association between the amplitude change and the Stroop bias change following HCF compared to NF images, for the participants altogether. (**a**) A topographic plot of the linear correlation magnitude between the food-cue induced LPPb amplitude and the Stroop bias changes. (**b**) Scatter plots of the same in the frontal and right posterior ROI. HCF: high-calorie food. NF: nonfood. LPP: late positive potential. ROI: region of interest.

4. Discussion

In the current study, we aimed to investigate psychobiological indices of food addiction in overweight and obesity. The results provide novel and unique neurobehavioral and psycho-cognitive markers characterizing FA versus no FA in overweight and obese participants. Our hypotheses were partially confirmed; during rest, the FAOB group showed greater left-brain asymmetry than that of the NFAOB group. This neurobiological signature is in line with approach motivation tendencies [82], indicating greater vulnerability to approach a motivationally salient cue relevant to the addictive condition [83]. The greater left-brain asymmetry at rest in the FAOB group may indicate a neuro-marker of repeatedly and compulsively approaching highly rewarding food, similarly to those observed in substance addiction [36]. Conversely, the greater right-brain dominance in the NFAOB group in comparison with the FAOB and (to a lesser extent) the H-C groups, may protect them from developing FA symptoms. The lack of addictive features in the NFAOB group may be reflected in their right frontal asymmetry. An extensive body of research points to inhibitory control deficits in obesity, which is associated with impulsivity and a lack of delayed discounting of food-related reward [84,85]. To overcome impulsivity toward food reward, the NFAOB may have developed heightened compensatory mechanisms in their right brain hemisphere, which is absent in healthy adults and the FAOB group, the latter who may be lacking the capacity to consistently control their food intake.

Our findings are in line with past research indicating left PFC asymmetry in chronic overeaters [31], in adults with high hedonic hunger (i.e., a drive for a food reward in the absence of hunger) [86], and in obesity [87]. In the current study, we also found left-brain asymmetry in the occipital ROI, and a general trend of reduced left alpha asymmetry in the whole hemisphere, possibly indicating a widespread left hemispheric dominance in the FAOB group, extending to brain areas in the sensory association cortex [77]. Specifically, brain responses with robust asymmetry to the left parietal and occipital brain areas may be related to strategic and tactical aspects of goal pursuit [88], such as reward valuation and integration [77,89], as well as hedonic valuation of food [77]. A widespread asymmetry that includes temporal and occipital electrodes has been observed in healthy individuals [76] and in psychiatric patients [90], but this is the first study that shows this electrophysiological marker in food addiction. Therefore, additional studies are needed to help uncover the significance of these findings.

In the Food Stroop task, participants viewed pictures of highly rewarding food, as well as nonfood items, before the Stroop word assignment. We hypothesized that the FAOB participants will show greater cue-reactivity in response to highly rewarding food cues, reflected in greater Stroop bias and heightened ERP responses, than that of NFAOB and H-C participants. This is based on previous literature indicating greater AB in the Food Stroop task in obese compared with lean participants [91,92]. Our results refuted previous findings; at earlier stages of cognitive processing (300–450 ms following picture presentation), the HCF pictures elicited heightened emotional reaction similarly in all groups. Thereafter, during the LPPb (at 450–495 ms) the FAOB group seems to have inhibited their emotional response to the HCF cues, differently from the NFAOB group, who displayed clear electrophysiological difference in response to HCF versus NF images. This neurobiological difference between the groups was behaviorally reflected in their differential performance on the Stroop word assignment, which started immediately thereafter. Indeed, in the Food Stroop task, the FAOB group showed a lower Stroop bias following images of HCF, suggesting that the lack of electrophysiological response to HCF images may result from an inhibitory process. The negative correlation between the neuronal response to HCF vs. NF images and the performance during the Stroop task further implies an inhibitory process of affective response, starting at the neuronal level and reducing attention bias on the Stroop task. The positive correlation between the occipital response to HCF vs. NF images and the performance during the Stroop task implies that increased sensory response to these images (without prefrontal inhibition) may induce the increased attention bias on the Stroop task.

The LPP component is commonly found in obesity research, and it indicates selective and motivated attention to rewarding cues [93], specifically highly rewarding food [94]. Greater LPP component response in addicted vs. non-addicted individuals, following a visual presentation of a cue associated with the addiction, has been shown in cannabis use and may be a neurobiological marker of addiction [36]. High LPP has also been found in response to an acute stressor, when cortisol levels are high [78], implying vigilant attention to a threat. In our sample, the strong inhibition of emotional-motivational reaction to the HCF cues in the FAOB group may point to the hypervigilance-avoidance hypothesis [95,96]. Research is pointing to an interaction between emotional valence and executive control demands in tasks involving attention and cognitive interference [44,97]. Attentional avoidance in a state of emotional vigilance has been observed in dieters who attempt to attentionally avoid pictures of the food they desire [64,98], and in addicted patients exposed to the stimuli they are trying to abstain from [99]. Accordingly, the HCF images in our study may have elicited in the FAOB group a strong affective response. At the early stages of information processing, the FAOB group, similarly to the other two groups, showed an initial heightened emotional response to the appetitive cues, which was thereafter extensively inhibited in the FAOB group, possibly when experiencing hypervigilance with triggers of an emotionally-laden problematic behavior associated with their condition.

The hypervigilance-avoidance hypothesis has been shown in adults with social phobia, who respond faster in a Stroop paradigm with images of socially challenging situations, following an anxiety-inducing task [100,101]. Similarly, overeating of highly rewarding food has been postulated to function as a relief from the physical tension associated with hypervigilance [102], suggesting an addictive cycle whereby compulsivity develops to relief from the psychophysiological tension associated with the condition. Following this hypothesis, individuals with overweight or obesity and FA may vigilantly detect highly rewarding food cues in their environment, propelling a negative affect and a negative urgency to impulsively consume that food. They may try to counteract their tendency to approach highly rewarding food (to relieve their hypervigilance) by exercising cognitive avoidance, up to the point where they disinhibit their restraint and lose their control over eating. At the point of disinhibition, consumption of the food may function to relieve the physical and emotional tension associated with generalized or cue-specific hypervigilance [84]. This hypothesis is in line with the theoretical understanding of impulsivity and loss of control of eating seeing in FA [5]. These behaviors and their neurobiological precursors may also be reflected in left-brain asymmetry [33].

The FAOB differed from the NFAOB (and the H–C) in both binge-eating (BE) and depressive symptoms and in symptoms of emotional and uncontrollable eating. These results are in-line with our hypotheses and replicate previous research comparing individuals with and without FA [103,104]. BED is related to FA [105], but research about distinctions and similarities between the two conditions is in its infancy. FAOB has been suggested to be an extreme form of BED [6]. However, in our sample, only 13 participants out of the 30 (i.e., 43%) in the FAOB group showed BE symptoms [106]. Moreover, BE symptom scores were not correlated with participants' performance on the Food Stroop task, nor with brain asymmetry scores or ERPs, and the three groups in our study differed in FA symptoms and BMI even when controlling for binge-eating scores. These are novel and important findings, pointing to FA as a unique clinical construct, characterized by distinct psycho-neurobiological markers, above and beyond BE symptoms. Future studies may employ the parameters addressed in the current study to directly compare two cohorts of overweight/obese adults: one with FA and the other with BE symptoms/BED.

BE and FA may escalate depressive symptoms [5,104], and there is ample evidence to support the co-occurrence of obesity and depression [2]. In our work, despite greater depressive symptoms in the FAOB group compared with the other two groups, symptoms level did not reach clinical significance but more of a melancholic state [107]. It is possible that the uncontrollable compulsion to eat, low self-esteem [19], self-inefficacy in controlling one's eating and weight [108], and the impairment in the quality of life in overweight/obesity with FA [109], contributed to greater depressive symptoms in the FAOB group.

The present study has strengths and limitations. This study is the first to find neurocognitive markers and psycho-behavioral correlations in overweight/obesity with FA, using brain asymmetry indices and ERP in a Food Stroop task. In the present study, participants' hunger and metabolism were carefully controlled for 24 h prior to, and on the day of, the study, to reduce the chance of confounding variables, such as metabolic hunger [47], biasing the results. Future work may regress these and other potential confounders on the neurocognitive parameters we applied in the current study; this was not performed here and may be a shortcoming of the present work. The present study has a limitation in terms of sample size, particularly in the NFAOB and H-C groups. Moreover, participants' recruitment in the present study poses several limitations to the generalizability of our findings; we did not recruit participants with the co-presence of obesity and SUD [110], and our study lacks a subgroup of FA who shows lean body mass [111], limiting the conclusions to FA in overweight/ obesity. Moreover, we used the original version of the YFAS, which is based on the DSM-IV, since participants' recruitment started prior to the publication of the most updated version, the YFAS version 2 [112]. Therefore, we did not distinguish between mild, moderate, and severe FA symptoms in participants' recruitment. We, therefore, suggest these limitations be addressed in future FA studies. Lastly, our research setting has possibly impacted the participants in the study. Future ecological momentary assessment studies may help examine overweight/obesity with FA in a different, more natural setting, to avoid possible confounding factors of conducting research in the lab.

5. Conclusions

Our study uniquely demonstrated that overweight and obese adults with FA show markers of left-brain asymmetry at rest, relative to overweight/obese adults without FA. In addition, the overweight/obese participants with FA show markers of a hypervigilant inhibition of emotional reaction to food triggers that may elicit excessive cravings, evident in a lower LPPb response to HCF images and reduced AB in a Food Stroop task. Our results are in line with psychobiological markers of SUD and behavioral addiction, and they introduce novel understandings of overweight/obesity with FA. Neurocognitive training and neuro-modulatory treatment, such as transcranial magnetic stimulation (TMS), may help rebalance hemispheric symmetry in obesity with FA. Future studies may also address potential therapies to help individuals with FA cope better with environmental stimuli relevant to their condition.

Author Contributions: Conceptualization, R.A.-F., U.A., A.Z., and L.K.; Methodology, R.A.-F., U.A., A.Z.; software, R.A.-F. and U.A.; validation, U.A.; formal analysis: R.A.-F., L.K., and U.A.; investigation, R.A.F, L.K. and G.B.; resources: A.Z.; Data curation, U.A. and L.K.; writing – initial draft preparation, R.A.-F.; writing - review and editing, U.A., L.K., G.B., and A.Z.; visualization, U.A.; supervision, A.Z.; project administration, R.A.-F., L.K. and G.B.; funding acquisition: U.A. and A.Z. All authors have read and agreed to the published version of the manuscript.

Funding: This research was funded by MAGNET program, Israel Innovation Authority, #61450.

Acknowledgments: We thank Noga Cohen at the Edmond J. Safra Brain Research Center for the Study of Learning Disabilities at the University of Haifa, Israel, for reviewing and commenting on this paper.

Conflicts of Interest: The authors declare no conflict of interest. The funders had no role in the design of the study; in the collection, analyses, or interpretation of data; in the writing of the manuscript, or in the decision to publish the results.

Appendix A

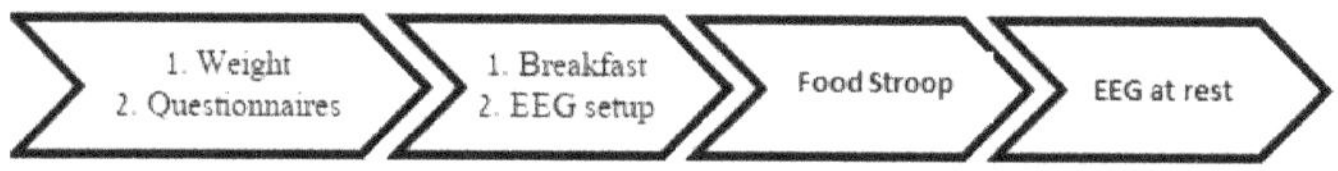

Figure A1. Procedures' timeline.

Table A1. Examples of YFAS items and scoring [45].

Item	Response Categories					Scoring		Criterion
In the past 12 months:	0	1	2	3	4	0	1	
1. I find that when I start eating certain foods, I end up eating much more than planned.	never	once a month	2–4 times a month	2–3 times a week	4 or more times or daily	0–3	4	Substance taken in larger amount and for longer period than intended
2. I find myself continuing to consume certain foods even though I am no longer hungry.	never	once a month	2–4 times a month	2–3 times a week	4 or more times or daily	0–3	4	Substance taken in larger amount and for longer period than intended
3. I eat to the point where I feel physically ill.	never	once a month	2–4 times a month	2–3 times a week	4 or more times or daily	0–2	3–4	Substance taken in larger amount and for longer period than intended
4. Not eating certain types of food or cutting down on certain types of food is something I worry about.	never	once a month	2–4 times a month	2–3 times a week	4 or more times or daily	0–3	4	Persistent desire or repeated unsuccessful attempt to quit
5. I spend a lot of time feeling sluggish or fatigued from overeating.	never	once a month	2–4 times a month	2–3 times a week	4 or more times or daily	0–2	3–4	Much time/activity to obtain, use, recover
6. I find myself constantly eating certain foods throughout the day.	never	once a month	2–4 times a month	2–3 times a week	4 or more times or daily	0–3	4	Much time/activity to obtain, use, recover
7. I find that when certain foods are not available, I will go out of my way to obtain them. For example, I will drive to the store to purchase certain foods even though I have other options available to me at home.	never	once a month	2–4 times a month	2–3 times a week	4 or more times or daily	0–2	3–4	Much time/activity to obtain, use, recover

Table A2. Number of participants included in the analysis.

	FAOB	NFAOB	H-C	Reasons for Drop-out
Stroop Reaction Time	30	15	16	Incomplete recording of RT (1) Failed to record covariate scores (4)
EEG at rest	18	15	16	Insufficient data quality
Event Related Potentials	22	15	15	Insufficient data quality
Food Stroop—ERP correlations	22	14	15	Insufficient data quality

FAOB: overweight and obese with food addiction. NFAOB: overweight and obese without food addiction. H-C: healthy controls. RT: reaction time. ERP: event-related potentials.

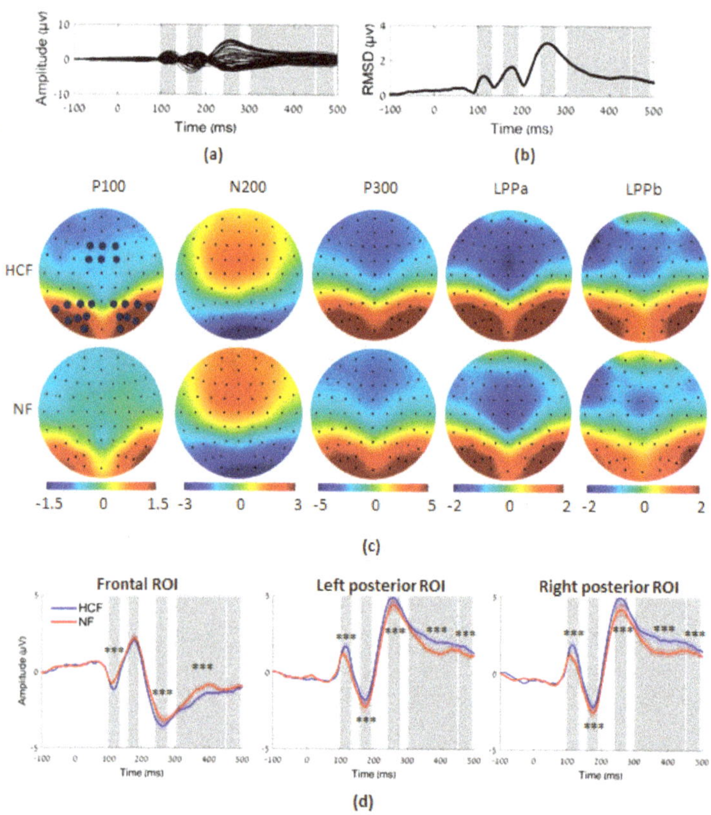

Figure A2. Event-related components in response to food (HCF) and nonfood (NF) images. Data is averaged for all participants together, regardless of group affiliation. (**a**) A butterfly plot of potential curves that include all electrodes. (**b**) Global mean field potential (GMFP) computed sample-wise as the root mean squared deviation (RMSD) from the mean amplitude of all electrodes. The different event-related components and time windows of interest (marked gray) were defined according to the peaks in GMFP and are in line with the literature [65,79]; from left to right: P100 (95–135 ms), N200 (155–195 ms), P300 (240–280 ms), and the Late Positive Potential (LPP) subdivided into LPPa (300–450 ms) and LPPb (450 495). (**c**) Topographic plots of the mean potentials during the different components following HCF and NF images. The frontal, right posterior, and left posterior regions of interest (electrodes marked in dark blue) were defined to capture differences in brain potentials induced by the food images. (**d**) Event-related potential curves in response to HCF and NF images. *** $p \leq 0.001$. HCF: high-calorie food. NF: nonfood.

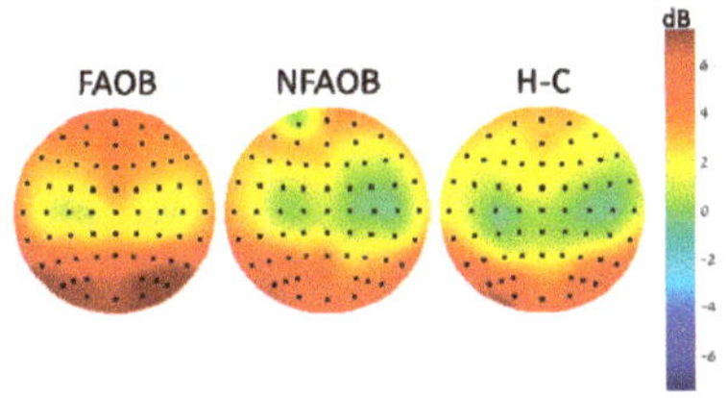

Figure A3. Topographical power maps of the raw resting state EEG activity. Topographical maps of the average resting state power in the experimental groups. No group differences were found using a nonparametric F test in any of the electrodes.

Appendix B

EDE-Q-I questionnaire's questions assessing BE symptoms [55].

"13. Over the past 28 days, how many times have you eaten what other people would regard as an unusually large amount of food (given the circumstances)?

14. On how many of these times did you have a sense of having lost control over your eating (at the time that you were eating)?

15. Over the past 28 days, how many DAYS have such episodes of overeating occurred (i.e., you have eaten an unusually large amount of food and have had a sense of loss of control at the time)?"

References

1. Jarolimova, J.; Tagoni, J.; Stern, T.A. Obesity: Its Epidemiology, Comorbidities, and Management. *Prim. Care Companion CNS Disord.* **2013**, *15*. [CrossRef] [PubMed]
2. Luppino, F.S.; De Wit, L.M.; Bouvy, P.F.; Stijnen, T.; Cuijpers, P.; Penninx, B.W.J.H.; Zitman, F.G. Overweight, Obesity, and Depression. *Arch. Gen. Psychiatry* **2010**, *67*, 220–229. [CrossRef] [PubMed]
3. Colditz, G.A. Economic costs of obesity and inactivity. *Med. Sci. Sports Exerc.* **1999**, *31* (Suppl. 11), S663–S667. [CrossRef]
4. Weiss, E.C.; Galuska, D.A.; Khan, L.K.; Gillespie, C.; Serdula, M.K. Weight Regain in U.S. Adults Who Experienced Substantial Weight Loss, 1999–2002. *Am. J. Prev. Med.* **2007**, *33*, 34–40. [CrossRef] [PubMed]
5. Davis, C.; Curtis, C.; Levitan, R.D.; Carter, J.C.; Kaplan, A.S.; Kennedy, J.L. Evidence that 'food addiction' is a valid phenotype of obesity. *Appetite* **2011**, *57*, 711–717. [CrossRef] [PubMed]
6. Davis, C. Compulsive Overeating as an Addictive Behavior: Overlap Between Food Addiction and Binge Eating Disorder. *Curr. Obes. Rep.* **2013**, *2*, 171–178. [CrossRef]
7. Carlier, N.; Marshe, V.S.; Cmorejova, J.; Davis, C.; Müller, D.J. Genetic Similarities between Compulsive Overeating and Addiction Phenotypes: A Case for "Food Addiction"? *Curr. Psychiatry Rep.* **2015**, *17*, 1–11. [CrossRef]
8. Ziauddeen, H.; Farooqi, I.S.; Fletcher, P.C. Obesity and the brain: how convincing is the addiction model? *Nat. Rev. Neurosci.* **2012**, *13*, 279–286. [CrossRef]
9. Fletcher, P.C.; Kenny, P.J. Food addiction: A valid concept? *Neuropsychopharmacology* **2018**, *43*, 2506–2513. [CrossRef]
10. Gearhardt, A.N.; Corbin, W.R.; Brownell, K.D. Preliminary validation of the Yale Food Addiction Scale. *Appetite* **2009**, *52*, 430–436. [CrossRef]
11. Meule, A.; Kübler, A. Food cravings in food addiction: The distinct role of positive reinforcement. *Eat. Behav.* **2012**, *13*, 252–255. [CrossRef] [PubMed]
12. Volkow, N.D.; Wise, R.A.; Baler, R. The dopamine motive system: implications for drug and food addiction. *Nat. Rev. Neurosci.* **2017**, *18*, 741. [CrossRef] [PubMed]
13. Jasinska, A.J.; Yasuda, M.; Burant, C.F.; Gregor, N.; Khatri, S.; Sweet, M.; Falk, E.B. Impulsivity and inhibitory control deficits are associated with unhealthy eating in young adults. *Appetite* **2012**, *59*, 738–747. [CrossRef] [PubMed]

14. Lubman, D.I.; Yücel, M.; Pantelis, C. Addiction, a condition of compulsive behaviour? Neuroimaging and neuropsychological evidence of inhibitory dysregulation. *Addiction* **2004**, *99*, 1491–1502. [CrossRef] [PubMed]
15. Lavagnino, L.; Arnone, D.; Cao, B.; Soares, J.C.; Selvaraj, S. Inhibitory control in obesity and binge eating disorder: A systematic review and meta-analysis of neurocognitive and neuroimaging studies. *Neurosci. Biobehav. Rev.* **2016**, *68*, 714–726. [CrossRef] [PubMed]
16. Batterink, L.; Yokum, S.; Stice, E. Body mass correlates inversely with inhibitory control in response to food among adolescent girls: An fMRI study. *Neuroimage* **2010**, *52*, 1696–1703. [CrossRef]
17. Moore, C.F.; Sabino, V.; Koob, G.F.; Cottone, P. Pathological Overeating: Emerging Evidence for a Compulsivity Construct. *Neuropsychopharmacology* **2017**, *42*, 1375–1389. [CrossRef]
18. Berkman, N.D.; Brownley, K.A.; Peat, C.M.; Lohr, K.N.; Cullen, K.E.; Morgan, L.X.; Bann, C.M.; Wallace, I.F.; Bulik, C.M. Table 1, DSM-IV and DSM-5 diagnostic criteria for binge-eating disorder. In *Management and Outcomes of Binge-Eating Disorder*; Agency for Healthcare Research and Quality: Rockville, MD, USA, 2015.
19. Gearhardt, A.N.; White, M.A.; Masheb, R.M.; Grilo, C.M. An examination of food addiction in a racially diverse sample of obese patients with binge eating disorder in primary care settings. *Compr. Psychiatry* **2013**, *54*, 500–505. [CrossRef]
20. Ms, A.N.G.; White, M.A.; Masheb, R.M.; Morgan, P.T.; Crosby, R.D.; Grilo, C.M. An examination of the food addiction construct in obese patients with binge eating disorder. *Int. J. Eat. Disord.* **2012**, *45*, 657–663.
21. Devoto, F.; Zapparoli, L.; Bonandrini, R.; Berlingeri, M.; Ferrulli, A.; Luzi, L.; Banfi, G.; Paulesu, E. Hungry brains: A meta-analytical review of brain activation imaging studies on food perception and appetite in obese individuals. *Neurosci. Biobehav. Rev.* **2018**, *94*, 271–285. [CrossRef]
22. Schulte, E.M.; Avena, N.M.; Gearhardt, A.N. Which foods may be addictive? The roles of processing, fat content, and glycemic load. *PLoS ONE* **2015**, *18*, e0117959. [CrossRef] [PubMed]
23. Pursey, K.M.; E Collins, C.; Stanwell, P.; Burrows, T.L. Foods and dietary profiles associated with 'food addiction' in young adults. *Addict. Behav. Rep.* **2015**, *2*, 41–48. [CrossRef] [PubMed]
24. Berridge, K.C.; Ho, C.-Y.; Richard, J.M.; DiFeliceantonio, A.G. The tempted brain eats: Pleasure and desire circuits in obesity and eating disorders. *Brain Res.* **2010**, *1350*, 43–64. [CrossRef] [PubMed]
25. Volkow, N.D.; Wang, G.J.; Fowler, J.S.; Tomasi, D.; Baler, R. Food and drug reward: Overlapping circuits in human obesity and addiction. *Curr. Top. Behav. Neurosci.* **2011**, 1–24. [CrossRef]
26. Nijs, I.M.T.; Franken, I.H.A. Attentional Processing of Food Cues in Overweight and Obese Individuals. *Curr. Obes. Rep.* **2012**, *1*, 106–113. [CrossRef]
27. Werthmann, J.; Jansen, A.; Roefs, A. Worry or craving? A selective review of evidence for food-related attention biases in obese individuals, eating-disorder patients, restrained eaters and healthy samples. In *Proceedings of the Nutrition Society*; Cambridge University Press: Cambridge, UK, 2015; pp. 99–114.
28. Castellanos, E.H.; Charboneau, E.; Dietrich, M.S.; Park, S.; Bradley, B.P.; Mogg, K.; Cowan, R.L. Obese adults have visual attention bias for food cue images: Evidence for altered reward system function. *Int. J. Obes.* **2009**, *33*, 1063–1073. [CrossRef]
29. Spielberg, J.M.; Stewart, J.L.; Levin, R.L.; Miller, G.A.; Heller, W. Prefrontal Cortex, Emotion, and Approach/Withdrawal Motivation. *Soc. Pers. Psychol. Compass* **2008**, *2*, 135–153. [CrossRef]
30. Alonso-Alonso, M.; Pascual-Leone, A. The Right Brain Hypothesis for Obesity. *JAMA* **2007**, *297*, 1819–1822. [CrossRef]
31. Ochner, C.N.; Green, D.; Van Steenburgh, J.J.; Kounios, J.; Lowe, M.R. Asymmetric prefrontal cortex activation in relation to markers of overeating in obese humans. *Appetite* **2009**, *53*, 44–49. [CrossRef]
32. Morys, F.; Janssen, L.K.; Cesnaite, E.; Beyer, F.; Garcia-Garcia, I.; Kube, J.; Kumral, D.; Liem, F.; Mehl, N.; Mahjoory, K.; et al. Hemispheric asymmetries in resting-state EEG and fMRI are related to approach and avoidance behaviour, but not to eating behaviour or BMI. *Hum. Brain Mapp.* **2019**, *41*, 1136–1152. [CrossRef]
33. McGeown, L.; Davis, R. Frontal EEG asymmetry moderates the association between attentional bias towards food and body mass index. *Biol. Psychol.* **2018**, *136*, 151–160. [CrossRef] [PubMed]
34. Karhunen, L.; Vanninen, E.J.; Kuikka, J.T.; Lappalainen, R.I.; Tiihonen, J.; Uusitupa, M.I. Regional cerebral blood flow during exposure to food in obese binge eating women. *Psychiatry Res. Neuroimaging* **2000**, *99*, 29–42. [CrossRef]

35. Finocchiaro, R.; Balconi, M. Left/Right Hemispheric "Unbalance" Model in Addiction. In *Addictive Substances and Neurological Disease: Alcohol, Tobacco, Caffeine, and Drugs of Abuse in Everyday Lifestyles*; Elsevier: Amsterdam, The Netherlands, 2017; pp. 383–391.
36. Shevorykin, A.; Ruglass, L.M.; Melara, R. Frontal Alpha Asymmetry and Inhibitory Control among Individuals with Cannabis Use Disorders. *Brain Sci.* **2019**, *9*, 219. [CrossRef] [PubMed]
37. Gordon, H.W. Laterality of Brain Activation for Risk Factors of Addiction. *Curr. Drug Abus. Rev.* **2016**, *9*, 1–18. [CrossRef] [PubMed]
38. Gearhardt, A.N. Neural Correlates of Food Addiction. *Arch. Gen. Psychiatry* **2011**, *68*, 808–816. [CrossRef] [PubMed]
39. McBride, D.; Barrett, S.P.; Kelly, J.T.; Aw, A.; Dagher, A. Effects of Expectancy and Abstinence on the Neural Response to Smoking Cues in Cigarette Smokers: An fMRI Study. *Neuropsychopharmacology* **2006**, *31*, 2728–2738. [CrossRef]
40. Garavan, H.; Pankiewicz, J.; Bloom, A.; Cho, J.-K.; Sperry, L.; Ross, T.J.; Salmeron, B.J.; Risinger, R.; Kelley, D.; Stein, E.A. Cue-Induced Cocaine Craving: Neuroanatomical Specificity for Drug Users and Drug Stimuli. *Am. J. Psychiatry* **2000**, *157*, 1789–1798. [CrossRef]
41. Grant, S.; London, E.D.; Newlin, D.B.; Villemagne, V.L.; Liu, X.; Contoreggi, C.; Phillips, R.L.; Kimes, A.S.; Margolin, A. Activation of memory circuits during cue-elicited cocaine craving. *Proc. Natl. Acad. Sci. USA* **1996**, *93*, 12040–12045. [CrossRef]
42. Wang, G.-J.; Volkow, N.D.; Fowler, J.S.; Cervany, P.; Hitzemann, R.J.; Pappas, N.R.; Wong, C.T.; Felder, C. Regional brain metabolic activation during craving elicited by recall of previous drug experiences. *Life Sci.* **1999**, *64*, 775–784. [CrossRef]
43. Davis, C.; Loxton, N.J.; Levitan, R.D.; Kaplan, A.S.; Carter, J.C.; Kennedy, J.L. 'Food addiction' and its association with a dopaminergic multilocus genetic profile. *Physiol. Behav.* **2013**, *118*, 63–69. [CrossRef]
44. Frayn, M.; Sears, C.R.; Von Ranson, K.M. A sad mood increases attention to unhealthy food images in women with food addiction. *Appetite* **2016**, *100*, 55–63. [CrossRef] [PubMed]
45. Meule, A.; Gearhardt, A.N. Five years of the Yale Food Addiction Scale: Taking stock and moving forward. *Curr. Addict. Rep.* **2014**, *1*, 193–205. [CrossRef]
46. Siep, N.; Roefs, A.; Roebroeck, A.; Havermans, R.; Bonte, M.L.; Jansen, A. Hunger is the best spice: An fMRI study of the effects of attention, hunger and calorie content on food reward processing in the amygdala and orbitofrontal cortex. *Behav. Brain Res.* **2009**, *198*, 149–158. [CrossRef] [PubMed]
47. Stockburger, J.; Schmälzle, R.; Flaisch, T.; Bublatzky, F.; Schupp, H.T. The impact of hunger on food cue processing: An event-related brain potential study. *Neuroimage* **2009**, *47*, 1819–1829. [CrossRef]
48. Loeber, S.; Grosshans, M.; Herpertz, S.; Kiefer, F.; Herpertz, S.C. Hunger modulates behavioral disinhibition and attention allocation to food-associated cues in normal-weight controls. *Appetite* **2013**, *71*, 32–39. [CrossRef]
49. Tapper, K.; Pothos, E.M.; Lawrence, A.D. Feast your eyes: Hunger and trait reward drive predict attentional bias for food cues. *Emotion* **2010**, *10*, 949–954. [CrossRef]
50. Duncan, N.W.; Northoff, G. Overview of potential procedural and participant- related confounds for neuroimaging of the resting state. *J. Psychiatry Neurosci.* **2013**, *38*, 84–96. [CrossRef]
51. Jasinska, A.J.; Stein, E.A.; Kaiser, J.; Naumer, M.J.; Yalachkov, Y. Factors modulating neural reactivity to drug cues in addiction: A survey of human neuroimaging studies. *Neurosci. Biobehav. Rev.* **2014**, *38*, 1–16. [CrossRef]
52. Lemeshow, A.R.; Gearhardt, A.N.; Genkinger, J.M.; Corbin, W.R. Assessing the psychometric properties of two food addiction scales. *Eat. Behav.* **2016**, *23*, 110–114. [CrossRef]
53. Shearin, E.N.; Russ, M.J.; Hull, J.W.; Clarkin, J.F.; Smith, G.P. Construct validity of the Three-Factor Eating Questionnaire: flexible and rigid control subscales. *Int. J. Eat. Disord.* **1994**, *16*, 187–198. [CrossRef]
54. Anglé, S.; Engblom, J.; Eriksson, T.; Kautiainen, S.; Saha, M.T.; Lindfors, P.; Lehtinen, M.; Rimpelä, A. Three factor eating questionnaire-R18 as a measure of cognitive restraint, uncontrolled eating and emotional eating in a sample of young Finnish females. *Int. J. Behav. Nutr. Phys. Act.* **2009**, *6*, 41. [CrossRef] [PubMed]
55. Goldfein, J.A.; Devlin, M.J.; Kamenetz, C. Eating Disorder Examination-Questionnaire with and without instruction to assess binge eating in patients with binge eating disorder. *Int. J. Eat. Disord.* **2005**, *37*, 107–111. [CrossRef] [PubMed]

56. Celio, A.A.; Wilfley, D.E.; Crow, S.; Mitchell, J.; Walsh, B.T. A comparison of the binge eating scale, questionnaire for eating and weight patterns-revised, and eating disorder examination questionnaire with instructions with the eating disorder examination in the assessment of binge eating disorder and its symptoms. *Int. J. Eat. Disord.* **2004**, *36*, 434–444. [CrossRef]
57. Beck, A.T.; Steer, R.A.; Carbin, M.G. Psychometric properties of the Beck Depression Inventory: Twenty-five years of evaluation. *Clin. Psychol. Rev.* **1988**, *8*, 77–100. [CrossRef]
58. Beck, A.T.; Ward, C.H.; Mendelson, M.; Mock, J.; Erbaugh, J. An Inventory for Measuring Depression. *Arch. Gen. Psychiatry* **1961**, *4*, 561–571. [CrossRef]
59. Watson, D.; Clark, L.A.; Tellegen, A. Development and Validation of Brief Measures of Positive and Negative Affect: The PANAS Scales. *J. Pers. Soc. Psychol.* **1988**, *54*, 1063–1070. [CrossRef]
60. De Carvalho, H.W.; Andreoli, S.B.; Lara, D.R.; Patrick, C.J.; Quintana, M.I.; Bressan, R.A.; De Melo, M.F.; Mari, J.D.J.; Jorge, M.R. Structural validity and reliability of the Positive and Negative Affect Schedule (PANAS): evidence from a large Brazilian community sample. *Rev. Bras. Psiquiatr.* **2013**, *35*, 169–172. [CrossRef]
61. Cserjési, R.; Luminet, O.; Poncelet, A.-S.; Lénárd, L. Altered executive function in obesity. Exploration of the role of affective states on cognitive abilities. *Appetite* **2009**, *52*, 535–539. [CrossRef]
62. Nijs, I.M.; Muris, P.; Euser, A.S.; Franken, I.H. Differences in attention to food and food intake between overweight/obese and normal-weight females under conditions of hunger and satiety. *Appetite* **2010**, *54*, 243–254. [CrossRef]
63. Cox, W.M.; Fadardi, J.S.; Pothos, E.M. The Addiction-Stroop test: Theoretical considerations and procedural recommendations. *Psychol. Bull.* **2006**, *132*, 443–476. [CrossRef]
64. Fadardi, J.S.; Bazzaz, M.M. A Combi-Stroop test for measuring food-related attentional bias. *Exp. Clin. Psychopharmacol.* **2011**, *19*, 371–377. [CrossRef] [PubMed]
65. Nijs, I.M.; Franken, I.H.; Muris, P. Food-related Stroop interference in obese and normal-weight individuals: Behavioral and electrophysiological indices. *Eat. Behav.* **2010**, *11*, 258–265. [CrossRef] [PubMed]
66. Forestell, C.A.; Lau, P.; Gyurovski, I.I.; Dickter, C.L.; Haque, S.S. Attentional biases to foods: The effects of caloric content and cognitive restraint. *Appetite* **2012**, *59*, 748–754. [CrossRef] [PubMed]
67. Werthmann, J.; Roefs, A.; Nederkoorn, C.; Mogg, K.; Bradley, B.P.; Jansen, A. Can(not) take my eyes off it: Attention bias for food in overweight participants. *Health Psychol.* **2011**, *30*, 561–569. [CrossRef] [PubMed]
68. Hume, D.J.; Howells, F.M.; Rauch, H.L.; Kroff, J.; Lambert, E.V. Electrophysiological indices of visual food cue-reactivity. Differences in obese, overweight and normal weight women. *Appetite* **2015**, *85*, 126–137. [CrossRef] [PubMed]
69. Blechert, J.; Lender, A.; Polk, S.; Busch, N.A.; Ohla, K. Food-Pics_Extended—An Image Database for Experimental Research on Eating and Appetite: Additional Images, Normative Ratings and an Updated Review. *Front. Psychol.* **2019**, *10*, 307. [CrossRef] [PubMed]
70. Bruce, G.; Jones, B.T. A pictorial Stroop paradigm reveals an alcohol attentional bias in heavier compared to lighter social drinkers. *J. Psychopharmacol.* **2004**, *18*, 527–533. [CrossRef]
71. Blechert, J.; Emeule, A.; Busch, N.A.; Eohla, K. Food-pics: an image database for experimental research on eating and appetite. *Front. Psychol.* **2014**, *5*, 617. [CrossRef]
72. Delorme, A.; Makeig, S. EEGLAB: an open source toolbox for analysis of single-trial EEG dynamics including independent component analysis. *J. Neurosci. Methods* **2004**, *134*, 9–21. [CrossRef]
73. Alyagon, U.; Shahar, H.; Hadar, A.; Barnea-Ygael, N.; Lazarovits, A.; Shalev, H.; Zangen, A. Alleviation of ADHD symptoms by non-invasive right prefrontal stimulation is correlated with EEG activity. *Neuroimage Clin.* **2020**, *26*, 102206. [CrossRef]
74. Oostenveld, R.; Fries, P.; Maris, E.; Schoffelen, J.-M. FieldTrip: Open Source Software for Advanced Analysis of MEG, EEG, and Invasive Electrophysiological Data. *Comput. Intell. Neurosci.* **2010**, *2011*, 1–9. [CrossRef] [PubMed]
75. Zibman, S.; Daniel, E.; Alyagon, U.; Etkin, A.; Zangen, A. Interhemispheric cortico-cortical paired associative stimulation of the prefrontal cortex jointly modulates frontal asymmetry and emotional reactivity. *Brain Stimul.* **2019**, *12*, 139–147. [CrossRef] [PubMed]
76. Ocklenburg, S.; Friedrich, P.; Schmitz, J.; Schlüter, C.; Genc, E.; Güntürkün, O.; Peterburs, J.; Grimshaw, G. Beyond frontal alpha: investigating hemispheric asymmetries over the EEG frequency spectrum as a function of sex and handedness. *Laterality Asymmetries Body Brain Cogn.* **2018**, *24*, 505–524. [CrossRef] [PubMed]

77. Van Bochove, M.E.; Ketel, E.; Wischnewski, M.; Wegman, J.; Aarts, E.; De Jonge, B.; Medendorp, W.P.; Schutter, D.J. Posterior resting state EEG asymmetries are associated with hedonic valuation of food. *Int. J. Psychophysiol.* **2016**, *110*, 40–46. [CrossRef] [PubMed]
78. AlOmari, R.A.; Fernandez, M.; Banks, J.B.; Acosta, J.; Tartar, J.L. Acute Stress Dysregulates the LPP ERP Response to Emotional Pictures and Impairs Sustained Attention: Time-Sensitive Effects. *Brain Sci.* **2015**, *5*, 201–219. [CrossRef]
79. Schupp, H.T.; Cuthbert, B.N.; Bradley, M.M.; Cacioppo, J.T.; Ito, T.; Lang, P.J. Affective picture processing: The late positive potential is modulated by motivational relevance. *Psychophysiology* **2000**, *37*, 257–261. [CrossRef]
80. Cuthbert, B.N.; Schupp, H.T.; Bradley, M.M.; Birbaumer, N.; Lang, P.J. Brain potentials in affective picture processing: covariation with autonomic arousal and affective report. *Biol. Psychol.* **2000**, *52*, 95–111. [CrossRef]
81. Becker, R.; Knock, S.; Ritter, P.; Jirsa, V. Relating Alpha Power and Phase to Population Firing and Hemodynamic Activity Using a Thalamo-cortical Neural Mass Model. *PLoS Comput. Biol.* **2015**, *11*, e1004352. [CrossRef]
82. Fetterman, A.K.; Ode, S.; Robinson, M.D. For which side the bell tolls: The laterality of approach-avoidance associative networks. *Motiv. Emot.* **2013**, *37*, 33–38. [CrossRef]
83. Tops, M.; Quirin, M.; Boksem, M.A.; Koole, S.L. Large-scale neural networks and the lateralization of motivation and emotion. *Int. J. Psychophysiol.* **2017**, *119*, 41–49. [CrossRef]
84. Lowe, M.R.; Van Steenburgh, J.; Ochner, C.; Coletta, M. Neural correlates of individual differences related to appetite. *Physiol. Behav.* **2009**, *97*, 561–571. [CrossRef] [PubMed]
85. Appelhans, B.M.; Woolf, K.; Pagoto, S.L.; Schneider, K.L.; Whited, M.C.; Liebman, R. Inhibiting Food Reward: Delay Discounting, Food Reward Sensitivity, and Palatable Food Intake in Overweight and Obese Women. *Obesity* **2011**, *19*, 2175–2182. [CrossRef] [PubMed]
86. Winter, S.; Feig, E.; Kounios, J.; Erickson, B.; Berkowitz, S.; Lowe, M. The relation of hedonic hunger and restrained eating to lateralized frontal activation. *Physiol. Behav.* **2016**, *163*, 64–69. [CrossRef] [PubMed]
87. Forcano, L.; Castellano, M.; Cuenca-Royo, A.; Goday-Arno, A.; Pastor, A.; Langohr, K.; Castañer, O.; Pérez-Vega, K.A.; Serra, C.; Ruffini, G.; et al. Prefrontal Cortex Neuromodulation Enhances Frontal Asymmetry and Reduces Caloric Intake in Patients with Morbid Obesity. *Obesity* **2020**, *28*, 696–705. [CrossRef]
88. Spielberg, J.M.; Heller, W.; Miller, G.A. Hierarchical Brain Networks Active in Approach and Avoidance Goal Pursuit. *Front. Hum. Neurosci.* **2013**, *7*. [CrossRef]
89. Hollmann, M.; Hellrung, L.; Pleger, B.; Schlogl, H.; Kabisch, S.; Stumvoll, M.; Villringer, A.; Horstmann, A. Neural correlates of the volitional regulation of the desire for food. *Int. J. Obes.* **2011**, *36*, 648–655. [CrossRef]
90. Gordon, E.; Palmer, D.M.; Cooper, N. EEG Alpha Asymmetry in Schizophrenia, Depression, PTSD, Panic Disorder, ADHD and Conduct Disorder. *Clin. EEG Neurosci.* **2010**, *41*, 178–183. [CrossRef]
91. Braet, C.; Crombez, G. Cognitive interference due to food cues in childhood obesity. *J. Clin. Child. Adolesc. Psychol.* **2003**, *32*, 32–39. [CrossRef]
92. Fagundo, A.B.; De La Torre, R.; Jiménez-Murcia, S.; Agüera, Z.; Granero, R.; Tárrega, S.; Botella, C.; Baños, R.; Fernández-Real, J.M.; Rodríguez, R.; et al. Executive Functions Profile in Extreme Eating/Weight Conditions: From Anorexia Nervosa to Obesity. *PLoS ONE* **2012**, *7*, e43382. [CrossRef]
93. Field, M.; Munafò, M.R.; Franken, I.H.A. A meta-analytic investigation of the relationship between attentional bias and subjective craving in substance abuse. *Psychol. Bull.* **2009**, *135*, 589–607. [CrossRef]
94. Carbine, K.A.; Rodeback, R.; Modersitzki, E.; Miner, M.; LeCheminant, J.D.; Larson, M.J. The utility of event-related potentials (ERPs) in understanding food-related cognition: A systematic review and recommendations. *Appetite* **2018**, *128*, 58–78. [CrossRef] [PubMed]
95. Hollitt, S.; Kemps, E.; Tiggemann, M.; Smeets, E.; Mills, J.S. Components of attentional bias for food cues among restrained eaters. *Appetite* **2010**, *54*, 309–313. [CrossRef] [PubMed]
96. Veenstra, E.M.; De Jong, P.J.; Koster, E.H.W.; Roefs, A. Attentional avoidance of high-fat food in unsuccessful dieters. *J. Behav. Exp. Psychiatry* **2010**, *41*, 282–288. [CrossRef] [PubMed]
97. Cohen, N.; Henik, A.; Mor, N. Can Emotion Modulate Attention? Evidence for Reciprocal Links in the Attentional Network Test. *Exp. Psychol.* **2011**, *58*, 171–179. [CrossRef]

98. Gearhardt, A.N.; Treat, T.A.; Hollingworth, A.; Corbin, W.R. The relationship between eating-related individual differences and visual attention to foods high in added fat and sugar. *Eat. Behav.* **2012**, *13*, 371–374. [CrossRef]
99. Field, M.; Cox, W.M. Attentional bias in addictive behaviors: A review of its development, causes, and consequences. *Drug Alcohol Depend.* **2008**, *97*, 1–20. [CrossRef]
100. Carrigan, M.H.; Randall, C.L. Self-medication in social phobia: a review of the alcohol literature. *Addict. Behav.* **2003**, *28*, 269–284. [CrossRef]
101. Beck, A.T.; Clark, D.A. An information processing model of anxiety: Automatic and strategic processes. *Behav. Res. Ther.* **1997**, *35*, 49–58. [CrossRef]
102. Root, M.P. Persistent, disordered eating as a gender-specific, post-traumatic stress response to sexual assault. *Psychother. Theory Res. Pract. Train.* **1991**, *28*, 96–102. [CrossRef]
103. Carter, J.C.; Van Wijk, M.; Rowsell, M. Symptoms of 'food addiction' in binge eating disorder using the Yale Food Addiction Scale version 2.0. *Appetite* **2019**, *133*, 362–369. [CrossRef]
104. Ivezaj, V.; White, M.A.; Grilo, C.M. Examining binge-eating disorder and food addiction in adults with overweight and obesity. *Obesity* **2016**, *24*, 2064–2069. [CrossRef] [PubMed]
105. White, M.A.; Potenza, M.N. Binge Eating Disorder and Food Addiction. *Curr. Drug Abus. Rev.* **2011**, *4*, 201–207.
106. Amianto, F.; Ottone, L.; Giovanni, A.D.; Fassino, S. Binge-eating disorder diagnosis and treatment: a recap in front of DSM-5. *BMC Psychiatry* **2015**, *15*, 70. [CrossRef] [PubMed]
107. Smarr, K.L.; Keefer, A.L. Measures of depression and depressive symptoms: Beck Depression Inventory-II (BDI-II), Center for Epidemiologic Studies Depression Scale (CES-D), Geriatric Depression Scale (GDS), Hospital Anxiety and Depression Scale (HADS), and Patient Health Questionna. *Arthritis Rheum.* **2011**, *63* (Suppl. 11), S454–S466. [CrossRef]
108. Burmeister, J.M.; Hinman, N.; Koball, A.; Hoffmann, D.A.; Carels, R.A. Food addiction in adults seeking weight loss treatment. Implications for psychosocial health and weight loss. *Appetite* **2013**, *60*, 103–110. [CrossRef] [PubMed]
109. Nunes-Neto, P.R.; Köhler, C.A.; Schuch, F.B.; Solmi, M.; Quevedo, J.; Maes, M.; Murru, A.; Vieta, E.; McIntyre, R.S.; McElroy, S.L.; et al. Food addiction: Prevalence, psychopathological correlates and associations with quality of life in a large sample. *J. Psychiatr. Res.* **2018**, *96*, 145–152. [CrossRef]
110. Sansone, R.A.; Sansone, L.A. Obesity and Substance Misuse: Is There a Relationship? *Innov. Clin. Neurosci.* **2013**, *10*, 30–35.
111. Schulte, E.M.; Gearhardt, A.N. Associations of Food Addiction in a Sample Recruited to Be Nationally Representative of the United States. *Eur. Eat. Disord. Rev.* **2018**, *26*, 112–119. [CrossRef]
112. Gearhardt, A.N.; Corbin, W.R.; Brownell, K.D. Development of the Yale Food Addiction Scale Version 2.0. *Psychol. Addict. Behav.* **2016**, *30*, 113–121. [CrossRef]

Publisher's Note: MDPI stays neutral with regard to jurisdictional claims in published maps and institutional affiliations.

nutrients

Review

The Melanocortin System behind the Dysfunctional Eating Behaviors

Emanuela Micioni Di Bonaventura [1,†], Luca Botticelli [1,†], Daniele Tomassoni [2], Seyed Khosrow Tayebati [1], Maria Vittoria Micioni Di Bonaventura [1,*,‡] and Carlo Cifani [1,‡]

1 School of Pharmacy, University of Camerino, 62032 Camerino, Italy; emanuela.micioni@unicam.it (E.M.D.B.); luca.botticelli@unicam.it (L.B.); khosrow.tayebati@unicam.it (S.K.T.); carlo.cifani@unicam.it (C.C.)
2 School of Bioscience and Veterinary Medicine, University of Camerino, 62032 Camerino, Italy; daniele.tomassoni@unicam.it
* Correspondence: mariavittoria.micioni@unicam.it; Tel.: +39-0737-403302
† These authors contributed equally to this work.
‡ These authors jointly supervised this work.

Received: 30 September 2020; Accepted: 11 November 2020; Published: 14 November 2020

Abstract: The dysfunction of melanocortin signaling has been associated with obesity, given the important role in the regulation of energy homeostasis, food intake, satiety and body weight. In the hypothalamus, the melanocortin-3 receptor (MC3R) and melanocortin-4 receptor (MC4R) contribute to the stability of these processes, but MC3R and MC4R are also localized in the mesolimbic dopamine system, the region that responds to the reinforcing properties of highly palatable food (HPF) and where these two receptors seem to affect food reward and motivation. Loss of function of the MC4R, resulting from genetic mutations, leads to overeating in humans, but to date, a clear understanding of the underlying mechanisms and behaviors that promote overconsumption of caloric foods remains unknown. Moreover, the MC4R demonstrated to be a crucial modulator of the stress response, factor that is known to be strictly related to binge eating behavior. In this review, we will explore the preclinical and clinical studies, and the controversies regarding the involvement of melanocortin system in altered eating patterns, especially binge eating behavior, food reward and motivation.

Keywords: melanocortin system; MC3R; MC4R; eating disorders; binge eating disorder; food reward; obesity; *MC4R* mutation; rs17782313; stress

1. Introduction

Nowadays, the increased consumption of food highly rich in fat, sugar and palatable components has fueled the so called Western diet, leading to excessive and non-homeostatic feeding behavior that impacts the quality of life [1,2]. The melanocortin system, known to be a key pathway in the regulation of food intake, body weight and energy balance [3–6], has been proposed as a possible underlying factor not only in obesity, in which there is evidence of a consistent relationship [7–14], but also in several dysfunctional eating patterns [15–20] that can lead to obesity, modulating the motivation for hedonic properties of food [21–23]. Among the altered feeding patterns, binge eating behavior is one of the most studied, due to the overlaps that exist with obesity [24–26], and melanocortin signaling can influence reward-related behaviors, given the presence of melanocortin receptors (MCRs) not only in the hypothalamus, but also in reward-related brain areas such as in the mesolimbic dopamine pathway [27,28].

Binge eating is a typical feature in eating disorders, in particular Bulimia Nervosa, binge/purging subtype of Anorexia Nervosa and Binge Eating Disorder (BED). A binge eating episode is characterized by an unusual consumption of a large amount of food that most people would not eat in the same discrete period of time, connected with the inability to stop overeating, accompanied by feelings of guilt, shame and regret [29]. Differently from Bulimia Nervosa and Anorexia Nervosa, BED is characterized by recurrent episodes of binge eating not followed by inappropriate compensatory behaviors, such as vomiting, prolonged fasting, or excessive exercise for controlling weight gain [29]. BED is the most prevalent eating disorder in adolescents and young women [24,25,30], and it is associated in some instances with overweight or obesity [31,32]. The subgroup of obese individuals that suffers also from BED seems to increase food-related impulsivity and reward sensitivity in comparison to obese people without BED [33,34]. Additionally, food craving is significantly higher under negative emotional states (including disappointment, anger, guilt, depressive symptoms) [35,36] and stress exposure [37–39] in obese binge eaters rather than obese. Thus, binge eating is a risk factor for obesity and, at the same time, overweight and obesity might enhance the possibility to engage binge eating behavior [40]. In light of these interconnected aberrant feeding patterns and the involvement of the MCRs in overeating and stress, the aim of this review is to revise the current literature on PubMed, regarding the role of the melanocortin system as a mutual underlying factor that may increase the susceptibility to develop aberrant eating behaviors. After a brief summary of the localization and the physiological functions of the melanocortin system, we will describe the role of melanocortin-3 receptor (MC3R) and melanocortin-4 receptor (MC4R) on food intake, focusing on their interaction with the brain reward system. Subsequently, we will highlight the impact of genetic mutations of MC4R on food consumption in humans. Finally, the melanocortin system, principally via MC4R, will be explored in stress response, considering stress as a key factor triggering altered feeding patterns.

2. An Overview of the Melanocortin Receptors in the Control of Food Intake

Pro-opiomelanocortin (POMC) is the precursor molecule of α-melanocyte-stimulating hormone (α-MSH), one of its proteolytic cleavage products, which has a regulatory role in feeding related behavior and satiety; the other active peptides are β-MSH, γ-MSH, adrenocorticotropic hormone (ACTH) and β-endorphin [3,4,9]. Localization of POMC neurons in the central nervous system (CNS) is in the arcuate nucleus of the hypothalamus (ARC) and in the nucleus of the tractus solitarius (NTS) of the brainstem, areas implicated in body weight loss, energy homeostasis and signaling of satiety, showing anorexigenic effects [4]. Adjacent to POMC cells, in the hypothalamic ARC, are localized agouti-related protein (AgRP) neurons and the neuropeptide Y (NPY) neurons producing, respectively, the endogenous antagonist of MCRs AgRP and the orexigenic neuropeptide NPY, both able to increase food intake [3,41,42]. In the 1990s, the first MCRs were initially cloned, and, subsequently, all five MCRs, members of the superfamily of G protein-coupled receptors, have been identified [43–46]. MC3R and MC4R are widely expressed in CNS, and, binding the endogenous MCRs agonist, α-MSH, are able to activate adenylate cyclase to elevate intracellular cAMP levels, generating an anorexigenic signal [41,47], regulating the homeostasis of energy intake and feeding behavior and suppressing food consumption [4,6,16]. Conversely, MC1R, MC2R and MC5R are primarily found in the periphery: the MC1R especially in the melanocytes, the MC2R in the adrenal cortex and MC5R in the exocrine glands [4,48].

The MC3R is predominantly expressed in the brain within the hypothalamus, mainly in the ARC and less in the dorsomedial portion of the ventromedial nucleus, anteroventral preoptic area, posterior hypothalamic area, the medial preoptic area and paraventricular nucleus (PVN) of the hypothalamus, but there is evidence of MC3R moderately localized also in the limbic system, in ventral tegmental area (VTA), central linear nucleus of raphe, in the lateral nucleus of the septum and in the medial habenula nucleus of the thalamus [43,46,48,49].

In contrast with MC3R, the MC4R has a more widespread expression in the CNS; indeed, the MC4R shows high prevalence in hypothalamic sites including PVN, the medial preoptic area, anterior hypothalamic nucleus, ventromedial nucleus of the hypothalamus, dorsomedial nucleus of the hypothalamus, tuberomammillary nucleus and other several hypothalamic areas, but it is also strongly expressed in the brainstem and moderately in the cortex, hippocampus, corpus striatum, amygdala, thalamus, spinal cord and also detected in the peripheral nervous system [27,43,48,50,51].

In the brain, the distinct localization, more widely for MC4R than for MC3R, also reflects a different binding with the peptides deriving from POMC cleavage: α-MSH and γ-MSH have high affinity for MC3R; meanwhile, MC4R is preferentially bound by α-MSH and less by γ-MSH [44,46,50,51]. Moreover, AgRP, endogenous antagonist of MCRs, has high affinity for both these receptors [41,42], reflecting a differential regulation of the metabolic response and food consumption [7,42,52]. Furthermore, in the hypothalamus, the melanocortin pathway interacts with other crucial hormones, such as leptin and insulin, which promote the processing of POMC to the anorexigenic α-MSH, signaling a decreased energy intake and contributing to the fed state (for details see ref. [12,53–55]). In addition, another functional interaction of the melanocortin system in the ARC nucleus is with the orexigenic neuropeptide Nociceptin/Orphanin FQ [56,57], which exerts an inhibitory influence on α-MSH cells [58], and is strictly involved in stress mechanisms [59,60] and binge eating behavior [56,61–64].

Preliminary information about the functions and physiological role related to feeding of MC3R and MC4R was provided by studies with the deletion of these MCRs in mice, which developed obesity, increased adipose mass, hyperphagia and lack of appetite control, in particular more pronounced in MC4R knockout (KO) mice rather than MC3R KO mice, even though mice lacking both receptors become significantly heavier than MC4R KO [11,16,65–68]. Additionally, all the previous effects, characteristic of severe obesity, are predominantly linked to *MC4R* mutations and defects in MC4R signaling in humans, compared to the alterations of MC3R, which frequently cause only moderate obesity or limited hyperphagia; to date, the role of MC3R remains an element that needs to be clarified [11,13,14,69–72]. Taking into account all these findings, it is interesting to explore the studies conducted so far regarding the association of MCRs with compulsive eating, food reward and motivation, and to support the possibility of their implication in binge eating behavior.

3. Melanocortin Receptors in Feeding

3.1. MC3R

3.1.1. MC3R: Preclinical Studies on Eating Behavior

The MC3R, compared to the MC4R subtype, exhibits a more limited distribution in the brain, being predominantly found in the hypothalamic nuclei and limbic regions, with dense expression in the ARC, ventromedial hypothalamus, VTA and medial habenula, structures in which it is supposed to regulate energy homeostasis and food seeking behavior [46,50,73–75]. MC3R and MC4R KO mice have been used to investigate the role of each receptor in regulating energy homeostasis, and many studies revealed that MC3Rs and MC4Rs might function independently, playing a complementary but non-redundant role in the regulation of energy balance [65,66,68,76]. Targeted deletion of the MC3R gene in mice promotes a modest obesity syndrome and increased accumulation of fat mass that is not related to hyperphagia, with a normal anorectic response to melanocortin agonists [65,66], suggesting that this receptor could be mostly involved in the regulation of energy homeostasis and metabolic processes, rather than in the control of feeding behavior. However, a study by Zhang et al. showed that MC3Rs and MC4Rs are of approximately equal importance in preventing weight gain during a high-fat chow diet, and that the absence of MC3Rs compromises leptin's ability to decrease food consumption [76], evidencing an altered anorectic response in MC3R null mice. Moreover, male MC3R KO mice, backcrossed onto the C57BL/6J background, showed a mild hyperphagia after exposure to a purified high-fat diet [6]. Sutton et al. demonstrated that obesity associated with MC3R deficiency is dependent on the dietary fat, considering that, if exposed to a low-fat diet, MC3R KO mice exhibited a modest increase in adiposity and a normal body weight, while during a high-fat diet, fat mass was comparable to that of MC4R KO littermates [68]. Additionally, MC3R KO mice were not hyperphagic under a low-fat diet, but showed a modest increase in food consumption under the high-fat diet, an effect that was gender specific, being mainly observed in male mice [68]. A recent experiment in mice with "humanized" MC3Rs further evidenced the role of the MC3R in appetite control: in this mouse model, the murine MC3R was replaced with the Wild Type (WT) human MC3R ($MC3R^{hWT/hWT}$) or the double-mutant C17A (Thr6Lys) + G241A (Val81Ile) human MC3R (MC3R$^{hDM/hDM}$) [77], characterized by a reduced receptor binding, signaling transduction and less protein expression, and associated with a greater risk of childhood obesity in human homozygous carriers [78–80]. Mutant homozygous mice with the double mutation (MC3R$^{hDM/hDM}$) had an increased adiposity and energy intake, compared to WT human MC3R (MC3R$^{hWT/hWT}$) littermates and were also hyperphagic [77], highlighting the contribution of MC3R signaling to energy homeostasis, metabolism and feeding behavior.

The behavioral phenotype linked to MC3R-deficiency may be also contextual and dependent on energy balance. In fact, MC3R-deficient mice appear to be less sensitive to the "pain" of hunger, and are not motivated to avoid unpleasant experiences associated with nutrient scarcity [81], as described by investigations using hypocaloric restricting feeding protocols. In this context, MC3Rs seem to be essential for entrainment of anticipatory behavior toward feeding time [81,82]. Food anticipatory behavior, consisting of a progressive rise of activity preceding food presentation, assessed using running wheels and measuring home cage activity, is attenuated in MC3R KO mice, compared to WT, under a restricting feeding protocol [73,82,83]. Moreover, the same mice did not exhibit the increased wakefulness generally coincident with food presentation and normally observed in non-mutant rodents [82]. Hypocaloric feeding protocols are known to promote binge-like eating behavior in WT mice, which reduce meal frequency but increase meal size and duration, with most of the food consumed within the first hour of presentation [21,83,84]. This behavioral phenotype is markedly attenuated in MC3R KO mice, without compensation in the feeding cycle later or changes in the meal structure, a finding that supports the essential role of an intact MC3R signaling in the compulsive eating response, observed after exposure to situations of poor nutrient availability and prolonged negative energy balance [83,84]. Additionally, the motivation to self-administer a food reward is markedly attenuated in MC3R-deficient mice, exposed to a caloric restriction protocol, while being normal if mice are fed in ad libitum conditions, reducing self-administration of chocolate flavored pellets [21]. The abnormal behavioral features associated with the deletion of MC3Rs could be partially explained by the neuroendocrine alterations found in the brain of MC3R-deficient mice, which failed to present the increase in the potent orexigenic neuropeptides AgRP and NPY during fasting and hypocaloric conditions [5,84,85]. Intriguingly, MC3R-deficient mice also exhibit altered responses of the hypothalamic-pituitary-adrenal (HPA) axis during caloric restriction, showing a lack of corticosterone serum increase in response to fasting, which instead is found in WT mice [84,85]. Furthermore, the dysregulation of fasting-induced corticosterone release was accompanied by a defect in the upregulation of hypothalamic corticotropin-releasing hormone (CRH) mRNA in the MC3R KO mice [85], indicating that both hypothalamic and adrenal functions are compromised by the absence of this receptor. Considering both the dysfunctional eating behaviors and the altered activity of the HPA axis observed in MC3R KO mice during fasting, future studies should be conducted to investigate if deletion or antagonism of the MC3R could reduce the compulsive-like eating in a preclinical model of binge eating, where a binge eating episode is elicited by the combination of food restriction and stress, trying to further characterize the role of this receptor in both homeostatic and non-homeostatic eating [86,87].

The information obtained from studies with MC3R KO mice are in accordance with the putative role of the MC3R as an inhibitory autoreceptor on POMC neurons [49,88,89], where α- and γ-MSH, released by POMC nerve terminals within the ARC, are supposed to regulate the activity of POMC neurons through activation of MC3R subtypes, and studies with selective MC3R agonists confirmed this observation. Indeed, Marks et al. found that stimulation of the MC3R, by peripheral administration of the selective MC3R agonist [D-trp8]-γ-MSH, results in the inhibition of POMC neuronal activity, which in turn leads to an increase in food intake in WT mice, while having no effect on feeding in MC3R KO littermates [90]. The suppression of POMC neuronal activity, after injection of [D-trp8]-γ-MSH, was demonstrated to be a consequence of an increased inhibitory synaptic transmission, due to the activation of GABAergic NPY neurons in the ARC, releasing GABA on POMC neurons [88,89].

Considering the highly potent orexigenic activity of NPY [91], this can explain the observed increase in food intake after stimulation of the MC3R [90], as reported in Figure 1.

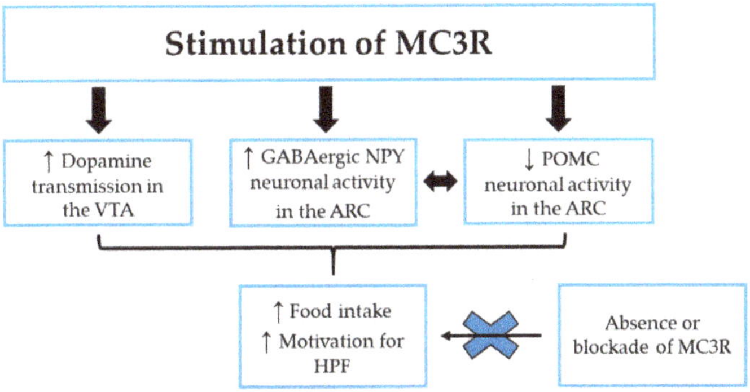

Figure 1. The potential MC3R mechanisms leading to increase food intake and motivation for highly palatable food (HPF) in preclinical studies. ↓: decrease; ↑: increase; ARC: Arcuate nucleus of the hypothalamus; HPF: Highly palatable food; MC3R: Melanocortin-3 receptor; NPY: neuropeptide Y; POMC: Pro-opiomelanocortin; VTA: Ventral Tegmental Area.

However, the study by Marks et al. had the limitation of investigating the effect of the MC3R agonist only by a peripheral administration, not clearly explaining if the results obtained were due to a peripheral or central action. Subsequently, Lee et al., using a rat model, obtained a similar finding, analyzing the effect of the same compound directly injected in the CNS, through intracerebroventricular (i.c.v.) injections, resulting in an increased food intake in treated rats, confirming a central mechanism of action [92]. Interestingly, it was examined if antagonism at MC3R would have the opposite effect, inhibiting feeding, but a strange result was obtained, observing that the MC3R antagonist PG-932, at a low dose, suppressed food intake, while at a higher dose significantly increased food consumption and body weight. These effects could be explained with the possible antagonism profile of PG-932 even at MC4R, when injected at high doses in rats [92].

Taken together, these studies confirm that the MC3Rs, despite their functions are still not completely understood, could represent important targets for the treatment of obesity and could also play a role in the aberrant feeding patterns that characterize eating disorders.

3.1.2. MC3R: Preclinical Studies in Food Reward

The melanocortin system interacts with several nuclei of the brain and neural circuits, among which, one of the most relevant in the control of food intake and body weight is the mesocorticolimbic dopamine system [93], connecting the VTA with the Nac, amygdala and PFC, regions particularly involved in reward, motivational processes and food consumption [94–96]. Dopamine has an essential role in food intake and reward, and thus, it is supposed that the melanocortin system can also influence feeding by modulating dopamine transmission in areas that are implicated in eating behaviors, satiety perception and reward processes. Indeed, α-MSH may affect food intake and reward, principally regulating dopamine neuronal activity in the VTA, which is part of the mesolimbic system that includes dopamine cells of the VTA projecting to the NAc [97], a key region for the reinforcing properties of highly palatable food (HPF). HPF, which consists of aliments rich in fat, sugar or both, is a potent reward and has been demonstrated to induce dopamine transmission in the NAc in both human and animal studies, increasing motivation to overconsume this type of food [94,95]. It is well documented that intra-VTA injections of α-MSH stimulate dopamine release in the NAc and dopamine-related behaviors, confirming that α-MSH increases dopamine neuronal activity in the VTA [98–101], and that POMC and AgRP neurons send projections to the VTA [102,103]. In this brain region, there is expression of both MC3Rs and MC4Rs in dopamine and non-dopamine neurons, but MC3Rs are expressed at a much higher level, compared to MC4Rs [22,46,74,104]. Conversely, the NAc shell shows a prominent concentration of MC4Rs that are found on both D1 and D2 receptor-expressing neurons [23], suggesting a differential action of MCRs on dopamine signaling in these brain areas. In light of the high expression of the MC3Rs in the VTA, the role of these receptors in the hedonic aspect of food intake was evaluated via activation of the reward circuitry. Accordingly, MC3R KO female mice, in a sucrose preference test, showed a significant reduction in the sucrose solution intake at all concentrations used (ranged from 1 to 2%), relative to WT littermates, and this was also accompanied by a decrease in sucrose preference at concentration of 1% [74]. Given the critical role of an intact VTA for sucrose preference and intake [105,106], and considering the high concentration of the MC3Rs in this region, it was hypothesized that the defect in sucrose intake in MC3R KO female mice was due to MC3R-related alterations in dopaminergic signaling in the VTA. Deletion of MC3Rs in mice was accompanied by changes in dopamine levels and its metabolites, DOPAC and homovanillic acid, in the VTA, but, interestingly, these parameters were restored in ovariectomized mice, suggesting an interaction between the melanocortin system and estrogens in the regulation of midbrain dopamine levels [74], a factor that could have an impact on food intake, taking into account the important relationship between ovarian hormones and emotional eating and binge eating, in both rodents and humans [107–110].

A following study, using MC3R^{tm1Butl} (MC3R$^{TB/TB}$) mice, the strain in which the expression of MC3R is suppressed by insertion of a loxP-flanked transcription blocker (TB) into the genes 5′ UTR [8], reported that the absence of MC3R signaling reduced self-administration of food reward (20 mg chocolate flavored food pellet) under a progressive ratio protocol in mice subjected to caloric restriction. The result of this study suggests that the motivation to obtain a food reward in MC3R$^{TB/TB}$ mice might be related to conditions of negative energy balance and nutrient scarcity, considering that this behavioral phenotype was not observed in mice with the same genotype, but with ad libitum access to food [21]. Moreover, acute refeeding after fasting did not induce neuronal activity (assessed by c-fos immunoreactivity) in the NAc of MC3R$^{TB/TB}$ mice, the region associated with reward and motivation for food [21], indicating the critical role performed by the MC3R in the appetitive responses to weight loss. Rescuing the expression of the endogenous MC3Rs in the VTA partially re-established the reduced motivation to work for food reward that characterized MC3R$^{TB/TB}$ mice, suggesting that MC3Rs expressed in the VTA could influence motivational responses to caloric restriction and have an important function in the defense of body weight during situations of poor nutrient availability [21].

Pandit et al. observed that pharmacological stimulation of the MC3Rs in the VTA increases the motivation to consume HPF, through a mechanism that involves dopaminergic transmission. Indeed, intra-VTA injection of the selective MC3R agonist γ-MSH increased response to sucrose in rats, evaluated under a progressive ratio schedule of reinforcement, an effect demonstrated by the increased number of active lever presses for sucrose. Conversely, when rats had free access to the sucrose pellet, the same treatment did not enhance free intake of both sucrose pellet or chow, indicating that MC3R stimulation selectively increases the incentive motivation for HPF and not its actual intake [22]. In the same study, i.c.v. administration of α-MSH, a MC3R/MC4R agonist, as expected, decreased the number of active lever presses, reducing response to sucrose, but when α-MSH was co-administrated with the MC4R antagonist HS014, motivation for sucrose was enhanced, supporting the role of MC3Rs in the motivation to obtain a food reward [22]. Interestingly, pretreatment with the dopamine receptors antagonist α-flupenthixol blocked the γ-MSH increased response to sucrose, and this confirms that MC3Rs in the VTA could affect food reward in a dopamine-dependent manner [22].

The result of this study is particularly interesting because it suggests that the melancortin system could fine tune motivation for HPF, depending on the type of MCR expression in different brain nuclei, considering that MC3R signaling in the VTA promotes the motivation-enhancing effects of food rewards (see Figure 1), while MC4R signaling in the NAc shell has the opposite effect, decreasing motivation for HPF [23].

3.2. MC4R

3.2.1. MC4R: Preclinical Studies on Food Preference and Motivation

As previously mentioned, the role of the MC4R in energy homeostasis and obesity is well established, and many preclinical and clinical studies investigated the implication of this receptor in preventing weight gain and regulating energy balance. However, it has been observed that MC4R could affect feeding behaviors also modulating the brain reward circuitry, in particular by influencing neural transmission in areas sensitive to reinforcing properties of HPF [28,111–113].

Indeed, central administration of the endogenous MCRs antagonist AgRP in rats has been demonstrated to preferentially increase intake of a high-fat diet, over a low-fat diet, with a mechanism involving opioid transmission, considering that Naloxone, an opioid receptor antagonist, was found to selectively counteract the consumption of high-fat pellets [112]. Additionally, a selective reduction in fat consumption was found in MC4R +/+ mice treated with intraperitoneally injection of melanotan II (MTII), a MC3R and MC4R agonist, without affecting the intake in MC4R −/− littermates and, in the same study, administration of the selective MC4R agonist (pentacyclo(D-K)-Asp-cis Apc-(D)Phe-Arg-Trp-Lys-NH2) had the same effect, suggesting that the MC4R is the necessary mediator for the reduction in fat intake [114]. When administered into the Central Amygdala, a region connected with hypothalamic areas that affect eating behavior, MTII strongly reduced the high-fat diet intake, but only moderately the low-fat or standard diet, conversely to injections of SHU-9119 and AgRP, antagonists of the MCRs, in the same brain area, that increased rat preference for the high-fat diet [115].

These findings were confirmed by the study of Tracy et al., in which rats, under operant and Pavlovian conditioning paradigms, after receiving i.c.v. injections of 1 nmol AgRP, enhanced active response to earn a peanut oil emulsion (100% fat) reinforcer, but not to obtain a sucrose (100% carbohydrate) reinforcer and increased responses to cues predictive of fat delivery [113]. These results extended previous evidence that melanocortins, via MC4Rs, are probably selective for the intake of high-fat food. Accordingly, Davis et al. observed that treatment with AgRP was able to support conditioned place preference for a high-fat diet compared to standard chow, while blocking the acquisition of place preference for sucrose pellets [111], indicating a selective reinforcement effect of melanocortin antagonism directed toward fat-rich food.

The ability of AgRP to modulate food intake is supposed to be mediated, at least in part, by its influence on dopaminergic signaling in the mesocorticolimbic dopamine circuitry, and central administration of AgRP promotes activation of c-fos immunoreactivity within tyrosine-hydroxylase midbrain dopamine neurons, indicating that melanocortin antagonists are able to elicit neuronal activation in these brain areas [111]. Furthermore, AgRP-treated rats increased dopamine turnover in the medial PFC, one of the major target of dopaminergic projections from the VTA, and it is known that dopaminergic neurons in the medial PFC respond to the positive hedonic aspect of HPF [111,116–118]. Activation of dopamine activity in the medial PFC could also be related to the AgRP ability to promote activation of orexin-A neurons in the lateral hypothalamus [119], strictly involved in the integration of rewarding stimuli, and orexin neurons in this area send projections to the VTA [120], which in turn could stimulate dopamine activity in the medial PFC. Orexin-A neurons are thought to principally regulate arousal, but also feeding and reward-related behaviors [120], and antagonism at the orexin-1 receptor has been demonstrated to block the compulsive-like eating episode in female rats, in a preclinical model of binge eating [121]. In light of these observations, the melanocortin system could be able to promote consumption of high-fat foods in a mechanism involving opioid, dopaminergic and orexin transmissions, and future investigation should be conducted to better understand how these neurotransmitter systems interact in order to facilitate the development of dysregulated eating behaviors.

Subsequent studies, testing MCRs agonists and antagonists, evaluated whether a direct injection of these compounds into the VTA was able to change feeding behavior, altering the activity of the mesolimbic dopamine system. Intra-VTA administration of MTII (a non-selective MC3R/MC4R agonist) dose-dependently suppressed the intake of standard chow in male rats, conversely to the MC3R/MC4R antagonist SHU-9119, which significantly stimulated 24-h food intake. Furthermore, a prolonged blockade of MCRs with the same MCRs antagonist, chronically injected for 5 days, increased total body weight, food intake and caloric efficiency, confirming that stimulation or blockade of MCRs might influence feeding behavior, by modulation of the mesolimbic dopamine transmission [122].

Taking into account this study, it was investigated if pharmacological stimulation of the MCRs in the VTA could also affect the intake of a rewarding sugar solution, under a two-bottle choice paradigm, a procedure in which rats had access to two identical drinking bottles, one containing normal water, and the other one filled with 1, 2 and 10% sucrose solutions. Intra-VTA administration of MTII dose-dependently decreased consumption of a 1 and 2% sucrose solutions, without affecting water intake in the 24-h prolonged access paradigm, while only the highest dose of MTII (50 pmol/side) reduced intake of the more appetizing 10% sucrose solution [123]. However, MTII treatment reduced not only sugar consumption in the two-bottle choice test, but also baseline 24-h food intake, raising the question of whether the effect of MCR stimulation in the mesolimbic pathway is specific or not to the hedonic aspect of food intake over the homeostatic level [123].

Additional studies have been performed to further investigate the role of the MC4Rs in the context of food reward, using self-administration paradigms, in order to evaluate if the melanocortin system could selectively affect food motivation. In light of the high expression of the MC4Rs in the NAc shell, α-MSH (0.2 nmol) and AgRP (0.1 nmol) were directly injected in this brain area, and they, respectively, decreased and increased food self-administration of 45 mg sucrose pellets, as indicated by the number of active lever presses and reinforcers earned in the operant conditioning chambers. This effect was demonstrated to be dopamine-dependent, considering that pretreatment with the dopamine receptors antagonist α-flupentixol, attenuated both active lever presses and reinforcers earned induced by AgRP [23]. Interestingly, α-MSH and AgRP, when administered in rats with free access to the sucrose pellets, did not influence feeding of the HPF, indicating that MC4Rs in the NAc shell are selectively involved in the motivation to obtain food reward [23]. A recent study, always using self-administration of sucrose pellets, under both a fixed and a progressive ratio schedule of reinforcement, obtained a similar result, considering that stimulation of MCRs with intra-VTA injections of MTII dose-dependently reduced sucrose self-administration on both schedules, while blockade of melanocortin signaling in the same area, with the MCRs antagonist SHU-9119, increased self-administration, but only under fixed ratio protocol [124].

These studies had the limit of using compounds that are not selective for MC3R or MC4R and are not in accordance with a recent finding by Pandit et al., who reported that the selective MC3R agonist γ-MSH increased sucrose self-administration when injected in the VTA [23]. This discrepancy can be explained by the activation of distinct pathways, depending on the selectivity of the agonist that activates MC3R or MC4R [124], confirming the different roles played by these MCRs in the mesolimbic dopamine system to regulate food reward and motivation, as shown in Table 1.

Finally, from these observations, it was demonstrated that the melanocortin system is able to affect different aspects of feeding behavior (from standard chow intake to self-administration of HPF) in light of its ability to interact with many other brain pathways implicated in the control of appetite and eating. Moreover, the identification of how this system is altered in aberrant eating patterns, including binge eating behavior, would be useful for a better understanding of these disorders and the discovery of new potential treatments.

Table 1. Summary of studies regarding MC3R and MC4R on food reward and motivation.

Species	Experiment	Result	Ref.
MC3R KO vs. WT mice	Sucrose preference test	↓ sucrose intake and preference in female MC3R KO mice	[74]
$MC3R^{tm1Butl}$ ($MC3R^{TB/TB}$) vs. WT mice	Food self-administration under fixed and progressive ratio protocols	↓ self-administration of a food reward in $MC3R^{TB/TB}$ mice exposed to caloric restriction	[21]
Rats	Food self-administration under fixed and progressive ratio protocols	↑ operant response, but not free access to sucrose after injection of the MC3R agonist γ-MSH	[22]
Rats	Consumption of a high-fat vs. low-fat diet	↑ intake of a high-fat diet vs. a low-fat diet after i.c.v. injection of AgRP	[112]
MC4R +/+ vs. MC4R −/− mice	Consumption of a three-choice diet (fat, protein, carbohydrate)	↓ fat intake in MC4R +/+, but not in MC4R −/− mice after injections of the MC3R/MC4R agonist MTII and the MC4R agonist (pentacyclo(D-K)-Asp-cis Apc-(D)Phe-Arg-Trp-Lys-NH2)	[114]
Rats	High-fat vs. low-fat diet paradigm	↓ the high-fat diet intake after injection of MTII in the CeA; ↑ high-fat diet consumption after the injection MCRs antagonists SHU-9119 and AgRP	[115]
Rats	Fat and sugar consumption under an operant conditioning paradigm	↑ active response to earn a peanut oil emulsion (100% fat) reinforcer, but not a sucrose (100% carbohydrate) reinforcer after i.c.v. injection of AgRP	[113]
Rats	Conditioned place preference for high-fat diet and sucrose pellets	AgRP supports conditioned place preference for a high-fat diet, while blocks the acquisition of place preference for sucrose pellets	[111]
Rats	Consumption of standard chow	↓ intake of standard chow after intra-VTA injection of MTII; ↑ 24-h food intake with SHU-9119	[122]
Rats	Two-bottle choice paradigm for a sucrose solution	↓ consumption of a 1 and 2% sucrose solutions with intra-VTA injections of MTII; ↓ intake of the more appetizing 10% sucrose solution only at the highest dose of MTII	[39]
Rats	Food self-administration under fixed and progressive ratio protocols	↓ operant response with α-MSH and ↑ operant response with AgRP injected in the NAc shell; no influence on free consumption of sucrose pellets	[23]
Rats	Food self-administration under fixed and progressive ratio protocols	↓ sucrose self-administration on both fixed and progressive ratio schedules with intra-VTA injections of MTII; ↑ self-administration, only under fixed ratio protocols with SHU-9119	[124]

↓: decrease; ↑: increase; AgRP: Agouti-related protein; α-MSH: α-melanocyte-stimulating hormone; γ-MSH: γ-melanocyte-stimulating hormone; CeA: Central Amygdala; i.c.v.: Intracerebroventricular; KO: Knock-out; MCRs: Melanocortin receptors; MC3R: Melanocortin-3 receptor; MC4R: Melanocortin-4 receptor; MTII: Melanotan II; NAc: Nucleus Accumbens; VTA: Ventral Tegmental Area; WT: Wild Type.

3.2.2. Clinical Studies on *MC4R* Mutations

Several human studies suggested that the dysfunction of the central melanocortin system, well established in the etiology of obesity, may be a potential mechanism underlying the development of altered eating patterns, due to its contribution to food seeking and consumption, appetite, hyperphagia and body weight control.

The majority of *MC4R* mutations [10], principally including missense and synonymous mutations, have demonstrated partial or complete no activity of MC4R through in vitro study [70], and this loss of function was associated with early-onset obesity in children, manifested particularly in homozygotes rather than heterozygotes, with a higher percentage of body fat mass, increased appetite and food seeking behavior during meals and hyperphagia [70]. Indeed, obese individuals, carriers of different *MC4R* mutations, compared with obese and normal weight participants without these variants, were diagnosed with BED through the completion of a validated questionnaire, thus resulting in the co-existence between obesity and BED [15,125]. In one of the first studies, despite a large number of obese children and adolescent carriers of *MC4R* gene mutations, only one girl met criteria for BED [126]. Conversely, Branson et al. found that obese individuals, carriers of *MC4R* gene mutations, met diagnostic criteria for BED, completing a validated eating disorder questionnaire [127] based on the fourth edition of Diagnostic and Statistical Manual of Mental Disorders (DMS-IV), defining BED as the major phenotype of MC4R genetic variants [15]. However, a significant controversy surrounded these findings [128], considering that other studies did not find an association between *MC4R* mutations and episodes of binge eating [129,130], and, in addition, no differences were detected in body mass index (BMI) or specific phenotype between adult carriers and non-carriers of the *MC4R* mutations [130]. In contrast to the study of Hebebrand et al., in which there were no strong associations between BED and *MC4R* mutations, Tao et al. identified BED in obese patients with specific mutations in this receptor (T11A, F51L, T112M and M200V), without being able to explain the possible pathogenesis of the development of this eating disorder in relation to *MC4R* mutations [131]. Additionally, variability of *MC4R* gene was also investigated in non-obese patients with binge eating behavior, showing a lower presence of *MC4R* mutation in this group in contrast to obese patients; however, the study was performed in a very small number of individuals with binge eating behavior and this limitation, together with the lack of a control group, might have affected the result [132].

The variants of *MC4R* were additionally considered for their possible association with the outcomes of bariatric surgery: in the study of Potoczna et al., obese patients, carriers of *MC4R* variants that presented an aggressive form of BED, were less responsive to weight loss after laparoscopic gastric banding treatment [125], while Vallette et al. did not find an influence of these genetic mutations in weight loss and body composition after the same surgical treatment [133]. A recent study evidenced that the presence of functional variants of *MC4R* significantly affected the efficacy of different laparoscopic operations in obese Swiss patients with BED, increasing the risk of reoperation due to a failure in postoperative weight loss [134].

These observations have encouraged further investigations of a possible involvement of *MC4R* mutations in different eating patterns, particularly in obese subjects, to explain and document the food attitudes leading to weight gain, hedonic overeating and behavioral addiction to obtain food rewards.

Valette et al. discussed how mutations could influence the choice and the preference for macronutrients: in obese adults, carriers of different functional mutations of *MC4R*, an increased carbohydrate intake compared to fat intake was reported. In the same study, using interviews with standardized questionnaires and binge eating scales, no statistical difference was found in eating behaviors in both carriers and non-carriers of *MC4R* mutations [135].

To investigate the impact of the complete loss of function of MC4R signaling on the brain response to anticipatory food reward, van der Klaauw et al. performed functional magnetic resonance imaging (fMRI) in a small group of obese individuals with heterozygous *MC4R* mutations and in obese and lean individuals without mutations in satiated state. After seeing images of HPF, surprisingly, no group difference was found in the amygdala or orbitofrontal cortex, but a hyporesponsivity to visual food cues was reported in the dorsal and ventral striatum in obese controls, compared to the response of MC4R-deficient obese patients and lean controls [136]. The result of this study is particularly relevant, knowing that dorsal striatum is a brain region involved in compulsive food seeking behavior and BED, even in a sated state [137,138]. Indeed, the understanding of how different brain responses and behavioral factors are involved in rewarding food cues may explain the reason for the development of HPF overconsumption.

3.2.3. The Polymorphism rs17782313 Nearby *MC4R* Gene and Eating Behavior

Recently, a genomewide association study (GWAS) identified several single nucleotide polymorphisms (SNPs) of the *MC4R* gene, associated with high BMI and the risk of the development of obesity [139]. Among them, the SNP rs17782313, mapping to a locus 188kb downstream from the coding sequence of *MC4R* gene region, has become increasingly relevant in relation to obesity and aberrant eating behaviors [140]. Additionally, this SNP rs17782313 seems to affect expression and function of the MC4R and it has been proposed, in several studies, as a factor leading to altered eating behavior patterns, increased vulnerability to higher BMI and changes in human brain regions, especially in women and children [141–144].

In the study of Qi et al., high preference and intake of nutrients rich in fat, saturated fat and partly protein, without any appetite deregulation regarding carbohydrate, were found in women carriers of this SNP compared to the non-carrier participants [142], leading to an elevated risk of severe obesity. Moreover, the following works have tried to clarify the possible implication of this *MC4R* SNP in feeding behavior: Stutzmann et al. revealed an excessive appetite in a large cohort of European populations, especially eating a large amount of food during meals with a higher frequency of snacking in children and teenagers carriers of this SNP, and a greater hunger in adults carriers of the same polymorphism [143]. Snacking is a particular dietary pattern, principally during childhood, in which energy-dense and nutrient-poor food is consumed between meals exhibiting a recurrent "snack episode", which can be translated into a bad feeding style and a risk factor for altered eating behavior and elevated BMI [145,146].

Furthermore, another study evidenced less postprandial satiation symptoms after a fully caloric satiating meal in obese individuals, carriers of rs17782313 polymorphism, promoting to eat more frequently and to increase the caloric intake in the subsequent meal, leading to higher BMI [147]. In addition, the presence of this genotypic variant demonstrated low satiety responsiveness scores, and high scores for enjoyment of food in Chilean obese children compared to the non-carrier participants [148], assessed through the Child Eating Behavior Questionnaire (CEBQ) [149] and 19-item Three-Factor Eating Questionnaire Parent (TFEQP-19) Chilean version of the TFEQ-R18 [150]. A following study, always conducted in a Chilean population, focused on obese children carriers of SNP rs17782313 revealing, in addition to lower satiety responsiveness and elevated enjoyment of food, even an overconsumption of snacks after a standard meal, using the Eating in the Absence of Hunger (EAH) Test and the CEBQ [17]. Moreover, in a three-generation Chilean family of obese women, the presence of a genetic variant of *MC4R*, generating an amino acid substitution Thr150Ile, characterized by a decreased activity of the MC4R, led to an elevated BMI and remarkable scores of cognitive restraint (CR), uncontrolled eating (UE) and emotional eating (EE), measured by TFEQ-R18 [19]. These three parameters indicate respectively: conscious lower consumption of HPF but higher intake of vegetables and proteins in order to control BMI; the tendency to eat unhealthy food more than usual in response to external stimuli with loss of control and hunger for extreme unstoppable appetite; and, finally, the inability to resist stress events, negative emotions and mood states, which often cause binge eating episodes [150–152]. These paradigms were also evaluated in Chilean adults, carriers of SNP rs17782313, presenting higher EE scores compared with non-carriers, while only women showed UE, evidencing a difference between women and men with this SNP [20].

Recent studies in obese, overweight and normal weight Chilean children extended the evidence about the SNP rs17782313, investigating how this genetic variant affects the ingestive behaviors related to reward properties of food [153]. Eating behavior scores were calculated from the EAH Test, CEBQ, TFEQ and Food Reinforcement Value Questionnaire (RVFQ), reporting differences between gender in eating patterns, but not in elevated BMI: in obese boys, carriers of the SNP, a significantly lower reinforcing value of food was observed compared to the non-carriers; meanwhile, obese girls, carriers of this polymorphism, showed lower satiety responsiveness, and UE with respect to obese girls without the SNP. These results are in accordance with the study of Vega et al., in which Chilean obese adults showed UE, suggesting the involvement of MC4R in dopamine pathways relating to food reward [153]. The hypothesis of a possible link between dopamine and melanocortin pathways has also been proposed by Yilmaz et al., underlying that this interaction could be responsible for the results of the study, in which, through the use of several questionnaires, significant EE, food craving, elevated BMI and depressive mood in European adult carriers of SNP rs17782313 were found [144]. Furthermore, the results of the case control comparisons with a group of female participants who had Anorexia or Bulimia Nervosa did not find any evidence that linked the genetic variant with these eating disorders [154].

The evidence concerning this specific polymorphism that might contribute to overweight and altered feeding patterns is not limited to the populations mentioned above, but it has been also investigated and found in subjects of different nationalities and ethnic origins [147,155–162], where the majority of these studies addressed the vulnerability of women and children to moderate and severe obesity and aberrant eating behaviors [141,157,161,163].

Horstmann et al. suggested that the genetic variation rs17782313 could affect reward mechanisms, showing that only women, homozygous carriers of the risk SNP, demonstrated EE and Disinhibition of Eating (loss of control over feeding, possibly due to external stimuli) measured by TFEQ-R18 and TFEQ-51. Moreover, through Magnetic Resonance Imaging (MRI), a sex-specific association was found between rs17782313 and an increased gray matter volume in the right amygdala, the anterior hippocampus, the medial orbitofrontal cortex and the left and the right PFC [141], crucial regions known to be involved in eating behavior [164,165].

All the studies discussed in this section (summarized in Table 2) highlighted that the partial or total loss of MC4R function, due to *MC4R* mutations, as well as the SNP rs17782313, are positively correlated with altered appetite and dysfunctional eating patterns, promoting obesity and elevated BMI.

Table 2. *MC4R* variant rs17782313 and manifestation of altered eating behavioral phenotype.

Subjects with the *MC4R* Variant rs17782313	Result	Ref.
Normal weight vs. obese children of both sexes	Obese children present high scores of Enjoyment of Food, Emotional Overeating, Food Responsiveness and lower Satiety Responsiveness	[17]
Adult participants	In both genders high scores of Emotional Eating associated with BMI were found, while only in women the Uncontrolled Eating scores were associated with BMI.	[20]
Healthy adult volunteers	Only women, especially homozygous carriers of *MC4R* variant rs17782313, demonstrated Emotional Eating and Disinhibition of Eating.	[141]
Adult women	Women had significantly higher intake of energy from fat, compared to carbohydrate.	[142]
Children, teenagers and adults	Children and teenagers presented snacking and eating large amounts of food during meals. Adults presented a greater hunger score	[143]
Adults between the ages of 24 and 50 years	Overeating behaviors, Emotional Eating and Food Cravings.	[144]
Overweight or obese participants	Less postprandial satiation symptoms after a fully caloric meal	[147]
Obese children	Low Satiety Responsiveness scores and high scores for the Enjoyment of Food	[148]
Obese, overweight and normal weight children	In obese girls were found significant lower scores of the Satiety Responsiveness and higher scores of the Uncontrolled Eating	[153]
Women	Association with increased BMI and obesity	[160,161,163]
Lean, overweight, and obese children	Association with increased BMI and obesity	[155,159]
Normal weight vs. obese adults	Association with increased BMI and obesity	[156,162]
Normal weight vs. obese adults	Association with increased BMI and obesity and a significant higher intake of energy from fat compared to carbohydrate.	[158]

BMI: Body Mass Index; MC4R: Melanocortin-4 receptor.

4. Melanocortin System and Stress Responses

Dieting, stress and negative affect are considered potential factors able to trigger binge eating episodes in patients with BED or Bulimia Nervosa [37,166,167]. Indeed, dieting periods are commonly observed in the history of binge eaters, but hunger alone appears to be non-sufficient to induce a compulsive-like eating, if not accompanied by conditions of stress or negative affect [168,169]. Stress has a central role in the etiology of binge eating, considering that obese individuals with BED, compared to those without, show a higher activity of the HPA axis and cortisol/corticosterone plasma level [170–173]. Additionally, higher cortisol levels, induced by stress, are able to promote a greater consumption of sweet foods [174], and are also positively correlated with the severity of binge eating [175].

The melanocortin system, principally via MC4R, has been demonstrated to play a central role in stress response and negative emotional states, including anxiety and depression [176,177], suggesting the MC4R as a possible target to treat these psychiatric conditions. In fact, MC4Rs are expressed in the limbic system, mainly in several nuclei of the amygdala, such as the central and basolateral nuclei, lateral septal nucleus, hippocampus and in the entorhinal cortex [50]; thus, the distribution of the MC4R in the brain indicates an important involvement of this receptor in promoting negative emotional states [176,177]. Moreover, the MC4R, contrary to MC3R, has been highly detected in the PVN of the hypothalamus, where it is supposed to regulate the activity of the HPA axis, via arginine vasopressin (AVP) and corticotropin releasing factor (CRF) neurons [50,176]. Initial evidence linking the MC4R and stress-related responses comes from studies in which the administration of α-MSH and ACTH in rats was able to increase grooming behavior [99–101,178,179], characterized by many activities directed to the animal body surface, such as face washing, body grooming, licking, scratching and genital grooming, and proposed as a rodent behavioral response to stress and novel environments [178,180]. The effect of α-MSH on grooming is principally due to its agonistic activity on MC4Rs, as demonstrated by Adan et al., who found that grooming behavior, induced by MCR agonists, was positively correlated with a greater affinity and potency for MC4R, rather than for MC3R. On the contrary, the antagonist SHU-9119 attenuated grooming induced by both melanocortins and by exposure to a novel environment [178]. This finding is further confirmed by the fact that the MC4R agonist MTII increased grooming in WT, but not in mutant rats deficient in MC4Rs, confirming that this behavior is principally mediated by MC4Rs, and not by MC3R subtypes [181].

Stress has been demonstrated to have profound effects on MC4R expression and activity in the brain. In fact, the exposure to electric foot shock stress in rats increased the expression of POMC and MC4R mRNA in the hypothalamus and in the amygdala [182], region implicated in the modulation of emotional- and fear-related behaviors [183] and binge eating episodes [184,185].

Furthermore, rats exposed to chronic restraint stress had increased MC4R mRNA expression in the ARC of the hypothalamus, compared to control rats, not exposed to stress [186]. The effect of stress on MC4R and on feeding behavior and appetite may be also dependent on the intensity and duration of the stressor, as supported by the study of Chagra et al., in which chronic exposure to a stress induced a significant decrease in c-fos- and MC4R-expressing cells in the ARC, indicating a shift toward more orexigenic behaviors, differently from control and acutely stressed rats [187].

Pharmacological stimulation of the MC4R is able to promote the activity of the HPA axis, as reported by the study of Von Frijtag et al., in which i.c.v. injection of ACTH1-24 (the N-terminal bioactive fragment of ACTH) in rats, significantly increased plasma concentrations of ACTH and corticosterone, an effect inhibited by pretreatment with the non-selective antagonist SHU-9119 and by the selective MC4R antagonist [D-Arg8] ACTH4-10 [188]. The influence of the melanocortin system on HPA axis tone and activity can be explained considering that MC4Rs are highly expressed in the parvocellular division of the PVN [50,189], the region in which CRF neurons are also predominantly localized and where they receive α-MSH neuronal terminals [190,191].

In fact, activation of MC4Rs by i.c.v. injections of α-MSH or MTII increases gene expression of CRF in the PVN [189,192] and enhances corticosterone plasma levels in rats, suggesting a functional interaction between CRF and the melanocortin system [189]. In the same study, the pretreatment with the CRF antagonist α-helical-CRH9–41 was able to prevent MTII-induced suppression of food intake, evidencing that the melanocortin system can alter endogenous CRF levels in order to modulate appetite [189].

A stress procedure that has been demonstrated to promote activation of the melanocortin neurons, and, consequently, of the HPA axis, is the acute restraint stress. Rats exposed to this stress had a robust c-fos mRNA expression in the medial amygdala (MeA) [18,193], a brain region with high levels of MC4Rs [50], and particularly sensitive to psychological stressors, characterized by an emotional component, such as restraint [194,195]. Lesions of the MeA result in a blunted response of the HPA axis to psychogenic stressors [196], conversely to pharmacological stimulation of the MC4R-expressing neurons in the MeA, which promotes corticosterone release [18]. Moreover, both stress-induced anorexia and corticosterone release, in response to the acute restraint stress, can be prevented by administration of a MC4R antagonist directly in the MeA [18,193]. The interaction between MC4Rs in the MeA and the CRF system is probably mediated by the efferents from the MeA to the Bed Nucleus of the Stria Terminalis (BNST), brain region enriched in CRF neurons [197], and involved in stress-induced emotional responses and activation of the HPA axis [198,199]. The BNST has been demonstrated to play a pivotal role in stress-induced binge eating for HPF, evoked by a combination of frustration stress and food restriction [200,201], and injection of a non-selective CRF receptor antagonist directly into the BNST was able to counteract this compulsive-like eating episode for HPF selectively in rats exposed to both stress and restriction [200]. These findings support the hypothesis that MC4R can also influence the activation of the HPA axis via extrahypothalamic sites, and thus could represent an important factor for the development of aberrant feeding behaviors in response to stress exposure. Consistently with this evidence, acute stress-induced release of ACTH and corticosterone, as well as neuronal activation in the PVN and MeA, were significantly attenuated in male rats with a *MC4R* mutation, producing a less functional receptor, compared to the WT littermates [202,203]. Intriguingly, it was observed that female rats with the same mutation revealed an unexpected and exaggerated acute stress-induced corticosterone release, contrary to mutant males, highlighting a difference in stress reactivity between male and female rats with the MC4R loss of function [202]. The result of this study suggests a sex-dependent responsivity in the basal HPA axis tone and acute stress-induced corticosterone in rodents with MC4R mutation [202]. Considering the heightened stress reactivity found in female rats with deficient MC4R activity [202], and that stress has been associated with EE [37,38] and binge eating behavior [38,204–206], it would be interesting in the future to evaluate the potential involvement of MC4R signaling in a female rat model of binge eating, in which the binge eating episode is elicited by a combination of food restriction plus stress [86,87] and by using HPF, in order to promote the aberrant feeding behavior and to increase the motivation to overconsume food [38,87,207–209].

5. Conclusions

The pivotal role played by melanocortin system in controlling feeding behavior, appetite, energy balance and motivation for rewarding properties of food can explain why dysfunction of this system, in both human and rodent studies, results in a breakdown of normal regulatory processes and in more vulnerability to the loss of control in food intake, possibly leading to altered eating patterns, as summarized in Figure 2. Further research needs to highlight the mechanisms driving the hyperphagia in melanocortin-associated obesity, evidencing whether the exaggerated food consumption is accompanied by the loss of behavioral control, food seeking and/or binge eating episodes. Recent studies concerning MC3R and MC4R revealed that melanocortin signaling can exert functional effects in reward-related behaviors, due the hedonic properties of HPF, which is a potent natural reinforcer, and it has been postulated that in humans, low melanocortin activity could predispose individuals to pathological overeating, developing obesity and altered feeding behavior. Finally, the consistent relationship between the MC4R and stress response can be considered an additional factor linking melanocortin signaling to binge eating episodes, given the key role of stress in the etiology of this compulsive behavior. More preclinical studies are needed to investigate the biological mechanisms underlying dysfunctional eating patterns and clarify the possible connection between MCRs and binge eating behavior.

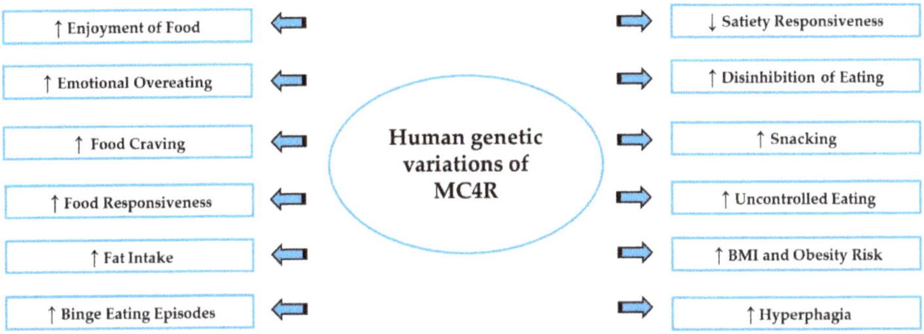

Figure 2. An overview of the altered eating patterns associated with the genetic variation of MC4R. ↓: decrease; ↑: increase; BMI: Body mass index; MC4R: Melanocortin-4 receptor.

Author Contributions: Conceptualization, E.M.D.B., L.B., M.V.M.D.B. and C.C.; writing—original draft preparation, E.M.D.B., L.B., M.V.M.D.B. and C.C.; writing—review and editing, E.M.D.B., L.B., D.T., S.K.T., M.V.M.D.B. and C.C.; visualization, E.M.D.B., L.B., D.T., S.K.T., M.V.M.D.B. and C.C.; supervision, M.V.M.D.B. and C.C.; funding acquisition, C.C. All authors have read and agreed to the published version of the manuscript.

Funding: The research was supported by grant PRIN2015KP7T2Y to C.C. of the Italian Ministry of Education, University and Research.

Conflicts of Interest: The authors declare no conflict of interest.

References

1. Cordain, L.; Eaton, S.B.; Sebastian, A.; Mann, N.; Lindeberg, S.; Watkins, B.A.; O'Keefe, J.H.; Brand-Miller, J. Origins and evolution of the Western diet: Health implications for the 21st century. *Am. J. Clin. Nutr.* **2005**, *81*, 341–354. [CrossRef] [PubMed]
2. Giudetti, A.M.; Micioni Di Bonaventura, M.V.; Ferramosca, A.; Longo, S.; Micioni Di Bonaventura, E.; Friuli, M.; Romano, A.; Gaetani, S.; Cifani, C. Brief daily access to cafeteria-style diet impairs hepatic metabolism even in the absence of excessive body weight gain in rats. *FASEB J.* **2020**, *34*, 9358–9371. [CrossRef] [PubMed]
3. Cone, R.D. The Central Melanocortin System and Energy Homeostasis. *Trends Endocrinol. Metab.* **1999**, *10*, 211–216. [CrossRef]
4. Cone, R.D. Studies on the physiological functions of the melanocortin system. *Endocr. Rev.* **2006**, *27*, 736–749. [CrossRef] [PubMed]
5. Girardet, C.; Butler, A.A. Neural melanocortin receptors in obesity and related metabolic disorders. *Biochim. Biophys. Acta* **2014**, *1842*, 482–494. [CrossRef] [PubMed]
6. Butler, A.A. The melanocortin system and energy balance. *Peptides* **2006**, *27*, 281–290. [CrossRef]
7. Fan, W.; Boston, B.A.; Kesterson, R.A.; Hruby, V.J.; Cone, R.D. Role of melanocortinergic neurons in feeding and the agouti obesity syndrome. *Nature* **1997**, *385*, 165–168. [CrossRef]
8. Beckers, S.; Zegers, D.; de Freitas, F.; Mertens, I.L.; Van Gaal, L.F.; Van Hul, W. Association study of MC4R with complex obesity and replication of the rs17782313 association signal. *Mol. Genet. Metab.* **2011**, *103*, 71–75. [CrossRef]
9. Benoit, S.; Schwartz, M.; Baskin, D.; Woods, S.C.; Seeley, R.J. CNS melanocortin system involvement in the regulation of food intake. *Horm. Behav.* **2000**, *37*, 299–305. [CrossRef]
10. Hinney, A.; Volckmar, A.L.; Knoll, N. Melanocortin-4 receptor in energy homeostasis and obesity pathogenesis. *Prog. Mol. Biol. Transl. Sci.* **2013**, *114*, 147–191. [CrossRef]
11. Huszar, D.; Lynch, C.A.; Fairchild-Huntress, V.; Dunmore, J.H.; Fang, Q.; Berkemeier, L.R.; Gu, W.; Kesterson, R.A.; Boston, B.A.; Cone, R.D.; et al. Targeted disruption of the melanocortin-4 receptor results in obesity in mice. *Cell* **1997**, *88*, 131–141. [CrossRef]
12. Kleinendorst, L.; van Haelst, M.M.; van den Akker, E.L.T. Genetics of Obesity. *Exp. Suppl.* **2019**, *111*, 419–441. [CrossRef] [PubMed]

13. Vaisse, C.; Clement, K.; Guy-Grand, B.; Froguel, P. A frameshift mutation in human MC4R is associated with a dominant form of obesity. *Nat. Genet.* **1998**, *20*, 113–114. [CrossRef] [PubMed]
14. Yeo, G.S.; Farooqi, I.S.; Aminian, S.; Halsall, D.J.; Stanhope, R.G.; O'Rahilly, S. A frameshift mutation in MC4R associated with dominantly inherited human obesity. *Nat. Genet.* **1998**, *20*, 111–112. [CrossRef]
15. Branson, R.; Potoczna, N.; Kral, J.G.; Lentes, K.U.; Hoehe, M.R.; Horber, F.F. Binge eating as a major phenotype of melanocortin 4 receptor gene mutations. *N. Engl. J. Med.* **2003**, *348*, 1096–1103. [CrossRef]
16. Butler, A.A.; Marks, D.L.; Fan, W.; Kuhn, C.M.; Bartolome, M.; Cone, R.D. Melanocortin-4 receptor is required for acute homeostatic responses to increased dietary fat. *Nat. Neurosci.* **2001**, *4*, 605–611. [CrossRef]
17. Ho-Urriola, J.; Guzman-Guzman, I.P.; Smalley, S.V.; Gonzalez, A.; Weisstaub, G.; Dominguez-Vasquez, P.; Valladares, M.; Amador, P.; Hodgson, M.I.; Obregon, A.M.; et al. Melanocortin-4 receptor polymorphism rs17782313: Association with obesity and eating in the absence of hunger in Chilean children. *Nutrition* **2014**, *30*, 145–149. [CrossRef]
18. Liu, J.; Garza, J.C.; Li, W.; Lu, X.Y. Melanocortin-4 receptor in the medial amygdala regulates emotional stress-induced anxiety-like behaviour, anorexia and corticosterone secretion. *Int. J. Neuropsychopharmacol.* **2013**, *16*, 105–120. [CrossRef]
19. Santos, J.L.; Amador, P.; Valladares, M.; Albala, C.; Martinez, J.A.; Marti, A. Obesity and eating behaviour in a three-generation Chilean family with carriers of the Thrl50Ile mutation in the melanocortin-4 receptor gene. *J. Physiol. Biochem.* **2008**, *64*, 205–210. [CrossRef]
20. Vega, J.A.; Salazar, G.; Hodgson, M.I.; Cataldo, L.R.; Valladares, M.; Obregon, A.M.; Santos, J.L. Melanocortin-4 Receptor Gene Variation Is Associated with Eating Behavior in Chilean Adults. *Ann. Nutr. Metab.* **2016**, *68*, 35–41. [CrossRef]
21. Mavrikaki, M.; Girardet, C.; Kern, A.; Faruzzi Brantley, A.; Miller, C.A.; Macarthur, H.; Marks, D.L.; Butler, A.A. Melanocortin-3 receptors in the limbic system mediate feeding-related motivational responses during weight loss. *Mol. Metab.* **2016**, *5*, 566–579. [CrossRef] [PubMed]
22. Pandit, R.; Omrani, A.; Luijendijk, M.C.; de Vrind, V.A.; Van Rozen, A.J.; Ophuis, R.J.; Garner, K.; Kallo, I.; Ghanem, A.; Liposits, Z.; et al. Melanocortin 3 Receptor Signaling in Midbrain Dopamine Neurons Increases the Motivation for Food Reward. *Neuropsychopharmacology* **2016**, *41*, 2241–2251. [CrossRef] [PubMed]
23. Pandit, R.; van der Zwaal, E.M.; Luijendijk, M.C.; Brans, M.A.; van Rozen, A.J.; Oude Ophuis, R.J.; Vanderschuren, L.J.; Adan, R.A.; la Fleur, S.E. Central melanocortins regulate the motivation for sucrose reward. *PLoS ONE* **2015**, *10*, e0121768. [CrossRef]
24. De Franca, G.V.; Gigante, D.P.; Olinto, M.T. Binge eating in adults: Prevalence and association with obesity, poor self-rated health status and body dissatisfaction. *Public Health Nutr.* **2014**, *17*, 932–938. [CrossRef]
25. Hudson, J.I.; Hiripi, E.; Pope, H.G., Jr.; Kessler, R.C. The prevalence and correlates of eating disorders in the National Comorbidity Survey Replication. *Biol. Psychiatry* **2007**, *61*, 348–358. [CrossRef] [PubMed]
26. Kessler, R.C.; Berglund, P.A.; Chiu, W.T.; Deitz, A.C.; Hudson, J.I.; Shahly, V.; Aguilar-Gaxiola, S.; Alonso, J.; Angermeyer, M.C.; Benjet, C.; et al. The prevalence and correlates of binge eating disorder in the World Health Organization World Mental Health Surveys. *Biol. Psychiatry* **2013**, *73*, 904–914. [CrossRef]
27. Kishi, T.; Aschkenasi, C.J.; Lee, C.E.; Mountjoy, K.G.; Saper, C.B.; Elmquist, J.K. Expression of melanocortin 4 receptor mRNA in the central nervous system of the rat. *J. Comp. Neurol.* **2003**, *457*, 213–235. [CrossRef]
28. Yoon, Y.R.; Baik, J.H. Melanocortin 4 Receptor and Dopamine D2 Receptor Expression in Brain Areas Involved in Food Intake. *Endocrinol. Metab.* **2015**, *30*, 576–583. [CrossRef] [PubMed]
29. American Psychiatric Association. *Diagnostic and Statistical Manual of Mental Disorders: DSM-5*; American Psychiatric Association: Arlington, VA, USA, 2013.
30. Guerdjikova, A.I.; Mori, N.; Casuto, L.S.; McElroy, S.L. Update on Binge Eating Disorder. *Med. Clin. N. Am.* **2019**, *103*, 669–680. [CrossRef]
31. Ulfvebrand, S.; Birgegard, A.; Norring, C.; Hogdahl, L.; von Hausswolff-Juhlin, Y. Psychiatric comorbidity in women and men with eating disorders results from a large clinical database. *Psychiatry Res.* **2015**, *230*, 294–299. [CrossRef] [PubMed]
32. Yanovski, S.Z. Binge eating disorder: Current knowledge and future directions. *Obes. Res.* **1993**, *1*, 306–324. [CrossRef] [PubMed]
33. Schag, K.; Schonleber, J.; Teufel, M.; Zipfel, S.; Giel, K.E. Food-related impulsivity in obesity and binge eating disorder—A systematic review. *Obes. Rev.* **2013**, *14*, 477–495. [CrossRef] [PubMed]

34. Manwaring, J.L.; Green, L.; Myerson, J.; Strube, M.J.; Wilfley, D.E. Discounting of Various types of rewards by women with and without binge eating Disorder: Evidence for general rather than specific Differences. *Psychol. Rec.* **2011**, *61*, 561–582. [CrossRef] [PubMed]
35. Moore, C.F.; Sabino, V.; Koob, G.F.; Cottone, P. Neuroscience of Compulsive Eating Behavior. *Front. Neurosci.* **2017**, *11*, 469. [CrossRef]
36. Zeeck, A.; Stelzer, N.; Linster, H.W.; Joos, A.; Hartmann, A. Emotion and eating in binge eating disorder and obesity. *Eur. Eat. Disord. Rev.* **2011**, *19*, 426–437. [CrossRef]
37. Adam, T.C.; Epel, E.S. Stress, eating and the reward system. *Physiol. Behav.* **2007**, *91*, 449–458. [CrossRef]
38. Micioni Di Bonaventura, M.V.; Micioni Di Bonaventura, E.; Polidori, C.; Cifani, C. Preclinical Models of Stress and Environmental Influences on Binge Eating. In *Binge Eating: A Transdiagnostic Psychopathology*; Frank, G.K.W., Berner, L.A., Eds.; Springer International Publishing: Cham, Switzerland, 2020; pp. 85–101. [CrossRef]
39. Rosenbaum, D.L.; White, K.S. The relation of anxiety, depression, and stress to binge eating behavior. *J. Health Psychol.* **2015**, *20*, 887–898. [CrossRef]
40. He, J.; Cai, Z.; Fan, X. Prevalence of binge and loss of control eating among children and adolescents with overweight and obesity: An exploratory meta-analysis. *Int. J. Eat. Disord.* **2017**, *50*, 91–103. [CrossRef]
41. Bagnol, D.; Lu, X.Y.; Kaelin, C.B.; Day, H.E.; Ollmann, M.; Gantz, I.; Akil, H.; Barsh, G.S.; Watson, S.J. Anatomy of an endogenous antagonist: Relationship between Agouti-related protein and proopiomelanocortin in brain. *J. Neurosci.* **1999**, *19*, RC26. [CrossRef]
42. Ollmann, M.M.; Wilson, B.D.; Yang, Y.K.; Kerns, J.A.; Chen, Y.; Gantz, I.; Barsh, G.S. Antagonism of central melanocortin receptors in vitro and in vivo by agouti-related protein. *Science* **1997**, *278*, 135–138. [CrossRef]
43. Gantz, I.; Konda, Y.; Tashiro, T.; Shimoto, Y.; Miwa, H.; Munzert, G.; Watson, S.J.; DelValle, J.; Yamada, T. Molecular cloning of a novel melanocortin receptor. *J. Biol. Chem.* **1993**, *268*, 8246–8250. [PubMed]
44. Gantz, I.; Miwa, H.; Konda, Y.; Shimoto, Y.; Tashiro, T.; Watson, S.J.; DelValle, J.; Yamada, T. Molecular cloning, expression, and gene localization of a fourth melanocortin receptor. *J. Biol. Chem.* **1993**, *268*, 15174–15179. [PubMed]
45. Mountjoy, K.G.; Robbins, L.S.; Mortrud, M.T.; Cone, R.D. The cloning of a family of genes that encode the melanocortin receptors. *Science* **1992**, *257*, 1248–1251. [CrossRef] [PubMed]
46. Roselli-Rehfuss, L.; Mountjoy, K.G.; Robbins, L.S.; Mortrud, M.T.; Low, M.J.; Tatro, J.B.; Entwistle, M.L.; Simerly, R.B.; Cone, R.D. Identification of a receptor for gamma melanotropin and other proopiomelanocortin peptides in the hypothalamus and limbic system. *Proc. Natl. Acad. Sci. USA* **1993**, *90*, 8856–8860. [CrossRef]
47. Adan, R.A.; Cone, R.D.; Burbach, J.P.; Gispen, W.H. Differential effects of melanocortin peptides on neural melanocortin receptors. *Mol. Pharmacol.* **1994**, *46*, 1182–1190.
48. Mountjoy, K.G. Distribution and function of melanocortin receptors within the brain. *Adv. Exp. Med. Biol.* **2010**, *681*, 29–48. [CrossRef]
49. Jegou, S.; Boutelet, I.; Vaudry, H. Melanocortin-3 receptor mRNA expression in pro-opiomelanocortin neurones of the rat arcuate nucleus. *J. Neuroendocrinol.* **2000**, *12*, 501–505. [CrossRef]
50. Mountjoy, K.G.; Mortrud, M.T.; Low, M.J.; Simerly, R.B.; Cone, R.D. Localization of the melanocortin-4 receptor (MC4-R) in neuroendocrine and autonomic control circuits in the brain. *Mol. Endocrinol.* **1994**, *8*, 1298–1308. [CrossRef]
51. Mountjoy, K.G.; Wild, J.M. Melanocortin-4 receptor mRNA expression in the developing autonomic and central nervous systems. *Dev. Brain Res.* **1998**, *107*, 309–314. [CrossRef]
52. Small, C.J.; Kim, M.S.; Stanley, S.A.; Mitchell, J.R.; Murphy, K.; Morgan, D.G.; Ghatei, M.A.; Bloom, S.R. Effects of chronic central nervous system administration of agouti-related protein in pair-fed animals. *Diabetes* **2001**, *50*, 248–254. [CrossRef]
53. Beckers, S.; Zegers, D.; Van Gaal, L.F.; Van Hul, W. The role of the leptin-melanocortin signalling pathway in the control of food intake. *Crit. Rev. Eukaryot. Gene Expr.* **2009**, *19*, 267–287. [CrossRef] [PubMed]
54. Fan, W.; Dinulescu, D.M.; Butler, A.A.; Zhou, J.; Marks, D.L.; Cone, R.D. The central melanocortin system can directly regulate serum insulin levels. *Endocrinology* **2000**, *141*, 3072–3079. [CrossRef] [PubMed]
55. Williams, K.W.; Scott, M.M.; Elmquist, J.K. Modulation of the central melanocortin system by leptin, insulin, and serotonin: Co-ordinated actions in a dispersed neuronal network. *Eur. J. Pharmacol.* **2011**, *660*, 2–12. [CrossRef] [PubMed]

56. Micioni Di Bonaventura, M.V.; Micioni Di Bonaventura, E.; Cifani, C.; Polidori, C. N/OFQ-NOP System in Food Intake. *Handb. Exp. Pharmacol.* **2019**, *254*, 279–295. [CrossRef] [PubMed]
57. Cifani, C.; Guerrini, R.; Massi, M.; Polidori, C. Chronic intracerebroventricular infusion of nociceptin/orphanin FQ increases food and ethanol intake in alcohol-preferring rats. *Peptides* **2006**, *27*, 2803–2810. [CrossRef] [PubMed]
58. Bomberg, E.M.; Grace, M.K.; Levine, A.S.; Olszewski, P.K. Functional interaction between nociceptin/orphanin FQ and alpha-melanocyte-stimulating hormone in the regulation of feeding. *Peptides* **2006**, *27*, 1827–1834. [CrossRef] [PubMed]
59. Filaferro, M.; Ruggieri, V.; Novi, C.; Calo, G.; Cifani, C.; Micioni Di Bonaventura, M.V.; Sandrini, M.; Vitale, G. Functional antagonism between nociceptin/orphanin FQ and corticotropin-releasing factor in rat anxiety-related behaviors: Involvement of the serotonergic system. *Neuropeptides* **2014**, *48*, 189–197. [CrossRef]
60. Vitale, G.; Filaferro, M.; Micioni Di Bonaventura, M.V.; Ruggieri, V.; Cifani, C.; Guerrini, R.; Simonato, M.; Zucchini, S. Effects of [Nphe(1), Arg(14), Lys(15)] N/OFQ-NH2 (UFP-101), a potent NOP receptor antagonist, on molecular, cellular and behavioural alterations associated with chronic mild stress. *J. Psychopharmacol.* **2017**, *31*, 691–703. [CrossRef]
61. Hardaway, J.A.; Jensen, J.; Kim, M.; Mazzone, C.M.; Sugam, J.A.; Diberto, J.F.; Lowery-Gionta, E.G.; Hwa, L.S.; Pleil, K.E.; Bulik, C.M.; et al. Nociceptin receptor antagonist SB 612111 decreases high fat diet binge eating. *Behav. Brain Res.* **2016**, *307*, 25–34. [CrossRef]
62. Micioni Di Bonaventura, M.V.; Ubaldi, M.; Liberati, S.; Ciccocioppo, R.; Massi, M.; Cifani, C. Caloric restriction increases the sensitivity to the hyperphagic effect of nociceptin/orphanin FQ limiting its ability to reduce binge eating in female rats. *Psychopharmacology* **2013**, *228*, 53–63. [CrossRef]
63. Pucci, M.; Micioni Di Bonaventura, M.V.; Giusepponi, M.E.; Romano, A.; Filaferro, M.; Maccarrone, M.; Ciccocioppo, R.; Cifani, C.; D'Addario, C. Epigenetic regulation of nociceptin/orphanin FQ and corticotropin-releasing factor system genes in frustration stress-induced binge-like palatable food consumption. *Addict. Biol.* **2016**, *21*, 1168–1185. [CrossRef]
64. Statnick, M.A.; Chen, Y.; Ansonoff, M.; Witkin, J.M.; Rorick-Kehn, L.; Suter, T.M.; Song, M.; Hu, C.; Lafuente, C.; Jimenez, A.; et al. A Novel Nociceptin Receptor Antagonist LY2940094 Inhibits Excessive Feeding Behavior in Rodents: A Possible Mechanism for the Treatment of Binge Eating Disorder. *J. Pharmacol. Exp. Ther.* **2016**, *356*, 493–502. [CrossRef] [PubMed]
65. Butler, A.A.; Kesterson, R.A.; Khong, K.; Cullen, M.J.; Pelleymounter, M.A.; Dekoning, J.; Baetscher, M.; Cone, R.D. A unique metabolic syndrome causes obesity in the melanocortin-3 receptor-deficient mouse. *Endocrinology* **2000**, *141*, 3518–3521. [CrossRef] [PubMed]
66. Chen, A.S.; Marsh, D.J.; Trumbauer, M.E.; Frazier, E.G.; Guan, X.M.; Yu, H.; Rosenblum, C.I.; Vongs, A.; Feng, Y.; Cao, L.; et al. Inactivation of the mouse melanocortin-3 receptor results in increased fat mass and reduced lean body mass. *Nat. Genet.* **2000**, *26*, 97–102. [CrossRef] [PubMed]
67. Ste Marie, L.; Miura, G.I.; Marsh, D.J.; Yagaloff, K.; Palmiter, R.D. A metabolic defect promotes obesity in mice lacking melanocortin-4 receptors. *Proc. Natl. Acad. Sci. USA* **2000**, *97*, 12339–12344. [CrossRef] [PubMed]
68. Sutton, G.M.; Trevaskis, J.L.; Hulver, M.W.; McMillan, R.P.; Markward, N.J.; Babin, M.J.; Meyer, E.A.; Butler, A.A. Diet-genotype interactions in the development of the obese, insulin-resistant phenotype of C57BL/6J mice lacking melanocortin-3 or -4 receptors. *Endocrinology* **2006**, *147*, 2183–2196. [CrossRef] [PubMed]
69. Begriche, K.; Girardet, C.; McDonald, P.; Butler, A.A. Melanocortin-3 receptors and metabolic homeostasis. *Prog. Mol. Biol. Transl. Sci.* **2013**, *114*, 109–146. [CrossRef] [PubMed]
70. Farooqi, I.S.; Keogh, J.M.; Yeo, G.S.; Lank, E.J.; Cheetham, T.; O'Rahilly, S. Clinical spectrum of obesity and mutations in the melanocortin 4 receptor gene. *N. Engl. J. Med.* **2003**, *348*, 1085–1095. [CrossRef]
71. Hinney, A.; Schmidt, A.; Nottebom, K.; Heibult, O.; Becker, I.; Ziegler, A.; Gerber, G.; Sina, M.; Gorg, T.; Mayer, H.; et al. Several mutations in the melanocortin-4 receptor gene including a nonsense and a frameshift mutation associated with dominantly inherited obesity in humans. *J. Clin. Endocrinol. Metab.* **1999**, *84*, 1483–1486. [CrossRef]
72. Schalin-Jantti, C.; Valli-Jaakola, K.; Oksanen, L.; Martelin, E.; Laitinen, K.; Krusius, T.; Mustajoki, P.; Heikinheimo, M.; Kontula, K. Melanocortin-3-receptor gene variants in morbid obesity. *Int. J. Obes. Relat. Metab. Disord.* **2003**, *27*, 70–74. [CrossRef]

73. Begriche, K.; Sutton, G.M.; Butler, A.A. Homeostastic and non-homeostatic functions of melanocortin-3 receptors in the control of energy balance and metabolism. *Physiol. Behav.* **2011**, *104*, 546–554. [CrossRef] [PubMed]
74. Lippert, R.N.; Ellacott, K.L.; Cone, R.D. Gender-specific roles for the melanocortin-3 receptor in the regulation of the mesolimbic dopamine system in mice. *Endocrinology* **2014**, *155*, 1718–1727. [CrossRef] [PubMed]
75. Duncan, A.; Heyer, M.P.; Ishikawa, M.; Caligiuri, S.P.B.; Liu, X.A.; Chen, Z.; Micioni Di Bonaventura, M.V.; Elayouby, K.S.; Ables, J.L.; Howe, W.M.; et al. Habenular TCF7L2 links nicotine addiction to diabetes. *Nature* **2019**, *574*, 372–377. [CrossRef]
76. Zhang, Y.; Kilroy, G.E.; Henagan, T.M.; Prpic-Uhing, V.; Richards, W.G.; Bannon, A.W.; Mynatt, R.L.; Gettys, T.W. Targeted deletion of melanocortin receptor subtypes 3 and 4, but not CART, alters nutrient partitioning and compromises behavioral and metabolic responses to leptin. *FASEB J.* **2005**, *19*, 1482–1491. [CrossRef] [PubMed]
77. Lee, B.; Koo, J.; Yun Jun, J.; Gavrilova, O.; Lee, Y.; Seo, A.Y.; Taylor-Douglas, D.C.; Adler-Wailes, D.C.; Chen, F.; Gardner, R.; et al. A mouse model for a partially inactive obesity-associated human MC3R variant. *Nat. Commun.* **2016**, *7*, 10522. [CrossRef] [PubMed]
78. Feng, N.; Young, S.F.; Aguilera, G.; Puricelli, E.; Adler-Wailes, D.C.; Sebring, N.G.; Yanovski, J.A. Co-occurrence of two partially inactivating polymorphisms of MC3R is associated with pediatric-onset obesity. *Diabetes* **2005**, *54*, 2663–2667. [CrossRef]
79. Lee, Y.S.; Poh, L.K.; Kek, B.L.; Loke, K.Y. The role of melanocortin 3 receptor gene in childhood obesity. *Diabetes* **2007**, *56*, 2622–2630. [CrossRef]
80. Savastano, D.M.; Tanofsky-Kraff, M.; Han, J.C.; Ning, C.; Sorg, R.A.; Roza, C.A.; Wolkoff, L.E.; Anandalingam, K.; Jefferson-George, K.S.; Figueroa, R.E.; et al. Energy intake and energy expenditure among children with polymorphisms of the melanocortin-3 receptor. *Am. J. Clin. Nutr.* **2009**, *90*, 912–920. [CrossRef]
81. Butler, A.A.; Girardet, C.; Mavrikaki, M.; Trevaskis, J.L.; Macarthur, H.; Marks, D.L.; Farr, S.A. A Life without Hunger: The Ups (and Downs) to Modulating Melanocortin-3 Receptor Signaling. *Front. Neurosci.* **2017**, *11*, 128. [CrossRef]
82. Sutton, G.M.; Perez-Tilve, D.; Nogueiras, R.; Fang, J.; Kim, J.K.; Cone, R.D.; Gimble, J.M.; Tschop, M.H.; Butler, A.A. The melanocortin-3 receptor is required for entrainment to meal intake. *J. Neurosci.* **2008**, *28*, 12946–12955. [CrossRef]
83. Begriche, K.; Marston, O.J.; Rossi, J.; Burke, L.K.; McDonald, P.; Heisler, L.K.; Butler, A.A. Melanocortin-3 receptors are involved in adaptation to restricted feeding. *Genes Brain Behav.* **2012**, *11*, 291–302. [CrossRef] [PubMed]
84. Girardet, C.; Mavrikaki, M.M.; Stevens, J.R.; Miller, C.A.; Marks, D.L.; Butler, A.A. Melanocortin-3 receptors expressed in Nkx2.1(+ve) neurons are sufficient for controlling appetitive responses to hypocaloric conditioning. *Sci. Rep.* **2017**, *7*, 44444. [CrossRef] [PubMed]
85. Renquist, B.J.; Murphy, J.G.; Larson, E.A.; Olsen, D.; Klein, R.F.; Ellacott, K.L.; Cone, R.D. Melanocortin-3 receptor regulates the normal fasting response. *Proc. Natl. Acad. Sci. USA* **2012**, *109*, E1489. [CrossRef]
86. Cifani, C.; Polidori, C.; Melotto, S.; Ciccocioppo, R.; Massi, M. A preclinical model of binge eating elicited by yo-yo dieting and stressful exposure to food: Effect of sibutramine, fluoxetine, topiramate, and midazolam. *Psychopharmacology* **2009**, *204*, 113–125. [CrossRef] [PubMed]
87. Hagan, M.M.; Wauford, P.K.; Chandler, P.C.; Jarrett, L.A.; Rybak, R.J.; Blackburn, K. A new animal model of binge eating: Key synergistic role of past caloric restriction and stress. *Physiol. Behav.* **2002**, *77*, 45–54. [CrossRef]
88. Cowley, M.A.; Cone, R.; Enriori, P.; Louiselle, I.; Williams, S.M.; Evans, A.E. Electrophysiological actions of peripheral hormones on melanocortin neurons. *Ann. N. Y. Acad. Sci.* **2003**, *994*, 175–186. [CrossRef]
89. Cowley, M.A.; Smart, J.L.; Rubinstein, M.; Cerdan, M.G.; Diano, S.; Horvath, T.L.; Cone, R.D.; Low, M.J. Leptin activates anorexigenic POMC neurons through a neural network in the arcuate nucleus. *Nature* **2001**, *411*, 480–484. [CrossRef]
90. Marks, D.L.; Hruby, V.; Brookhart, G.; Cone, R.D. The regulation of food intake by selective stimulation of the type 3 melanocortin receptor (MC3R). *Peptides* **2006**, *27*, 259–264. [CrossRef]
91. Gehlert, D.R. Role of hypothalamic neuropeptide Y in feeding and obesity. *Neuropeptides* **1999**, *33*, 329–338. [CrossRef]

92. Lee, M.; Kim, A.; Conwell, I.M.; Hruby, V.; Mayorov, A.; Cai, M.; Wardlaw, S.L. Effects of selective modulation of the central melanocortin-3-receptor on food intake and hypothalamic POMC expression. *Peptides* **2008**, *29*, 440–447. [CrossRef]
93. West, K.S.; Lu, C.; Olson, D.P.; Roseberry, A.G. Alpha-melanocyte stimulating hormone increases the activity of melanocortin-3 receptor-expressing neurons in the ventral tegmental area. *J. Physiol.* **2019**, *597*, 3217–3232. [CrossRef] [PubMed]
94. Kenny, P.J. Reward mechanisms in obesity: New insights and future directions. *Neuron* **2011**, *69*, 664–679. [CrossRef] [PubMed]
95. Volkow, N.D.; Wang, G.J.; Baler, R.D. Reward, dopamine and the control of food intake: Implications for obesity. *Trends Cogn. Sci.* **2011**, *15*, 37–46. [CrossRef] [PubMed]
96. Nair, S.G.; Navarre, B.M.; Cifani, C.; Pickens, C.L.; Bossert, J.M.; Shaham, Y. Role of dorsal medial prefrontal cortex dopamine D1-family receptors in relapse to high-fat food seeking induced by the anxiogenic drug yohimbine. *Neuropsychopharmacology* **2011**, *36*, 497–510. [CrossRef]
97. Wise, R.A. Roles for nigrostriatal—Not just mesocorticolimbic—Dopamine in reward and addiction. *Trends Neurosci.* **2009**, *32*, 517–524. [CrossRef]
98. Lindblom, J.; Opmane, B.; Mutulis, F.; Mutule, I.; Petrovska, R.; Klusa, V.; Bergstrom, L.; Wikberg, J.E. The MC4 receptor mediates alpha-MSH induced release of nucleus accumbens dopamine. *Neuroreport* **2001**, *12*, 2155–2158. [CrossRef]
99. Sanchez, M.S.; Barontini, M.; Armando, I.; Celis, M.E. Correlation of increased grooming behavior and motor activity with alterations in nigrostriatal and mesolimbic catecholamines after alpha-melanotropin and neuropeptide glutamine-isoleucine injection in the rat ventral tegmental area. *Cell. Mol. Neurobiol.* **2001**, *21*, 523–533. [CrossRef]
100. Torre, E.; Celis, M.E. Alpha-MSH injected into the substantia nigra or intraventricularly alters behavior and the striatal dopaminergic activity. *Neurochem. Int.* **1986**, *9*, 85–89. [CrossRef]
101. Torre, E.; Celis, M.E. Cholinergic mediation in the ventral tegmental area of alpha-melanotropin induced excessive grooming: Changes of the dopamine activity in the nucleus accumbens and caudate putamen. *Life Sci.* **1988**, *42*, 1651–1657. [CrossRef]
102. Dietrich, M.O.; Bober, J.; Ferreira, J.G.; Tellez, L.A.; Mineur, Y.S.; Souza, D.O.; Gao, X.B.; Picciotto, M.R.; Araujo, I.; Liu, Z.W.; et al. AgRP neurons regulate development of dopamine neuronal plasticity and nonfood-associated behaviors. *Nat. Neurosci.* **2012**, *15*, 1108–1110. [CrossRef]
103. King, C.M.; Hentges, S.T. Relative number and distribution of murine hypothalamic proopiomelanocortin neurons innervating distinct target sites. *PLoS ONE* **2011**, *6*, e25864. [CrossRef] [PubMed]
104. Liu, H.; Kishi, T.; Roseberry, A.G.; Cai, X.; Lee, C.E.; Montez, J.M.; Friedman, J.M.; Elmquist, J.K. Transgenic mice expressing green fluorescent protein under the control of the melanocortin-4 receptor promoter. *J. Neurosci.* **2003**, *23*, 7143–7154. [CrossRef] [PubMed]
105. Martinez-Hernandez, J.; Lanuza, E.; Martinez-Garcia, F. Selective dopaminergic lesions of the ventral tegmental area impair preference for sucrose but not for male sexual pheromones in female mice. *Eur. J. Neurosci.* **2006**, *24*, 885–893. [CrossRef] [PubMed]
106. Shibata, R.; Kameishi, M.; Kondoh, T.; Torii, K. Bilateral dopaminergic lesions in the ventral tegmental area of rats influence sucrose intake, but not umami and amino acid intake. *Physiol. Behav.* **2009**, *96*, 667–674. [CrossRef] [PubMed]
107. Culbert, K.M.; Racine, S.E.; Klump, K.L. Hormonal Factors and Disturbances in Eating Disorders. *Curr. Psychiatry Rep.* **2016**, *18*, 65. [CrossRef] [PubMed]
108. Edler, C.; Lipson, S.F.; Keel, P.K. Ovarian hormones and binge eating in bulimia nervosa. *Psychol. Med.* **2007**, *37*, 131–141. [CrossRef]
109. Micioni Di Bonaventura, M.V.; Lutz, T.A.; Romano, A.; Pucci, M.; Geary, N.; Asarian, L.; Cifani, C. Estrogenic suppression of binge-like eating elicited by cyclic food restriction and frustrative-nonreward stress in female rats. *Int. J. Eat. Disord.* **2017**, *50*, 624–635. [CrossRef]
110. Alboni, S.; Micioni Di Bonaventura, M.V.; Benatti, C.; Giusepponi, M.E.; Brunello, N.; Cifani, C. Hypothalamic expression of inflammatory mediators in an animal model of binge eating. *Behav. Brain Res.* **2017**, *320*, 420–430. [CrossRef]

111. Davis, J.F.; Choi, D.L.; Shurdak, J.D.; Krause, E.G.; Fitzgerald, M.F.; Lipton, J.W.; Sakai, R.R.; Benoit, S.C. Central melanocortins modulate mesocorticolimbic activity and food seeking behavior in the rat. *Physiol. Behav.* **2011**, *102*, 491–495. [CrossRef]
112. Hagan, M.M.; Rushing, P.A.; Benoit, S.C.; Woods, S.C.; Seeley, R.J. Opioid receptor involvement in the effect of AgRP- (83-132) on food intake and food selection. *Am. J. Physiol. Regul. Integr. Comp. Physiol.* **2001**, *280*, R814–R821. [CrossRef]
113. Tracy, A.L.; Clegg, D.J.; Johnson, J.D.; Davidson, T.L.; Benoit, S.C. The melanocortin antagonist AgRP (83-132) increases appetitive responding for a fat, but not a carbohydrate, reinforcer. *Pharmacol. Biochem. Behav.* **2008**, *89*, 263–271. [CrossRef] [PubMed]
114. Samama, P.; Rumennik, L.; Grippo, J.F. The melanocortin receptor MCR4 controls fat consumption. *Regul. Pept.* **2003**, *113*, 85–88. [CrossRef]
115. Boghossian, S.; Park, M.; York, D.A. Melanocortin activity in the amygdala controls appetite for dietary fat. *Am. J. Physiol. Regul. Integr. Comp. Physiol.* **2010**, *298*, R385–R393. [CrossRef] [PubMed]
116. Bassareo, V.; Di Chiara, G. Differential influence of associative and nonassociative learning mechanisms on the responsiveness of prefrontal and accumbal dopamine transmission to food stimuli in rats fed ad libitum. *J. Neurosci.* **1997**, *17*, 851–861. [CrossRef]
117. Botticelli, L.; Micioni Di Bonaventura, E.; Del Bello, F.; Giorgioni, G.; Piergentili, A.; Romano, A.; Quaglia, W.; Cifani, C.; Micioni Di Bonaventura, M.V. Underlying Susceptibility to Eating Disorders and Drug Abuse: Genetic and Pharmacological Aspects of Dopamine D4 Receptors. *Nutrients* **2020**, *12*, 2288. [CrossRef]
118. Cifani, C.; Koya, E.; Navarre, B.M.; Calu, D.J.; Baumann, M.H.; Marchant, N.J.; Liu, Q.R.; Khuc, T.; Pickel, J.; Lupica, C.R.; et al. Medial prefrontal cortex neuronal activation and synaptic alterations after stress-induced reinstatement of palatable food seeking: A study using c-fos-GFP transgenic female rats. *J. Neurosci.* **2012**, *32*, 8480–8490. [CrossRef]
119. Zheng, H.; Corkern, M.M.; Crousillac, S.M.; Patterson, L.M.; Phifer, C.B.; Berthoud, H.R. Neurochemical phenotype of hypothalamic neurons showing Fos expression 23 h after intracranial AgRP. *Am. J. Physiol. Regul. Integr. Comp. Physiol.* **2002**, *282*, R1773–R1781. [CrossRef]
120. Stuber, G.D.; Wise, R.A. Lateral hypothalamic circuits for feeding and reward. *Nat. Neurosci.* **2016**, *19*, 198–205. [CrossRef]
121. Piccoli, L.; Micioni Di Bonaventura, M.V.; Cifani, C.; Costantini, V.J.; Massagrande, M.; Montanari, D.; Martinelli, P.; Antolini, M.; Ciccocioppo, R.; Massi, M.; et al. Role of orexin-1 receptor mechanisms on compulsive food consumption in a model of binge eating in female rats. *Neuropsychopharmacology* **2012**, *37*, 1999–2011. [CrossRef]
122. Roseberry, A.G. Altered feeding and body weight following melanocortin administration to the ventral tegmental area in adult rats. *Psychopharmacology* **2013**, *226*, 25–34. [CrossRef]
123. Yen, H.H.; Roseberry, A.G. Decreased consumption of rewarding sucrose solutions after injection of melanocortins into the ventral tegmental area of rats. *Psychopharmacology* **2015**, *232*, 285–294. [CrossRef] [PubMed]
124. Shanmugarajah, L.; Dunigan, A.I.; Frantz, K.J.; Roseberry, A.G. Altered sucrose self-administration following injection of melanocortin receptor agonists and antagonists into the ventral tegmental area. *Psychopharmacology* **2017**, *234*, 1683–1692. [CrossRef] [PubMed]
125. Potoczna, N.; Branson, R.; Kral, J.G.; Piec, G.; Steffen, R.; Ricklin, T.; Hoehe, M.R.; Lentes, K.U.; Horber, F.F. Gene variants and binge eating as predictors of comorbidity and outcome of treatment in severe obesity. *J. Gastrointest. Surg.* **2004**, *8*, 971–981; discussion 972–981. [CrossRef] [PubMed]
126. Sina, M.; Hinney, A.; Ziegler, A.; Neupert, T.; Mayer, H.; Siegfried, W.; Blum, W.F.; Remschmidt, H.; Hebebrand, J. Phenotypes in three pedigrees with autosomal dominant obesity caused by haploinsufficiency mutations in the melanocortin-4 receptor gene. *Am. J. Hum. Genet.* **1999**, *65*, 1501–1507. [CrossRef]
127. Spitzer, R.L.; Yanovski, S.; Wadden, T.; Wing, R.; Marcus, M.D.; Stunkard, A.; Devlin, M.; Mitchell, J.; Hasin, D.; Horne, R.L. Binge eating disorder: Its further validation in a multisite study. *Int. J. Eat. Disord.* **1993**, *13*, 137–153.
128. Gotoda, T. Binge eating as a phenotype of melanocortin 4 receptor gene mutations. *N. Engl. J. Med.* **2003**, *349*, 606–609. [CrossRef]

129. Hebebrand, J.; Geller, F.; Dempfle, A.; Heinzel-Gutenbrunner, M.; Raab, M.; Gerber, G.; Wermter, A.K.; Horro, F.F.; Blundell, J.; Schafer, H.; et al. Binge-eating episodes are not characteristic of carriers of melanocortin-4 receptor gene mutations. *Mol. Psychiatry* **2004**, *9*, 796–800. [CrossRef]
130. Lubrano-Berthelier, C.; Dubern, B.; Lacorte, J.M.; Picard, F.; Shapiro, A.; Zhang, S.; Bertrais, S.; Hercberg, S.; Basdevant, A.; Clement, K.; et al. Melanocortin 4 receptor mutations in a large cohort of severely obese adults: Prevalence, functional classification, genotype-phenotype relationship, and lack of association with binge eating. *J. Clin. Endocrinol. Metab.* **2006**, *91*, 1811–1818. [CrossRef]
131. Tao, Y.X.; Segaloff, D.L. Functional analyses of melanocortin-4 receptor mutations identified from patients with binge eating disorder and nonobese or obese subjects. *J. Clin. Endocrinol. Metab.* **2005**, *90*, 5632–5638. [CrossRef]
132. Gamero-Villarroel, C.; Rodriguez-Lopez, R.; Jimenez, M.; Carrillo, J.A.; Garcia-Herraiz, A.; Albuquerque, D.; Flores, I.; Gervasini, G. Melanocortin-4 receptor gene variants are not associated with binge-eating behavior in nonobese patients with eating disorders. *Psychiatr. Genet.* **2015**, *25*, 35–38. [CrossRef]
133. Valette, M.; Poitou, C.; Le Beyec, J.; Bouillot, J.L.; Clement, K.; Czernichow, S. Melanocortin-4 receptor mutations and polymorphisms do not affect weight loss after bariatric surgery. *PLoS ONE* **2012**, *7*, e48221. [CrossRef] [PubMed]
134. Bonnefond, A.; Keller, R.; Meyre, D.; Stutzmann, F.; Thuillier, D.; Stefanov, D.G.; Froguel, P.; Horber, F.F.; Kral, J.G. Eating Behavior, Low-Frequency Functional Mutations in the Melanocortin-4 Receptor (MC4R) Gene, and Outcomes of Bariatric Operations: A 6-Year Prospective Study. *Diabetes Care* **2016**, *39*, 1384–1392. [CrossRef] [PubMed]
135. Valette, M.; Poitou, C.; Kesse-Guyot, E.; Bellisle, F.; Carette, C.; Le Beyec, J.; Hercberg, S.; Clement, K.; Czernichow, S. Association between melanocortin-4 receptor mutations and eating behaviors in obese patients: A case-control study. *Int. J. Obes.* **2014**, *38*, 883–885. [CrossRef]
136. Van der Klaauw, A.A.; von dem Hagen, E.A.; Keogh, J.M.; Henning, E.; O'Rahilly, S.; Lawrence, A.D.; Calder, A.J.; Farooqi, I.S. Obesity-associated melanocortin-4 receptor mutations are associated with changes in the brain response to food cues. *J. Clin. Endocrinol. Metab.* **2014**, *99*, E2101–E2106. [CrossRef] [PubMed]
137. Furlong, T.M.; Jayaweera, H.K.; Balleine, B.W.; Corbit, L.H. Binge-like consumption of a palatable food accelerates habitual control of behavior and is dependent on activation of the dorsolateral striatum. *J. Neurosci.* **2014**, *34*, 5012–5022. [CrossRef]
138. Wang, G.J.; Geliebter, A.; Volkow, N.D.; Telang, F.W.; Logan, J.; Jayne, M.C.; Galanti, K.; Selig, P.A.; Han, H.; Zhu, W.; et al. Enhanced striatal dopamine release during food stimulation in binge eating disorder. *Obesity* **2011**, *19*, 1601–1608. [CrossRef]
139. Thorleifsson, G.; Walters, G.B.; Gudbjartsson, D.F.; Steinthorsdottir, V.; Sulem, P.; Helgadottir, A.; Styrkarsdottir, U.; Gretarsdottir, S.; Thorlacius, S.; Jonsdottir, I.; et al. Genome-wide association yields new sequence variants at seven loci that associate with measures of obesity. *Nat. Genet.* **2009**, *41*, 18–24. [CrossRef]
140. Loos, R.J.; Lindgren, C.M.; Li, S.; Wheeler, E.; Zhao, J.H.; Prokopenko, I.; Inouye, M.; Freathy, R.M.; Attwood, A.P.; Beckmann, J.S.; et al. Common variants near MC4R are associated with fat mass, weight and risk of obesity. *Nat. Genet.* **2008**, *40*, 768–775. [CrossRef]
141. Horstmann, A.; Kovacs, P.; Kabisch, S.; Boettcher, Y.; Schloegl, H.; Tonjes, A.; Stumvoll, M.; Pleger, B.; Villringer, A. Common genetic variation near MC4R has a sex-specific impact on human brain structure and eating behavior. *PLoS ONE* **2013**, *8*, e74362. [CrossRef]
142. Qi, L.; Kraft, P.; Hunter, D.J.; Hu, F.B. The common obesity variant near MC4R gene is associated with higher intakes of total energy and dietary fat, weight change and diabetes risk in women. *Hum. Mol. Genet.* **2008**, *17*, 3502–3508. [CrossRef]
143. Stutzmann, F.; Cauchi, S.; Durand, E.; Calvacanti-Proenca, C.; Pigeyre, M.; Hartikainen, A.L.; Sovio, U.; Tichet, J.; Marre, M.; Weill, J.; et al. Common genetic variation near MC4R is associated with eating behaviour patterns in European populations. *Int. J. Obes.* **2009**, *33*, 373–378. [CrossRef] [PubMed]
144. Yilmaz, Z.; Davis, C.; Loxton, N.J.; Kaplan, A.S.; Levitan, R.D.; Carter, J.C.; Kennedy, J.L. Association between MC4R rs17782313 polymorphism and overeating behaviors. *Int. J. Obes.* **2015**, *39*, 114–120. [CrossRef] [PubMed]
145. Blaine, R.E.; Kachurak, A.; Davison, K.K.; Klabunde, R.; Fisher, J.O. Food parenting and child snacking: A systematic review. *Int. J. Behav. Nutr. Phys. Act.* **2017**, *14*, 146. [CrossRef] [PubMed]

146. Potter, M.; Vlassopoulos, A.; Lehmann, U. Snacking Recommendations Worldwide: A Scoping Review. *Adv. Nutr.* **2018**, *9*, 86–98. [CrossRef] [PubMed]
147. Acosta, A.; Camilleri, M.; Shin, A.; Carlson, P.; Burton, D.; O'Neill, J.; Eckert, D.; Zinsmeister, A.R. Association of melanocortin 4 receptor gene variation with satiation and gastric emptying in overweight and obese adults. *Genes Nutr.* **2014**, *9*, 384. [CrossRef]
148. Valladares, M.; Dominguez-Vasquez, P.; Obregon, A.M.; Weisstaub, G.; Burrows, R.; Maiz, A.; Santos, J.L. Melanocortin-4 receptor gene variants in Chilean families: Association with childhood obesity and eating behavior. *Nutr. Neurosci.* **2010**, *13*, 71–78. [CrossRef] [PubMed]
149. Wardle, J.; Guthrie, C.A.; Sanderson, S.; Rapoport, L. Development of the Children's Eating Behaviour Questionnaire. *J. Child Psychol. Psychiatry Allied Discip.* **2001**, *42*, 963–970. [CrossRef]
150. De Lauzon, B.; Romon, M.; Deschamps, V.; Lafay, L.; Borys, J.M.; Karlsson, J.; Ducimetiere, P.; Charles, M.A.; Fleurbaix Laventie Ville Sante Study, G. The Three-Factor Eating Questionnaire-R18 is able to distinguish among different eating patterns in a general population. *J. Nutr.* **2004**, *134*, 2372–2380. [CrossRef]
151. Turton, R.; Chami, R.; Treasure, J. Emotional Eating, Binge Eating and Animal Models of Binge-Type Eating Disorders. *Curr. Obes. Rep.* **2017**, *6*, 217–228. [CrossRef]
152. Van Strien, T.; Cebolla, A.; Etchemendy, E.; Gutierrez-Maldonado, J.; Ferrer-Garcia, M.; Botella, C.; Banos, R. Emotional eating and food intake after sadness and joy. *Appetite* **2013**, *66*, 20–25. [CrossRef]
153. Obregon, A.M.; Oyarce, K.; Santos, J.L.; Valladares, M.; Goldfield, G. Association of the melanocortin 4 receptor gene rs17782313 polymorphism with rewarding value of food and eating behavior in Chilean children. *J. Physiol. Biochem.* **2017**, *73*, 29–35. [CrossRef] [PubMed]
154. Yilmaz, Z.; Kaplan, A.S.; Tiwari, A.K.; Levitan, R.D.; Piran, S.; Bergen, A.W.; Kaye, W.H.; Hakonarson, H.; Wang, K.; Berrettini, W.H.; et al. The role of leptin, melanocortin, and neurotrophin system genes on body weight in anorexia nervosa and bulimia nervosa. *J. Psychiatr. Res.* **2014**, *55*, 77–86. [CrossRef] [PubMed]
155. Abadi, A.; Peralta-Romero, J.; Suarez, F.; Gomez-Zamudio, J.; Burguete-Garcia, A.I.; Cruz, M.; Meyre, D. Assessing the effects of 35 European-derived BMI-associated SNPs in Mexican children. *Obesity* **2016**, *24*, 1989–1995. [CrossRef] [PubMed]
156. Cheung, C.Y.; Tso, A.W.; Cheung, B.M.; Xu, A.; Ong, K.L.; Fong, C.H.; Wat, N.M.; Janus, E.D.; Sham, P.C.; Lam, K.S. Obesity susceptibility genetic variants identified from recent genome-wide association studies: Implications in a Chinese population. *J. Clin. Endocrinol. Metab.* **2010**, *95*, 1395–1403. [CrossRef] [PubMed]
157. El Hajj Chehadeh, S.; Osman, W.; Nazar, S.; Jerman, L.; Alghafri, A.; Sajwani, A.; Alawlaqi, M.; AlObeidli, M.; Jelinek, H.F.; AlAnouti, F.; et al. Implication of genetic variants in overweight and obesity susceptibility among the young Arab population of the United Arab Emirates. *Gene* **2020**, *739*, 144509. [CrossRef]
158. Khalilitehrani, A.; Qorbani, M.; Hosseini, S.; Pishva, H. The association of MC4R rs17782313 polymorphism with dietary intake in Iranian adults. *Gene* **2015**, *563*, 125–129. [CrossRef]
159. Lv, D.; Zhang, D.D.; Wang, H.; Zhang, Y.; Liang, L.; Fu, J.F.; Xiong, F.; Liu, G.L.; Gong, C.X.; Luo, F.H.; et al. Genetic variations in SEC16B, MC4R, MAP2K5 and KCTD15 were associated with childhood obesity and interacted with dietary behaviors in Chinese school-age population. *Gene* **2015**, *560*, 149–155. [CrossRef]
160. Rana, S.; Rahmani, S.; Mirza, S. MC4R variant rs17782313 and manifestation of obese phenotype in Pakistani females. *RSC Adv.* **2018**, *8*, 16957–16972. [CrossRef]
161. Shi, J.; Long, J.; Gao, Y.T.; Lu, W.; Cai, Q.; Wen, W.; Zheng, Y.; Yu, K.; Xiang, Y.B.; Hu, F.B.; et al. Evaluation of genetic susceptibility loci for obesity in Chinese women. *Am. J. Epidemiol.* **2010**, *172*, 244–254. [CrossRef]
162. Zobel, D.P.; Andreasen, C.H.; Grarup, N.; Eiberg, H.; Sorensen, T.I.; Sandbaek, A.; Lauritzen, T.; Borch-Johnsen, K.; Jorgensen, T.; Pedersen, O.; et al. Variants near MC4R are associated with obesity and influence obesity-related quantitative traits in a population of middle-aged people: Studies of 14,940 Danes. *Diabetes* **2009**, *58*, 757–764. [CrossRef]
163. Orkunoglu-Suer, F.E.; Harmon, B.T.; Gordish-Dressman, H.; Clarkson, P.M.; Thompson, P.D.; Angelopoulos, T.J.; Gordon, P.M.; Hubal, M.J.; Moyna, N.M.; Pescatello, L.S.; et al. MC4R variant is associated with BMI but not response to resistance training in young females. *Obesity* **2011**, *19*, 662–666. [CrossRef] [PubMed]
164. Stevenson, R.J.; Francis, H.M. The hippocampus and the regulation of human food intake. *Psychol. Bull.* **2017**, *143*, 1011–1032. [CrossRef] [PubMed]

165. Uher, R.; Murphy, T.; Brammer, M.J.; Dalgleish, T.; Phillips, M.L.; Ng, V.W.; Andrew, C.M.; Williams, S.C.; Campbell, I.C.; Treasure, J. Medial prefrontal cortex activity associated with symptom provocation in eating disorders. *Am. J. Psychiatry* **2004**, *161*, 1238–1246. [CrossRef] [PubMed]
166. Freeman, L.M.; Gil, K.M. Daily stress, coping, and dietary restraint in binge eating. *Int. J. Eat. Disord.* **2004**, *36*, 204–212. [CrossRef]
167. Cifani, C.; Di Bonaventura, M.V.M.; Ciccocioppo, R.; Massi, M. Binge eating in female rats induced by yo-yo dieting and stress. In *Animal Models of Eating Disorders*; Springer: Totowa, NJ, USA, 2013; pp. 27–49.
168. Polivy, J.; Zeitlin, S.B.; Herman, C.P.; Beal, A.L. Food restriction and binge eating: A study of former prisoners of war. *J. Abnorm. Psychol.* **1994**, *103*, 409–411. [CrossRef]
169. Woods, A.M.; Racine, S.E.; Klump, K.L. Examining the relationship between dietary restraint and binge eating: Differential effects of major and minor stressors. *Eat. Behav.* **2010**, *11*, 276–280. [CrossRef]
170. Gluck, M.E.; Geliebter, A.; Hung, J.; Yahav, E. Cortisol, hunger, and desire to binge eat following a cold stress test in obese women with binge eating disorder. *Psychosom. Med.* **2004**, *66*, 876–881. [CrossRef]
171. Gluck, M.E.; Geliebter, A.; Lorence, M. Cortisol stress response is positively correlated with central obesity in obese women with binge eating disorder (BED) before and after cognitive-behavioral treatment. *Ann. N. Y. Acad. Sci.* **2004**, *1032*, 202–207. [CrossRef]
172. Cifani, C.; Micioni Di, B.M.; Vitale, G.; Ruggieri, V.; Ciccocioppo, R.; Massi, M. Effect of salidroside, active principle of Rhodiola rosea extract, on binge eating. *Physiol. Behav.* **2010**, *101*, 555–562. [CrossRef]
173. Micioni Di Bonaventura, M.V.; Vitale, G.; Massi, M.; Cifani, C. Effect of Hypericum perforatum Extract in an Experimental Model of Binge Eating in Female Rats. *J. Obes.* **2012**, *2012*, 956137. [CrossRef]
174. Epel, E.; Lapidus, R.; McEwen, B.; Brownell, K. Stress may add bite to appetite in women: A laboratory study of stress-induced cortisol and eating behavior. *Psychoneuroendocrinology* **2001**, *26*, 37–49. [CrossRef]
175. Coutinho, W.F.; Moreira, R.O.; Spagnol, C.; Appolinario, J.C. Does binge eating disorder alter cortisol secretion in obese women? *Eat. Behav.* **2007**, *8*, 59–64. [CrossRef] [PubMed]
176. Chaki, S.; Okubo, T. Melanocortin-4 receptor antagonists for the treatment of depression and anxiety disorders. *Curr. Top. Med. Chem.* **2007**, *7*, 1145–1151. [CrossRef] [PubMed]
177. Chaki, S.; Okuyama, S. Involvement of melanocortin-4 receptor in anxiety and depression. *Peptides* **2005**, *26*, 1952–1964. [CrossRef] [PubMed]
178. Adan, R.A.; Szklarczyk, A.W.; Oosterom, J.; Brakkee, J.H.; Nijenhuis, W.A.; Schaaper, W.M.; Meloen, R.H.; Gispen, W.H. Characterization of melanocortin receptor ligands on cloned brain melanocortin receptors and on grooming behavior in the rat. *Eur. J. Pharmacol.* **1999**, *378*, 249–258. [CrossRef]
179. De Barioglio, S.R.; Lezcano, N.; Celis, M.E. Alpha MSH-induced excessive grooming behavior involves a GABAergic mechanism. *Peptides* **1991**, *12*, 203–205. [CrossRef]
180. Spruijt, B.M.; van Hooff, J.A.; Gispen, W.H. Ethology and neurobiology of grooming behavior. *Physiol. Rev.* **1992**, *72*, 825–852. [CrossRef]
181. Mul, J.D.; van Boxtel, R.; Bergen, D.J.; Brans, M.A.; Brakkee, J.H.; Toonen, P.W.; Garner, K.M.; Adan, R.A.; Cuppen, E. Melanocortin receptor 4 deficiency affects body weight regulation, grooming behavior, and substrate preference in the rat. *Obesity* **2012**, *20*, 612–621. [CrossRef]
182. Yamano, Y.; Yoshioka, M.; Toda, Y.; Oshida, Y.; Chaki, S.; Hamamoto, K.; Morishima, I. Regulation of CRF, POMC and MC4R gene expression after electrical foot shock stress in the rat amygdala and hypothalamus. *J. Vet. Med Sci.* **2004**, *66*, 1323–1327. [CrossRef]
183. Davis, M.; Shi, C. The extended amygdala: Are the central nucleus of the amygdala and the bed nucleus of the stria terminalis differentially involved in fear versus anxiety? *Ann. N. Y. Acad. Sci.* **1999**, *877*, 281–291. [CrossRef]
184. Blasio, A.; Iemolo, A.; Sabino, V.; Petrosino, S.; Steardo, L.; Rice, K.C.; Orlando, P.; Iannotti, F.A.; Di Marzo, V.; Zorrilla, E.P.; et al. Rimonabant precipitates anxiety in rats withdrawn from palatable food: Role of the central amygdala. *Neuropsychopharmacology* **2013**, *38*, 2498–2507. [CrossRef] [PubMed]
185. Micioni Di Bonaventura, M.V.; Pucci, M.; Giusepponi, M.E.; Romano, A.; Lambertucci, C.; Volpini, R.; Micioni Di Bonaventura, E.; Gaetani, S.; Maccarrone, M.; D'Addario, C.; et al. Regulation of adenosine A2A receptor gene expression in a model of binge eating in the amygdaloid complex of female rats. *J. Psychopharmacol.* **2019**. [CrossRef] [PubMed]

186. Karami Kheirabad, M.; Namavar Jahromi, B.; Tamadon, A.; Ramezani, A.; Ahmadloo, S.; Sabet Sarvestan, F.; Koohi-Hosseinabadi, O. Expression of Melanocortin-4 Receptor mRNA in Male Rat Hypothalamus During Chronic Stress. *Int. J. Mol. Cell. Med.* **2015**, *4*, 182–187. [PubMed]
187. Chagra, S.L.; Zavala, J.K.; Hall, M.V.; Gosselink, K.L. Acute and repeated restraint differentially activate orexigenic pathways in the rat hypothalamus. *Regul. Pept.* **2011**, *167*, 70–78. [CrossRef]
188. Von Frijtag, J.C.; Croiset, G.; Gispen, W.H.; Adan, R.A.; Wiegant, V.M. The role of central melanocortin receptors in the activation of the hypothalamus-pituitary-adrenal-axis and the induction of excessive grooming. *Br. J. Pharmacol.* **1998**, *123*, 1503–1508. [CrossRef]
189. Lu, X.Y.; Barsh, G.S.; Akil, H.; Watson, S.J. Interaction between alpha-melanocyte-stimulating hormone and corticotropin-releasing hormone in the regulation of feeding and hypothalamo-pituitary-adrenal responses. *J. Neurosci.* **2003**, *23*, 7863–7872. [CrossRef]
190. Hwang, B.H.; Guntz, J.M. Downregulation of corticotropin-releasing factor mRNA, but not vasopressin mRNA, in the paraventricular hypothalamic nucleus of rats following nutritional stress. *Brain Res. Bull.* **1997**, *43*, 509–514. [CrossRef]
191. Liposits, Z.; Sievers, L.; Paull, W.K. Neuropeptide-Y and ACTH-immunoreactive innervation of corticotropin releasing factor (CRF)-synthesizing neurons in the hypothalamus of the rat. An immunocytochemical analysis at the light and electron microscopic levels. *Histochemistry* **1988**, *88*, 227–234. [CrossRef]
192. Sarkar, S.; Legradi, G.; Lechan, R.M. Intracerebroventricular administration of alpha-melanocyte stimulating hormone increases phosphorylation of CREB in TRH- and CRH-producing neurons of the hypothalamic paraventricular nucleus. *Brain Res.* **2002**, *945*, 50–59. [CrossRef]
193. Liu, J.; Garza, J.C.; Truong, H.V.; Henschel, J.; Zhang, W.; Lu, X.Y. The melanocortinergic pathway is rapidly recruited by emotional stress and contributes to stress-induced anorexia and anxiety-like behavior. *Endocrinology* **2007**, *148*, 5531–5540. [CrossRef]
194. Ulrich-Lai, Y.M.; Herman, J.P. Neural regulation of endocrine and autonomic stress responses. *Nat. Rev. Neurosci.* **2009**, *10*, 397–409. [CrossRef] [PubMed]
195. Bedse, G.; Romano, A.; Tempesta, B.; Lavecchia, M.A.; Pace, L.; Bellomo, A.; Duranti, A.; Micioni Di Bonaventura, M.V.; Cifani, C.; Cassano, T.; et al. Inhibition of anandamide hydrolysis enhances noradrenergic and GABAergic transmission in the prefrontal cortex and basolateral amygdala of rats subjected to acute swim stress. *J. Neurosci. Res.* **2015**, *93*, 777–787. [CrossRef] [PubMed]
196. Dayas, C.V.; Buller, K.M.; Day, T.A. Neuroendocrine responses to an emotional stressor: Evidence for involvement of the medial but not the central amygdala. *Eur. J. Neurosci.* **1999**, *11*, 2312–2322. [CrossRef]
197. Roozendaal, B.; McEwen, B.S.; Chattarji, S. Stress, memory and the amygdala. *Nat. Rev. Neurosci.* **2009**, *10*, 423–433. [CrossRef] [PubMed]
198. Choi, D.C.; Furay, A.R.; Evanson, N.K.; Ostrander, M.M.; Ulrich-Lai, Y.M.; Herman, J.P. Bed nucleus of the stria terminalis subregions differentially regulate hypothalamic-pituitary-adrenal axis activity: Implications for the integration of limbic inputs. *J. Neurosci.* **2007**, *27*, 2025–2034. [CrossRef]
199. Forray, M.I.; Gysling, K. Role of noradrenergic projections to the bed nucleus of the stria terminalis in the regulation of the hypothalamic-pituitary-adrenal axis. *Brain Res. Brain Res. Rev.* **2004**, *47*, 145–160. [CrossRef] [PubMed]
200. Micioni Di Bonaventura, M.V.; Ciccocioppo, R.; Romano, A.; Bossert, J.M.; Rice, K.C.; Ubaldi, M.; St Laurent, R.; Gaetani, S.; Massi, M.; Shaham, Y.; et al. Role of bed nucleus of the stria terminalis corticotrophin-releasing factor receptors in frustration stress-induced binge-like palatable food consumption in female rats with a history of food restriction. *J. Neurosci.* **2014**, *34*, 11316–11324. [CrossRef] [PubMed]
201. Micioni Di Bonaventura, M.V.; Ubaldi, M.; Giusepponi, M.E.; Rice, K.C.; Massi, M.; Ciccocioppo, R.; Cifani, C. Hypothalamic CRF1 receptor mechanisms are not sufficient to account for binge-like palatable food consumption in female rats. *Int. J. Eat. Disord.* **2017**, *50*, 1194–1204. [CrossRef]
202. Chaffin, A.T.; Fang, Y.; Larson, K.R.; Mul, J.D.; Ryan, K.K. Sex-dependent effects of MC4R genotype on HPA axis tone: Implications for stress-associated cardiometabolic disease. *Stress* **2019**, *22*, 571–580. [CrossRef]
203. Ryan, K.K.; Mul, J.D.; Clemmensen, C.; Egan, A.E.; Begg, D.P.; Halcomb, K.; Seeley, R.J.; Herman, J.P.; Ulrich-Lai, Y.M. Loss of melanocortin-4 receptor function attenuates HPA responses to psychological stress. *Psychoneuroendocrinology* **2014**, *42*, 98–105. [CrossRef]

204. Cottone, P.; Sabino, V.; Roberto, M.; Bajo, M.; Pockros, L.; Frihauf, J.B.; Fekete, E.M.; Steardo, L.; Rice, K.C.; Grigoriadis, D.E.; et al. CRF system recruitment mediates dark side of compulsive eating. *Proc. Natl. Acad. Sci. USA* **2009**, *106*, 20016–20020. [CrossRef] [PubMed]
205. Iemolo, A.; Blasio, A.; St Cyr, S.A.; Jiang, F.; Rice, K.C.; Sabino, V.; Cottone, P. CRF-CRF1 receptor system in the central and basolateral nuclei of the amygdala differentially mediates excessive eating of palatable food. *Neuropsychopharmacology* **2013**, *38*, 2456–2466. [CrossRef] [PubMed]
206. Romano, A.; Micioni Di Bonaventura, M.V.; Gallelli, C.A.; Koczwara, J.B.; Smeets, D.; Giusepponi, M.E.; De Ceglia, M.; Friuli, M.; Micioni Di Bonaventura, E.; Scuderi, C.; et al. Oleoylethanolamide decreases frustration stress-induced binge-like eating in female rats: A novel potential treatment for binge eating disorder. *Neuropsychopharmacology* **2020**, *45*, 1931–1941. [CrossRef] [PubMed]
207. Cifani, C.; Micioni Di Bonaventura, E.; Botticelli, L.; Del Bello, F.; Giorgioni, G.; Pavletic, P.; Piergentili, A.; Quaglia, W.; Bonifazi, A.; Schepmann, D.; et al. Novel Highly Potent and Selective Sigma1 Receptor Antagonists Effectively Block the Binge Eating Episode in Female Rats. *ACS Chem. Neurosci.* **2020**, *11*, 3107–3116. [CrossRef] [PubMed]
208. Pucci, M.; Micioni Di Bonaventura, M.V.; Zaplatic, E.; Bellia, F.; Maccarrone, M.; Cifani, C.; D'Addario, C. Transcriptional regulation of the endocannabinoid system in a rat model of binge-eating behavior reveals a selective modulation of the hypothalamic fatty acid amide hydrolase gene. *Int. J. Eat. Disord.* **2018**, *52*, 51–60. [CrossRef] [PubMed]
209. Micioni Di Bonaventura, M.V.; Cifani, C.; Lambertucci, C.; Volpini, R.; Cristalli, G.; Massi, M. A2A adenosine receptor agonists reduce both high-palatability and low-palatability food intake in female rats. *Behav. Pharmacol.* **2012**, *23*, 567–574. [CrossRef] [PubMed]

nutrients

MDPI

Review

Negative Affectivity and Emotion Dysregulation as Mediators between ADHD and Disordered Eating: A Systematic Review

Sarah El Archi [1], Samuele Cortese [2,3,4,5], Nicolas Ballon [6,7], Christian Réveillère [1], Arnaud De Luca [8,9], Servane Barrault [1,10,†] and Paul Brunault [1,6,7,*,†]

1 Qualipsy, EE 1901, Université de Tours, 37041 Tours, France; sarah.elarchi@univ-tours.fr (S.E.A.); christian.reveillere@univ-tours.fr (C.R.); servane.barrault@univ-tours.fr (S.B.)
2 Center for Innovation in Mental Health, Academic Unit of Psychology, Clinical and Experimental Sciences (CNS and Psychiatry), Faculty of Medicine, University of Southampton, Southampton SO17 1BJ, UK; samuele.cortese@gmail.com
3 Solent NHS Trust, Southampton SO19 8BR, UK
4 New York University Child Study Center, New York, NY 10016, USA
5 Division of Psychiatry and Applied Psychology, School of Medicine, University of Nottingham, Nottingham NG72UH, UK
6 CHRU de Tours, Service d'Addictologie Universitaire, Équipe de Liaison et de Soins en Addictologie, 37044 Tours, France; nicolas.ballon@univ-tours.fr
7 UMR 1253, iBrain, Université de Tours, INSERM, 37032 Tours, France
8 CHRU de Tours, Centre Spécialisé de l'Obésité, 37044 Tours, France; a.deluca@chu-tours.fr
9 UMR 1069, Nutrition, Croissance et Cancer, Université de Tours, INSERM, 37032 Tours, France
10 CHRU de Tours, Service d'Addictologie Universitaire, Centre de Soins d'Accompagnement et de Prévention en Addictologie d'Indre-et-Loire (CSAPA-37), 37000 Tours, France
* Correspondence: paul.brunault@univ-tours.fr; Tel.: +33-2-18-37-05-81
† These authors contributed equally to this work.

Received: 30 September 2020; Accepted: 23 October 2020; Published: 27 October 2020

Abstract: Attention-Deficit/Hyperactivity Disorder (ADHD) is associated with disordered eating, especially addictive-like eating behavior (i.e., binge eating, food addiction, loss of control overeating). The exact mechanisms underlying this association are unclear. ADHD and addictive-like eating behavior are both associated with negative affectivity and emotion dysregulation, which we hypothesized are mediators of this relationship. The purpose of this systematic review was to review the evidence related to this hypothesis from studies assessing the relationship between childhood or adulthood ADHD symptomatology, negative affectivity, emotion dysregulation and addictive-like eating behavior. The systematic review followed the Preferred Reporting Items for Systematic reviews and Meta-Analyses (PRISMA) recommendations. The literature search was conducted in PubMed and PsycINFO (publication date: January 2015 to August 2020; date of search: 2 September 2020). Out of 403 potentially relevant articles, 41 were retained; 38 publications reported that ADHD and disordered eating or addictive-like eating behavior were significantly associated, including 8 articles that suggested a mediator role of negative affectivity or emotion dysregulation. Sixteen publications reported that the association between ADHD symptomatology and disordered eating or addictive-like eating behavior differed according to gender, eating behavior and ADHD symptoms (hyperactivity, impulsivity and inattention). We discuss the practical implications of these findings and directions future research.

Keywords: food addiction; addictive-like eating; binge eating; eating disorders; loss of control overeating; Attention-Deficit/Hyperactivity Disorder; emotion self-regulation; negative mood

1. Introduction

Attention-Deficit/Hyperactivity Disorder (ADHD) is a neurodevelopmental disorder characterized by impairing levels of inattention and/or hyperactivity-impulsivity, which is thought to begin generally in childhood (before the age of 12) and significantly interferes with social, academic, and/or occupational functioning. Childhood ADHD prevalence is estimated to be between 5 and 7% [1–3]. Current evidence indicates that impairing symptoms of the disorder persist in adulthood in 50 to 60% of cases [4]. The prevalence of adult ADHD is between 1.4 and 3.6% [1]. The treatment for individuals with ADHD includes pharmacologic [5] and non-pharmacologic [6] options. It has been demonstrated that both childhood and adult ADHD is associated with higher prevalence and risk of a large number of medical and psychiatric comorbidities. According to Kooij and colleagues (2019), 60–80% of individuals with ADHD show life-time comorbidities such as anxiety disorder (34%), mood disorder (22%), behavioral disorder (15%) and substance use disorders (11%). One of the most prevalent medical comorbidities is obesity; meta-analytic evidence indicates a 70% increased risk of obesity in adults with ADHD compared to those without ADHD [7,8]. ADHD has also been found to be significantly associated with eating disorders (EDs) (i.e., anorexia nervosa [AN], bulimia nervosa [BN], and binge eating disorder [BED]) [9]. In addition, ADHD is associated more generally with addictive-like eating behavior, even when no ED is diagnosed, notably loss of control overeating [10], binge eating (i.e., recurrent consumption of unusually large amounts of food during a discrete period of time while experiencing loss of control over food intake), and food addiction (FA) (i.e., addictive-like eating behaviors in relation to specific foods high in fat and/or refined carbohydrates, including craving, loss of control overeating, harm related to the behavior, and maintenance of the behavior despite negative consequences) [11–13].

An important research area related to addictive-like eating behavior focuses on the "food addiction" phenotype. According to Gearhardt and colleagues (2009) [12], this can be measured by applying the Diagnostic and Statistical Manual of mental disorders (DSM) criteria for substance dependence to highly palatable foods. FA has been assessed in the general population, among individuals with obesity or ED [14], and with impulse control disorders and psychiatric disorders, including major depressive disorder [15], substance use disorders [16], post-traumatic stress disorder [17], and ADHD [11]. Although FA is not part of the DSM-5 [18] and remains a hotly debated topic, a growing body of literature demonstrates that the "food addiction" phenotype shares some risk factors with other addictive behaviors and could improve our understanding of disordered eating behavior. On the one hand, FA shares neurobiological and clinical features with substance use disorder, such as reward system involvement, loss of control over intake, experience of craving and high impulsivity. On the other hand, it shares features with binging type ED, such as eating a large amount of food in a discrete period of time, and a sense of lack of control overeating during this episode [18]. In fact, FA is over-represented among EDs, especially the binging/overeating types (BN, BED and binging subtype AN [19,20]), but can also be present when no ED is diagnosed. According to Maxwell, Gardiner and Loxton (2020), FA and binge eating are associated with impulsivity, and "there seems to be a pattern emerging regarding overconsumption of food, task effort and lack of inhibition control, specifically that FA is associated with an inability to put the "brakes" on behavior" [21].

Different explanations have been proposed to explain the association between adult ADHD and addictive-like eating behavior. One hypothesis is that the impulsivity dimension of ADHD symptoms may explain the co-occurrence of ADHD and addictive-like eating behavior, such as binge eating [22]. The impulsivity associated with ADHD may increase the overall risk of sensation seeking and addictive disorders, including both substance-use disorders and behavioral addictions [23,24]. Urgency, defined as the tendency to commit rash or regrettable actions as a result of intense negative affect [25], has been hypothesized to be one of the main facets of impulsivity explaining the association between ADHD and addictive disorders [26,27]. As reported by Van Emmerik-Van Oortmerssen and colleagues (2012) in their meta-analysis [28], 23.1% of individuals with a substance-use disorder meet DSM criteria for ADHD. In addition, Anker, Bendiksen and Heir (2018) found that the prevalence of substance-use disorder among the ADHD population ranged from 4% to 23.6%, depending on gender or the substance used [29]. Similarly, addictive disorders are over-represented among people with ADHD [30–32]. Some publications [31–33] report that inattention and hyperactivity/impulsivity are related to the severity of addictive behavior, notably in gambling disorder and symptoms of internet addiction as assessed by the Internet Addiction Test [34]. They also posit that emotion self-regulation may be an important mediator in the association between ADHD and addictive disorders, highlighting the need for a systematic review in this field.

Another hypothesis regarding the relationship between adult ADHD and disordered eating behavior concerns the emotional self-regulation difficulties observed in both groups. Emotion regulation refers to conscious and unconscious processes regulating emotions. "Because emotions are multicomponential processes that unfold overtime, emotion regulation involves changes in emotion dynamics, or the latency, rise time, magnitude, duration, and offset of responses in behavioral, experiential, or physical domains" [35]. Five types of emotion regulation strategies have been described: situation selection (selecting situations that avoid uncomfortable emotions), situation modification (modifying situation features that lead to uncomfortable emotions), attentional deployment (distracting oneself from the attention-grabbing features of an emotional situation), cognitive change (reappraising the emotional meaning of a situation in non-emotional terms) and response modulation (modulating the behavioral, experiential, or physical aspect of the emotional response) [36]. Disruption of these processes leads to difficulties in generating and controlling emotions, associated with inappropriate behavior. Emotion regulation difficulties are encountered in some disorders, including ADHD [37], substance-use disorder [38] and disordered eating [39]. Masi and colleagues (2020) found that emotional dysregulation was a predictor of the persistence of ADHD symptoms after 4 weeks of pharmacological treatment. Higher levels of emotional dysregulation at the baseline assessment predicted higher levels of overall symptoms of ADHD at follow-up [40].

The hypothesis of a mediating role of emotion dysregulation in the association between ADHD and disordered eating is supported by the strong association found between emotion dysregulation and ED [39]. Emotion dysregulation affects up to 70% of adults with ADHD and substantially worsens the psychosocial outcomes of the disorder [41]. Moreover, the DSM-5 highlights emotion dysregulation as a feature supporting the diagnosis of ADHD [18]. According to the systematic review of ADHD-associated emotion dysregulation conducted by Beheshti, Chavanon and Christiansen (2020), the persistence of ADHD inattention symptoms in older age correlates with impaired situation identification, which requires attention processes, whereas hyperactive symptoms are associated more with impaired capacity to inhibit emotional responses. Additionally, emotional lability and negative emotional responses might play a key role in the emotion dysregulation-associated psychopathology of adults with ADHD [42]. Emotion dysregulation has been identified as a mediator between ADHD symptoms and several disorders such as depressive symptoms [43]. Emotion regulation difficulties particularly concern negative affect. Negative affectivity has been shown to be higher in individuals with ADHD and to be associated with a negative impact on ADHD experience and medication adherence, and increased risk of suicidal ideation and behavior, or various comorbid disorders [44–46]. Individuals with ADHD also show lack of emotion regulation strategies. As hypothesized for persons with a substance-use disorder [47], individuals who are less likely to use coping strategies to deal with or express emotions may resort to more problematic behavior. We can hypothesize that substance-use disorder and addictive disorders may provide immediate pleasure and/or a dissociative-like state to individuals with ADHD, offering psychological escape from the offending reality [48], and thus constitute a dysfunctional coping strategy to regulate negative affect.

The role of emotion dysregulation in the association between ADHD and addictive behavior has also been investigated in gambling disorder. For example, Mestre-Bach and colleagues (2019) found people with gambling disorder and ADHD symptomatology had greater emotion regulation difficulties than those without ADHD. The authors found that individuals with ADHD-gambling disorder comorbidity had higher rates of the following emotion regulation difficulties: non-acceptance of emotional responses, difficulty pursuing goal-directed behaviors when experiencing negative emotions, difficulty controlling impulsive behaviors when experiencing negative emotions, limited access to emotion regulation strategies, and lack of emotional clarity [32]. Their results are in line with the mediating role of emotion regulation in the relationship between ADHD symptomatology and addictive disorders in patients with gambling disorder. However, to our knowledge, no systematic review has been conducted to assess the mediating role of emotion regulation in ADHD symptoms and ED/addictive-like eating behavior (i.e., FA, binge eating, loss of control overeating).

To fill this gap, the aim of this study was to conduct a systematic review of studies investigating the association between childhood/adult ADHD, negative affectivity, emotion regulation, and disordered eating, with a specific focus on addictive-like eating behavior (i.e., binge eating, FA, loss of control overeating). We investigated negative affectivity, a common term involving many negative emotions such as anxiety, depression, negative urgency, stress. To this end, we first explored the characteristics of studies conducted in this field of research. In order to investigate the association between ADHD and disordered eating, we examined the prevalence of ADHD and disordered eating comorbidity within different populations. Next, we assessed negative affectivity and emotion regulation in individuals with ADHD, and finally we examined the involvement of these features in the relationship between ADHD symptomatology and addictive-like eating behavior. Due to potential difference in these relationships between children/adolescents and adults, we investigated both populations.

We hypothesized that: (1) individuals with disordered eating would show more ADHD symptoms; (2) individuals with ADHD symptoms would have higher levels of disordered eating; (3) ADHD symptoms would be associated with severity of addictive-like eating behavior; (4) the level of ADHD symptoms would be associated with high levels of negative affectivity and emotion regulation difficulties; (5) negative affectivity and emotion regulation difficulties may be mediators in the relationship between ADHD symptoms and addictive-like eating behavior.

2. Materials and Methods

This systematic review included publications investigating the association between ADHD and addictive-like eating behavior such as loss of control overeating, binge eating, and preoccupation with food, which are the main FA symptoms, and some DSM-5 EDs (eating disorders mentioned in the Diagnostic and Statistical Manual of Mental Disorders, Fifth edition) such as BN and BED, which show high FA prevalence [19,20].

This review was undertaken according to the quality standards of the Preferred Reporting Items for Systematic reviews and Meta-Analyses (PRISMA; Figure 1).

2.1. Literature Search

We conducted the literature search on 2 September 2020; the systematic literature review methodology included analysis of the electronic databases PsycINFO and PubMed. In order to identify all relevant publications on the association between [ADHD] and [FA symptoms and/or disordered eating], we used the following key words: ["ADHD" OR "attention-deficit hyperactivity disorder"] AND ["food addiction" OR "binge eating" OR "eating disorder" OR "bulimia" OR "obesity" OR "obese" OR "overweight"]. We included studies that used these keywords in their abstract (criterion I1, see Table 1). We focused on articles published from January 2015 to August 2020 (criterion I2) in peer-reviewed journals (criterion I3). Moreover, as we did not have funding for translation, we only included publications written in English or French (criterion I4). Based on these inclusion criteria, we excluded book chapters, letters to the editor and articles published before January 2015 and not written in English or French (criteria E1–E4). After removing duplicates, 403 article abstracts were identified for "abstract screening".

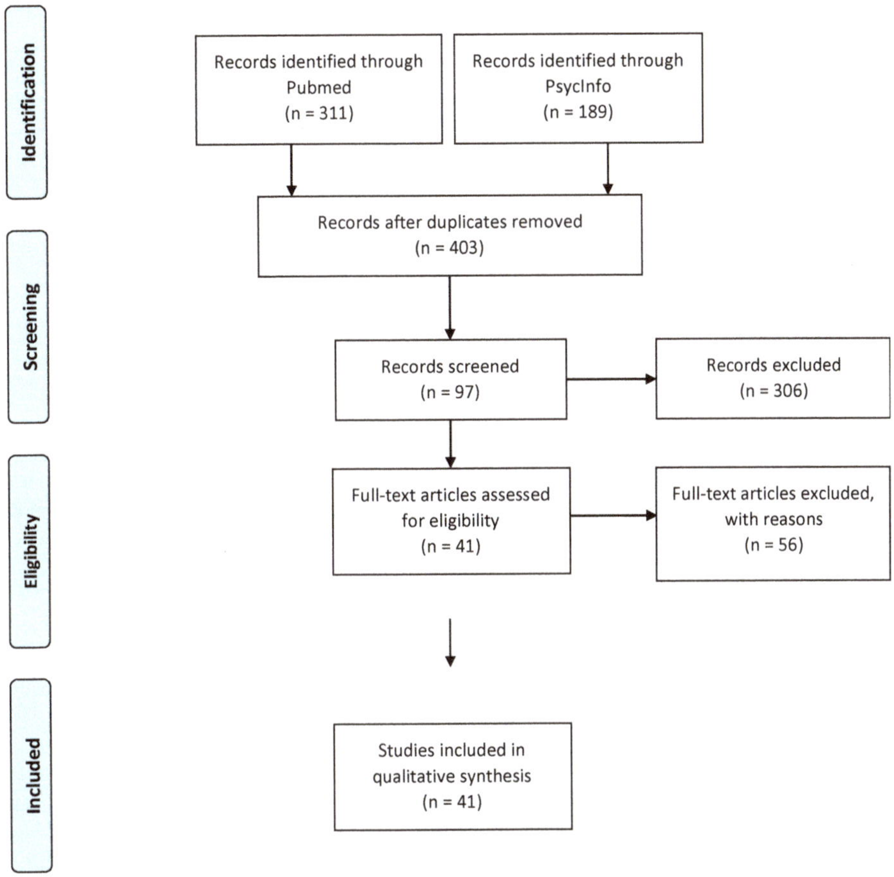

Figure 1. Study selection flow chart.

Table 1. Inclusion and exclusion criteria.

	Inclusion Criteria		Exclusion Criteria
I1	Key words cited in the abstract	E1	Key words not cited in the title/abstract
I2	Date of publication: January 2015 to June 2020	E2	Publication before January 2015
I3	Journal article with peer-review	E3	Book chapter, letter to the editor or other non-empirical type of publications
I4	Written in English or French	E4	Paper not written in English or French
I5	Empirical research	E5	Review and meta-analysis papers
I6	Focus on the association between ADHD and eating behavior	E6	Focus on treatment, medical imaging, genetics
I7	ADHD and disordered eating symptoms in the same individual	E7	Focus on the impact of parents' disordered eating or BMI on their child's ADHD symptoms
I8	Assessment of ADHD and disordered eating symptoms		

Note: ADHD: Attention-Deficit Hyperactivity Disorder; BMI: Body Mass Index.

Careful reading of these abstracts allowed us to select articles with an empirical approach (criterion I5), concerned directly or indirectly with ADHD and eating behavior (criterion I6) and investigating ADHD and disordered eating symptoms in the same individual (criterion I7). These inclusion criteria led to exclusion of review and meta-analysis articles (criterion E5), publications which did not address ADHD and eating behavior directly or indirectly, or focused on ADHD treatment or medical imaging (criterion E6). We also excluded all publications that investigated the impact of parents' disordered eating or body mass index (BMI) on their child's ADHD symptoms (criterion E7).

The papers thus retained were then read in full and appraised. We did not use a specific tool to appraise the quality of these studies, but they were checked for all the inclusion criteria and selection errors. We also checked that all the studies assessed ADHD and eating behavior using a validated instrument such as self-administered questionnaires or clinical interviews (criterion I8).

Regarding the characteristics of the populations studied, as our aim was to provide an overview of the association between ADHD and disordered eating, we did not consider age or gender as exclusion criteria.

2.2. Data Extraction

To investigate the characteristics of the publications, the following data were extracted: author names, country and year of publication, source, sample characteristics (age, gender, size, recruitment method and place), study design. We also extracted data about the prevalence of ADHD in individuals with disordered eating and the prevalence of disordered eating in individuals with ADHD. We thus identified the ADHD assessment tools used, the use of medication especially for individuals with ADHD, the type of eating behavior and the tools used to assess it. Finally, we examined the main results and conclusions about disordered eating and ADHD comorbidity. In this way, we extracted data regarding the association between ADHD and disordered eating, especially addictive-like eating symptoms and the involvement of negative affectivity and emotion self-regulation.

It should be noted that we use the word "symptom" to describe features of disordered eating and ADHD assessed only through self-administered questionnaires, and "diagnosis" or "severity" when assessment was through clinical interviews. Moreover, we use the word "eating disorder" (or ED) only for DSM disorders such as BN, BED and AN, and the word "disordered eating" as a generic word to include all pathological eating behaviors/symptoms such as binge eating, food addiction, loss of control overeating, strong desire for food, preoccupation with food, bulimic symptoms . . . .

3. Results

We initially identified 403 articles, of which 97 were screened and selected for full-text reading. After full-text reading, 56 publications were excluded for the following reasons:

- No data about behavioral features of eating (n = 38), including 30 publications which focused on the association between ADHD and BMI [8,49–85]
- Non-representative sample, e.g., autism spectrum disorder (n = 4) [86–89]
- Previous selection errors (n = 10) [22,90–98]
- Investigations did not include ADHD-disordered eating association (n = 3) [99–101]
- No access to full text (n = 1) [102]

Thus, 41 publications were included in this systematic literature review for qualitative synthesis (see Figure 1 for the study flow chart).

3.1. Article Characteristics

3.1.1. Country of Investigation

The majority of these studies were conducted in the USA (n = 10, 25.6% of the included publications). Others were conducted in Sweden (n = 5, 12.8%), France, Canada (n = 4, 10.3% for each), the UK (n = 3,

7.7%), Spain, Brazil (n = 2, 5.1% for each), Norway, Australia, Israel, Korea, Switzerland, Greece, Iran, Germany and China (n = 1, 2.6% for each). One study did not specify the country of recruitment.

3.1.2. Year of Publication

Included articles were published between January 2015 and August 2020. Eleven articles were published in 2017, 10 before 2017 and 20 after 2017. See Figure 2.

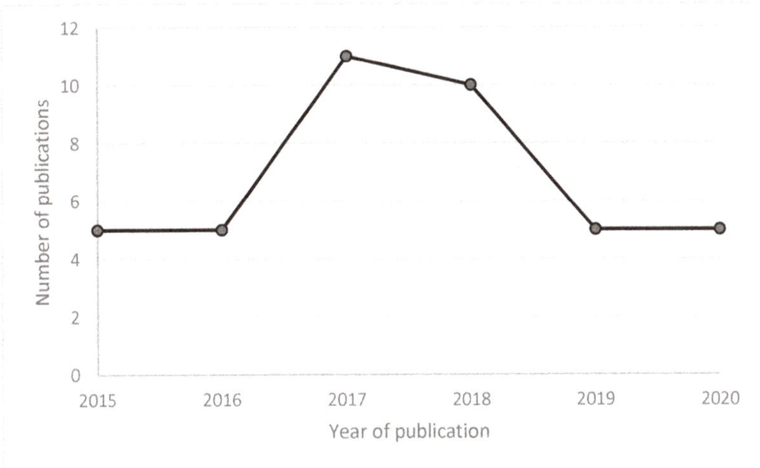

Figure 2. Number of publications from 2015 to 2020.

3.1.3. Study Design

Among the 41 publications, 80.5% were cross-sectional (n = 33), and 19.5% were prospective longitudinal studies (n = 8).

3.1.4. Age of Interest

Nineteen studies were conducted with children and/or adolescents (46.3%) and 24 with adults (58.5%). Two studies had a mixed adolescent-adult sample (4.9%).

3.1.5. Population

Twenty-two studies were conducted with participants from the general population (53.7%), and 46.3% (n = 19) involved clinical populations: patients with severe obesity recruited in obesity and centers or prior to bariatric surgery (n = 8), patients with disordered eating (n = 6), ADHD outpatients (n = 2), or patients recruited in psychiatric departments (n = 3).

3.1.6. ADHD Assessment and Medication

ADHD was assessed through clinical interviews (including semi-structured interviews) in 21 studies (51.2%), and through self-administered questionnaires in 20 studies (48.8%).

For children and adolescents, the main assessment tool for ADHD was the Kiddie Schedule for Affective Disorders and Schizophrenia (KSADS; 26.3% of the 19 studies conducted with children or adolescents) and the ADHD Rating Scale (ADHD-RS; 15.8%). For adults, ADHD was mainly assessed with DSM-IV or DSM5 semi-structured interviews using the Composite International Diagnostic Interview (CIDI), the Diagnostisch Interview Voor ADHD bij volwassenen (DIVA 2.0), or the Structured Clinical Interview for DSM Disorders (SCID). The main self-administered questionnaire was the Adult

ADHD Self-Report Scale (ASRS; 41.7% of the 24 studies conducted with adults). It should be noted that some studies used the ASRS, a screening scale, as a diagnostic tool.

Fifty-four percent of the studies with adults included a retrospective assessment of childhood ADHD symptoms (n = 13), included in the diagnostic tool or additionally reported mainly through the Wender Utah Rating Scale (WURS) (n = 3).

Despite the known influence of ADHD pharmacological treatment on eating behavior [103], only 10 studies specified the ADHD medication status (25.6%). Three of them were conducted in medication-naïve populations, the remainder reported the rate of ADHD participants on medication.

3.1.7. Disordered Eating Assessment Tools

Among the studies of children-adolescents, 7 (36.8%) assessed eating behavior through interviews (including semi-structured interviews), 10 (52.6%) through self-administered questionnaire, and 2 (10.5%) used both interviews and self-administered questionnaires. Various tools were used to assess disordered eating behavior, including the following self-administered questionnaires: the Eating Disorder Inventory-2 (EDI-2) (n = 3), the Children's Eating Attitude Test (ChEAT), the Child Eating Behavior Questionnaire (CEBQ), the Eating Disorder Examination Questionnaire (EDE-Q) (n = 3 for each), and the Child Eating Disorder Examination (ChEDE), which specifically assesses loss of control overeating (n = 3). None of the studies used the Yale Food Addiction Scale for Children.

For adults, 14 studies (58.3%) used professional interviews (including semi-structured interviews), 14 publications (58.3%) were based on self-administered questionnaires investigating disordered eating, and 4 (16.7%) assessed disordered eating through both interviews and self-administered questionnaires. The main ED diagnostic tools used during clinical interview were the Mini International Neuropsychiatric Interview (MINI) and the SCID (n = 3 for each). The main self-administered questionnaires were the Binge Eating Scale (BES) to assess binge eating (n = 5), the original (DSM-IV-TR based) Yale Food Addiction Scale (YFAS) and the YFAS 2.0 (DSM-5 based) to assess FA (n = 2), the EDE-Q and the EDI-2 to assess disordered eating (n = 4 for each), and the Bulimic Investigatory Test Edinburgh (BITE) to assess bulimic symptoms (n = 4).

3.2. *Association between ADHD and Disordered Eating*

3.2.1. Prevalence of Disordered Eating in Individuals with ADHD

Children and Adolescents

Four studies focused on the association between disordered eating and addictive-like eating behavior among children with ADHD symptoms (Table 2). They showed divergent results depending on the type of population. Wentz and colleagues (2019) [104], who assessed children recruited in an obesity clinic found no significant difference between individuals with and without ADHD diagnosis in terms of loss of control overeating. However, a study conducted in the general non-clinical population found a higher prevalence of loss of control overeating in children with than without ADHD diagnosis (70.5% vs. 20%; $p < 0.001$). The odds of loss of control overeating were increased 12.68 times for children with ADHD (95% Confidence Interval (CI): 3.11–51.64; $p < 0.001$) after adjusting for age, sex and race [105]. Another study with children attending psychiatric outpatient clinics found a higher prevalence of binge eating in individuals with ADHD than in controls (26% vs. 2%; $p < 0.001$) [103]. Moreover, in a longitudinal study by Bisset and colleagues (2019) [106], adolescents who screened positive for ADHD symptoms at age 12–13 tended to have a higher risk of objective binge eating at age 14–15 than adolescents without ADHD symptomatology (3.7% vs. 1.3%; Odds Ratio (OR) = 2.9, 95% CI: 0.9–8.6). Interestingly, this association was significant only for boys (2.9% vs. 0.3%; OR = 9.4, 95% CI: 1.7–52.8) and not for girls (6.5% vs. 2.2%; OR = 3.1, 95% CI: 0.7–14.0). The authors found no difference in terms of BN and BED symptoms (even partial syndromes) between adolescents with and without ADHD symptoms.

Table 2. Disordered eating prevalence among individuals with Attention-Deficit Hyperactivity Disorder symptomatology.

Population	Country	Children-Adolescents	Adults	ADHD Diagnosis Instruments	N	Age Mean (SD) (Years)	Gender Female (%)	Disordered Eating	ADHD Symptoms		Non-ADHD Symptoms		Statistics		
									n	Disordered Eating Prevalence (%)	*n*	Disordered Eating Prevalence (%)	Odds Ratio	95% Confidence Intervals	*p*
General population															
[105]	USA	x		K-SADS PL and CRPS	79	11.0 (1.9)	48.1	LOC-E	31	70.5	7	20	12.68	3.11–51.64 [a]	
[106]	Australia	x		SDQ or parent-reported ADHD diagnosis or medication	2672	14.9 (0.3)	0	Regular objective BE		2.9		0.3	9.4	1.7–52.8	
								Partial syndrome BN		1.6		1.5	1.0	0.1–8.0	
								Partial syndrome BED		1.3		0.2	6.2	0.6–61.1	
							100	Regular objective BE		6.5		2.2	3.1	0.7–14.0	
								Partial syndrome BN		6.5		3.6	1.9	0.4–8.2	
								Partial syndrome BED		0		0.6	-	-	
[107]	Sweden		x	DSM-IV criteria	18,029	33.6 (7.6)	55.6	BE behavior	113	7.17		-	3.65	2.72–4.91	
								DSM-5 BE behavior	58	3.72		-	3.01	2.09–4.35	
								DSM-5 BED	7	0.45		-	2.55	1.11–5.86	
								DSM-5 BN	48	3.11		-	3.09	2.09–4.56	
[108]	UK		x	ASRS	7403	46.3 (18.6)	51.4	Self-sick for feeling full		8.5		2.7	2.79	1.76–4.42 [b]	
													1.26	0.63–2.51 [c]	
								Uncontrolled eating		22.8		6.4	3.94	2.94–5.28 [b]	
													1.67	1.14–2.46 [c]	
								Possible ED		19.2		5.7	3.48	2.56–4.72 [b]	
													1.32	0.82–2.13 [c]	
[13]	France		x	WURS-25 + ASRS	1517	20.6 (3.6)	68.2	Food addiction	12	14.1	57	4.0	2.27	1.05–4.88 [d]	
								Any ED	28	32.9	249	17.4	1.33	0.76–2.33 [d]	
[109]	USA		x	DIS-IV (childhood) + adult ACDS	4719	31 (DNS)	52.1	Past 12-month any ED					9.74	4.23–22.40 [b]	
													2.84	1.22–6.63 [c]	
								Past 12-month BED					4.53	1.82–11.24 [b]	
													1.65	0.67–4.04 [c]	
								Past 12-month BN					28.24	6.33–126.01 [b]	
													5.04	1.15–22.08 [c]	
								Past 12-month subthreshold BED					5.55	1.90–16.24 [b]	
													3.83	0.94–15.67 [c]	

Table 2. *Cont.*

Population	Country	Children-Adolescents	Adults	ADHD Diagnosis Instruments	N	Age Mean (SD) (Years)	Gender Female (%)	Disordered Eating	ADHD Symptoms		Non-ADHD Symptoms		Statistics		
									n	Disordered Eating Prevalence (%)	*n*	Disordered Eating Prevalence (%)	Odds Ratio	95% Confidence Intervals	*p*
Psychiatry outpatients															
[103]	USA	x		DSM-IV criteria	252	10.8 (3.7)	47.2	BE	28	26	3	2.0	-	-	***
[110]	USA		x	SCID-IV	1134	39.7 (14.4)	58	Any ED	19	9.3	35	3.8	2.67	1.45–4.80	
Patients with obesity															
[104]	Sweden	x		Medical records or DAWBA	40	12.4 (3.0)	48.7	LOC-E	5	55.6	21	67.7	-	-	>0.05
[11]	France		x	DIVA 2.0	105	46.5 (10.7)	86.7	Food addiction	8	28.6	7	9.1	4.00	1.29–12.40	
								Significant distress in relation to food	9	32.1	9	11.7	3.58	1.25–10.30	
ADHD outpatients															
[111]	France		x	Children-MINI adapted for adults	81	34.8 (11.6)	37	Bulimia nervosa	7	8.6	-	-	-	-	-
[29]	Norway		x	DSM-IV criteria	533	36.2 (11.3)	100	Any ED	36	13.0	-	-	-	-	-
						37.4 (10.7)	0	Any ED	3	1.1	-	-	-	-	-

Note: N or n: group size; SD: Standard Deviation; ADHD: Attention-Deficit Hyperactivity Disorder; DSM-IV: Diagnostic and Statistical Manual of mental disorders, fouth edition; LOC-E: Loss Of Control overEating; BE: Binge Eating; BED: Binge Eating Disorder; BN: Bulimia Nervosa; ED: Eating Disorder; K-SADS PL: Schedule for Affective Disorders and Schizophrenia for school-age children-Present and Lifetime Version; CRPS: Conners-3 Parent Rating Scale-Revised; SDQ: Strengths and Difficulties Questionnaire; ASRS: Adult ADHD Self-Report Scale; WURS: Wender Utah Rating Scale; DIS-IV: Diagnostic Interview Schedule for DSM-IV; ACDS: ADHD Clinical Diagnostic Scale; SCID-IV: The Structured Clinical Interview for DSM-IV; DAWBA: Development and Well-Being Assessment; DIVA: Diagnostische Interview Voor ADHD; MINI: Mini International Neuropsychiatric Interview; DNS: data not specified; [a]: model adjusted for age, sex, race, body mass index z score; [b]: model adjusted for age, race, sex; [c]: model adjusted for age, race, sex and lifetime diagnosis of psychiatric comorbidities; [d]: model adjusted on universities (place of recruitment), cursus and financial difficulties; *** $p < 0.001$.

Adults

Within adult population, eight studies assessing disordered eating prevalence among individuals with ADHD symptomatology.

Two of these studies, with no control group, found a prevalence of 8.6% for BN [111], and 1.1% and 13% for any ED in ADHD patient men and women respectively [29]. Four studies with a general non-clinical population examined ED prevalence; ADHD-ED association odds ratio ranged from 1.32 (95% CI: 0.82–2.13) to 28.24 (95% CI: 6.33–126.01) [13,107–109]. These associations were particularly strong for BN (up to OR = 28.24, 95% CI: 6.33–126.01) [107,109]. Three of these studies found that ADHD symptoms were associated with an increased risk of ED. However, the odds ratio was significant after adjusting for age, sex and race, but not after adjusting for age, sex, race and psychiatric comorbidities, especially for BED (details in Table 2) [108,109]. Among psychiatric outpatients, Gorlin and colleagues (2016) [110] found higher ED prevalence for individuals diagnosed with ADHD (9.3% vs. 3.8%, $p < 0.01$), especially for the inattentive subtype (inattentive subtype: 10.3% individuals with an ED; OR = 3.01, 95% CI: 1.30–6.34; combined subtype: 8.1%, OR = 2.17, 95% CI: 0.90–4.68).

All publications assessing addictive-like eating symptoms in individuals with ADHD symptoms ($n = 4$) reported that ADHD was associated with a higher risk of addictive-like eating symptomatology: food addiction, binge eating, uncontrolled eating, significant distress in relation to food, and made him/herself be sick because he/she felt uncomfortably full [11,13,107,108] (details in Table 2). The FA prevalence rate was higher in patients with ADHD symptoms or diagnosis. In a study conducted in a non-clinical student population, FA prevalence was observed in 14.1% of the sample with ADHD symptoms compared to only 4% of those without ADHD symptoms (OR = 2.27, 95% CI: 1.05–4.88) [13]. In a sample of patients with severe obesity, FA prevalence was higher in those with than without ADHD diagnosis (28.6% vs. 9.1%; OR = 4.00, 95% CI: 1.29–12.40) [11]. Moreover, in a sample of adults with severe obesity, FA was associated with a retrospective assessment of childhood ADHD (24.3% vs. 8.8% without childhood ADHD symptoms, OR: 3.32, 95% CI: 1.08–10.23, $p = 0.034$) [11].

3.2.2. Prevalence of ADHD in Individuals with Disordered Eating

Children and Adolescents

Three studies of overweight or obese children assessed ADHD prevalence (Table 3). One study with a non-clinical sample by Gowey and colleagues (2017) [112] found a rate of clinical levels of ADHD of 5% and subclinical levels of 5.91%, similar to the prevalence in the normal weight population. However, other studies conducted in clinical populations of children with obesity found higher rates of ADHD, ranging from 11% [113] to 18.4% [104]. Reinblatt and colleagues (2015) [105] found that the odds of children with obesity and loss of control overeating having an ADHD diagnosis was 7.3 times higher (95% CI: 1.88–28.17) than obese children without loss of control overeating, and 10.44 times higher (95% CI: 2.96–36.75) than children without obesity. These results were observed for both inattentive and hyperactivity/impulsivity ADHD subtypes.

Rojo-Moreno and colleagues (2015) [114] and Mohammadi and colleagues (2019) [115] assessed ADHD and eating disorder in general non-clinical populations. They found higher rates of ADHD in children with than without eating disorders ([114]: 31.4% vs. 8.4%, $p < 0.05$; [115]: 7.6% vs. 3.9%, $p = 0.026$). Furthermore, Kim and colleagues (2018) [116] found that 21.1% of children presenting with addictive-like eating behavior such as every-day overeating had a high risk of ADHD (see Table 4).

Table 3. Attention-Deficit Hyperactivity Disorder prevalence among population with overweight or obesity.

Population	Country	Children-Adolescents	Adults	N	Age Mean (SD) (Years)	Gender Female (%)	Mean BMI or zBMI (SD)	ADHD Instruments	Childhood ADHD Prevalence (%)	Adult ADHD Prevalence
Non clinical population										
[112]	USA	x		220	10.3 (1.4)	53.6	2.19 (0.38)	CBCL	5.0	
Clinical obesity population										
[113]	USA	x		385	10.9 (2.3)	63	2.26 (0.35)	CBCL	11.0	
[104]	Sweden	x		76	12.4 (3.0)	48.7	3.40 (0.50)	Medical records or DAWBA	18.4	
[9]	Brazil		x	106	39.0 (10.7)	100	39.21 (5.29)	K-SADS adapted for adults, DSM IV		28.3 [a]
[117]	Germany		x	120	41.0 (11.5)	79.2	47.76 (7.41)	WURS-k + CAARS-S:S	17.5 [b]	8.3 [a]
[11]	France		x	105	46.4 (10.7)	86.7	46.90 (7.80)	DIVA 2.0	35.2 [b]	26.7 [a]

Note: N: group size; SD: Standard Deviation; BMI: Body Mass Index; zBMI: Body Mass Index z score; ADHD: Attention-Deficit Hyperactivity Disorder; CBCL: Child Behavior Checklist; DAWBA: Development and Well-Being Assessment; K-SADS: Schedule for Affective Disorders and Schizophrenia for school-age children; WURS-k: Wender Utah Rating Scale Short Version; CAARS-S:S: Conners' Adult ADHD Rating Scale-Self-Report: Short Version; DIVA: Diagnostische Interview Voor ADHD; DSM IV: Diagnostic and Statistical Manual of mental disorders, Fourth edition [a]: ADHD symptomatology since childhood as expected by DSM criteria; [b]: retrospectively estimated.

Table 4. Attention-Deficit Hyperactivity Disorder prevalence among disordered eating.

Population	Country	Children-Adolescents	Adults	ADHD Instruments	N	Mean Age (SD) (Years)	Gender Female (%)	Disordered Eating	Disordered Eating		Non Disordered Eating		Statistics		
									n	ADHD Symptoms Prevalence (%)	*n*	ADHD Symptoms Prevalence (%)	Odds Ratio	95% Confidence Intervals	*p*
General population															
[114]	Spain	x		K-SADS	962	DNS (12–16)	47.8	ED	11	31.4	80	8.4	5.03	2.37–10.64	
[116]	Korea	x		K-ARS	16,831	9.29 (1.71)	50.2	Every day overeating	68	21.1	-	-			
[115]	Iran	x		K-SADS PL	27,111	DNS (6–18)	48.6	Lifetime ED	-	7.6	-	3.9			0.026
[118]	USA		x	CIDI	1686	DNS (18–44)	100	Lifetime any ED	18	21.9	75	5.7	4.51	2.01–10.15 [b]	
								Past 12-month any ED	10	30.6	83	6.1	7.11	2.61–19.39 [b]	
								Lifetime BED	8	17.1	85	6.3	3.01	1.14–7.95 [b]	
								Past 12-month BED	4	19.3	89	6.5	3.57	1.06–12.09 [b]	
								Lifetime BN	10	33.2	83	6.1	7.93	2.75–22.85 [b]	
								Past 12-month BN	6	56.7	87	6.3	21.15	3.76–118-98 [b]	
								Lifetime any binge	16	18.7	77	5.9	3.66	1.71–7.87 [b]	
								Past 12-month any binge	8	19.4	85	6.3	3.71	1.68–8.20 [b]	
							0	Lifetime any ED	6	21.3	85	9.7	2.23	0.81–6.13 [b]	
								Past 12-month any ED	4	45.9	87	9.7	6.48	1.33–31.60 [b]	
								Lifetime BED	6	25.4	85	9.6	2.93	0.98–8.76 [b]	
								Past 12-month BED	4	45.9	87	9.7	6.47	1.33–31.61 [b]	
								Lifetime BN	1	66.9	90	9.9	18.18	1.39–238.40 [b]	
								Lifetime any binge	11	19.4	80	9.5	2.39	1.17–4.91 [b]	
								Past 12-month any binge	7	38.9	84	9.5	5.02	1.90–13.28 [b]	
Mood disorder outpatients [a]															
[119]	Canada		x	MINI	631	37.8–40.0 (12.0–12.4)	59.0	BE	26	20.8	61	12.5			0.018

Table 4. *Cont.*

Population	Country	Children-Adolescents	Adults	ADHD Instruments	N	Mean Age (SD) (Years)	Gender Female (%)	Disordered Eating	Disordered Eating		Non Disordered Eating		Statistics		
									n	ADHD Symptoms Prevalence (%)	*n*	ADHD Symptoms Prevalence (%)	Odds Ratio	95% Confidence Intervals	*p*
ED patients															
[120]	Sweden		x	ASRS	1094	27.7 (8.7)	100	Any ED	346	31.6	-	-			
								BED	25	27.5	-	-			
								BN	156	37.1	-	-			
								AN-BP	13	35.1	-	-			
								AN-R	12	17.6	-	-			
								EDNOS-BP	102	31.0	-	-			
								EDNOS-R	38	25.7	-	-			
[121]	Sweden		x	ASRS	443	27.5 (8.5)	100	Any ED	45	10.2	-	-			
[122]	Israel	x	x	K-SADS PL	168	DNS (15–28)	100	BE	-	16.6	-	13.6			0.392
								BN	-	12.0	-	-			
								AN-BP	-	28.0	-	-			
								AN-R	-	9.0	-	-			
[123]	Canada		x	ASRS	500	27.6 (10.6)	95.2	Any ED	-	49.8					
[124]	France		x	WURS + BAADS	73	28.1 (7.3)	100	AN-R	3	8.1	-	-			
								AN-BP	9	32.1	-	-			
								BN	1	12.5	-	-			

Note: N: group size; SD: Standard Deviation; ADHD: Attention-Deficit Hyperactivity Disorder; K-SADS PL: Schedule for Affective Disorders and Schizophrenia for school-age children-Present and Lifetime Version; K-ARS: Korean version of the ADHD rating scale; CIDI: Composite International Diagnostic Interview; MINI: Mini International Neuropsychiatric Interview; ASRS: Adult ADHD Self-Report Scale; WURS: Wender Utah Rating Scale; BAADS: Brown Attention Deficit Disorder Scale; ED: eating disorder; BE: binge eating; BN: bulimia nervosa; BED: Binge Eating Disorder; AN: Anorexia Nervosa; BP: Binging/purging type; R: restrictive type; EDNOS: Eating Disorders Not Otherwise Specified; DNS: data not specified; [a]: major depressive disorder or bipolar disorder; [b]: logistic regression models adjusted for age and race/ethnicity.

Adults

Three studies conducted in adults with severe obesity, recruited in obesity hospital departments, reported the prevalence of ADHD (Table 3). Nielsen and colleagues (2017) [117] estimated that 8.3% of bariatric surgery patients screened positive for ADHD on both the WURS (childhood ADHD symptoms scale) and the CAARS (adult ADHD symptoms scale). Based on adult ADHD DSM-IV criteria (including ADHD symptoms before the age of seven years), Brunault and colleagues (2019) [11] and Nazar and colleagues (2016) [9] found prevalence rates of 26.7% and 28.3% respectively in semi-structured diagnostic interviews. Looking only at current ADHD symptomatology, the prevalence rates of inattention, hyperactivity, and impulsivity were 23.3%, 12.5% and 21.7%, respectively [117]. Retrospective childhood ADHD was estimated at 35.2% [11] and 17.5% [117].

Five studies assessed ADHD in clinical populations of women with ED. High ADHD prevalence was found, especially among women with ED involving binging/purging behavior (AN-BP, EDNOS-BP, BN): from 10.2% to 49.8% [120–124]. However, Halevy-Yosef and colleagues (2019) [122] observed no significant difference in terms of ADHD prevalence between ED patients with BE (16.6%) and those without BE (13.6%) ($p = 0.392$).

After assessing disordered eating in a general non-clinical population, Brewerton & Duncan (2016) [118] found that the prevalence of ADHD was significantly higher in adults with lifetime or past 12-month disordered eating (BED, BN and binge eating), except for men diagnosed with lifetime disordered eating, and especially BED (see details Table 4). Similarly, in a sample of adults with major depressive or bipolar disorder, Woldeyohannes and colleagues (2015) [119] found an ADHD diagnosis rate of 20.8% among those with binge-eating behavior compared to 12.5% among those who did not binge ($p = 0.018$).

3.2.3. ADHD and Disordered Eating

Children and Adolescents

Twelve studies explored the association between ADHD and addictive-like eating in children or adolescents.

Kim and colleagues (2018) [116] found that children with overeating had higher scores on the K-ARS (Korean version of the ADHD rating scale assessing ADHD symptom severity), increasing with frequency of overeating. Egbert and colleagues (2018) and Halevy-Yosef and colleagues (2019) conducted studies with individuals with clinical obesity and clinical ED respectively, and found that ADHD scale scores (Child Behavior Checklist, CBCL and ADHD-RS respectively) were higher in groups with dysregulated eating (56.17, Standard Deviation (SD) = 8.26 vs. 54.42, SD = 6.18, $p < 0.05$) [113] or binge eating (22.92, SD = 9.78 vs. 19.86, SD = 10.48, $p < 0.001$) [122]. In the clinical ED sample, further investigations found that severity of ADHD inattention symptoms was greater among binge-eating than non-binge eating individuals and controls (Bonferroni corrected $p = 0.0003$), and that severity of ADHD hyperactivity/impulsivity symptoms was greater in binge-eating and non-binge eating individuals than in controls (Bonferroni corrected $p < 0.01$). Patients who reported binging/purging behavior scored higher on both inattentive and hyperactivity/impulsivity ADHD subscales [122]. Kurz and colleagues (2017) [125] used a laboratory test meal and found no difference between individuals with ADHD and controls in loss of control overeating, liking for food and desire to eat.

Two studies conducted with non-clinical samples of children found that ADHD symptoms [126] and ADHD diagnosis [127] were related to emotional overeating. One of these studies [127] with 4-year-old children found a positive association between ADHD scale scores and eating behaviors, especially food responsiveness and emotional overeating. Moreover, children who scored in the medium and highest tertiles of the responsiveness scale and in the highest tertile of the emotional eating scale scored higher on the ADHD scales. In girls, food responsiveness was significantly associated only with impulsivity symptoms; in boys, it was significantly associated with inattentive and hyperactivity symptoms, while emotional overeating was significantly associated only with hyperactivity symptoms.

Some studies corroborated these results through correlation analysis. They found that ADHD severity was positively correlated with objective overeating ($r = 0.10$, $p < 0.05$), objective binge eating ($r = 0.17$, $p < 0.01$) [113], BN symptoms ($r = 0.19$, $p < 0.0001$), emotional overeating ($r = 0.31$, $p < 0.0001$) and emotional undereating ($r = 0.28$, $p < 0.0001$) [126], and with disordered eating as assessed on scales including the EAT-26 (ED severity, $r = 0.53$, $p < 0.0001$), EDE-Q (disordered eating behavior, $r = 0.48$, $p < 0.0001$), EDI-2 (impulse regulation and interoceptive awareness subscales, $r = 0.65$, $p < 0.001$ and $r = 0.66$, $p < 0.001$ respectively) [122].

Four studies conducted regression analyses and found a significant association between ADHD and disordered eating, and more specifically addictive-like eating behavior. These studies showed that ADHD symptoms were associated with loss of control overeating and binge eating [113], food preoccupation and oral control (i.e., self-control of eating and pressure from others to eat) [112]. Similarly, ADHD diagnosis was associated with loss of control overeating [105] and binge eating [103]. Egbert and colleagues (2018) [113] demonstrated that ADHD symptoms were positively associated with frequency of objective binge eating and objective overeating (respectively 6% and 5% increase in frequency of objective binge eating and objective overeating for every one-point increase in ADHD symptoms, $\chi^2(1) = 16.61$, $p < 0.001$; $\chi^2(1) = 10.64$, $p < 0.01$), but not subjective binge eating ($\chi^2(1) = 1.30$, $p = 0.25$).

Further investigations involving mediation analyses highlighted the mediator role of loss of control overeating and binge eating in the relation between ADHD and BMI [103,105].

Four longitudinal studies found a positive association between ADHD symptoms during early-childhood and addictive-like eating behavior in later childhood or adolescence [128–130]. One of these studies [128] found a significant effect of ADHD symptoms on change in eating behaviors from early childhood (around 4 years old) to later childhood (around 7 years). They found that ADHD symptomatology was associated with changes in food responsiveness and emotional overeating when attention symptoms occurred, and only in emotional overeating when hyperactivity symptoms occurred. Conversely, the effect of eating behaviors on changes in ADHD symptomatology from early childhood to later childhood was not significant [128]. According to Sonneville and colleagues (2015) [130], mid- and late-childhood hyperactivity/impulsivity symptoms were correlated with mid- and late-childhood overeating and late-childhood BMI, leading to strong desire for food in early adolescence, correlated with binge eating in mid-adolescence. These results suggest that ADHD hyperactivity/impulsivity symptoms may lead indirectly to binge eating through overeating and desire for food. Similarly, Zhang and colleagues (2020) [131] found that ADHD symptoms at 14 predicted the development of binge eating (OR: 1.27, 95% CI: 1.03–1.57, $p = 0.024$) and purging (OR: 1.35, 95% CI: 1.12–1.64, $p = 0.0016$) behaviors at 16 or 19. However, Yilmaz and colleagues (2017) [129] found that only high inattention combined with high hyperactivity/impulsivity throughout childhood and adolescence predicted disordered eating, such as bulimia nervosa, in late adolescence ($p < 0.01$).

Adults

Thirteen studies focused on the association between ADHD and disordered eating in adults.

In a study with mood disorder outpatients, Woldeyoannes and colleagues (2015) [119] found no association between BE and childhood or adult ADHD (OR = 1.33, 95% CI: 0.40–4.49; OR = 1.05, 95% CI: 0.43–2.58 respectively). However, individuals with both BE and bipolar disorder had significantly higher scores on the WURS (retrospective childhood ADHD scale) and the ASRS (current adult ADHD scale; $p = 0.007$ and $p < 0.001$, respectively). Nazar (2018) [132] found no difference in binge eating between students with and without ADHD ($p = 0.07$), but greater binge eating among those with comorbid ADHD-ED ($p < 0.001$). In individuals with ADHD diagnosis, there was no difference between individuals with and without ED comorbidity in terms of inattentive and hyperactivity/impulsivity symptomatology ($p = 0.53$ and $p = 0.75$ respectively). Van der Oord and colleagues (2017) [133] assessed individuals with severe obesity and found that only comorbid BE was associated with an increase in ADHD symptomatology, mainly inattentive symptoms ($p < 0.01$). In this population, ADHD diagnosis was associated with bulimic symptoms, greater binge eating and higher FA scores [9,11]. Similar results were found when childhood ADHD was retrospectively assessed [11].

Six publications involved samples of individuals with ED. They found that ADHD symptomatology and diagnosis were associated with ED, especially binging/purging behaviors such as BN and AN binge/purge subtype, which were related to inattentive symptoms [122–124]. However, Halevy-Yosef and colleagues (2019) [122] found no differences in ASRS scores between ED with and without binging/purging behavior after Bonferroni correction. ED symptoms related to ADHD symptomatology were mostly addictive-like eating behaviors such as binge eating, purging and loss of control overeating [120,122]. Individuals diagnosed with ED scored higher on disordered eating scales if they also had ADHD. Ferre and colleagues (2017) [134] and Sala and colleagues (2018) [124] reported higher scores on the EAT-40 (assessing disordered eating) and BITE-symptomatology subscale (assessing binge eating symptomatology) among ED patients with than without comorbid ADHD symptomatology. However, while Ferre and colleagues (2017) [134] found similar results for binge-eating severity on the BITE-severity subscale, Sala and colleagues (2018) [124] found no significant difference between individuals with and without ADHD diagnosis. Carlucci and colleagues (2017) reported significant small multivariate effect of ED diagnosis on ASRS-total score ($F(4992) = 2.43$, $p = 0.046$), which was not found for either inattentive or hyperactivity-impulsivity factors ($p = 0.06$ and $p = 0.016$ respectively) [123]. Finally, a high baseline ASRS-total score (>18) was associated with a lower rate of ED recovery at 1 year follow-up (72.1% vs. 46.7%, $p = 0.001$), especially for binging (75.1% vs. 48.5%, $p = 0.003$), purging (74.0% vs. 47.6%, $p = 0.001$) and loss of control overeating (75.6% vs. 47.4%, $p < 0.001$) symptoms. This association remained significative only with ASRS inattentive factor, especially for binging and loss of control overeating. Regression analyses confirmed the predictive role of high ASRS scores on the persistence of disordered eating (OR = 2.59, 95% CI: 1.36–4.91) [121].

Among the six studies that analyzed the correlations between ADHD symptomatology and disordered eating, three were conducted with a student population and found positive correlations between ADHD and bulimic symptoms ($r = 0.34$, $p < 0.001$) [135] and binge eating ([132]: $r = 0.43$, $p < 0.001$; [136]: $r = 0.21$, $p < 0.001$). Similar results were found for patients with ED [122,123] or severe obesity [117], for both inattentive ($r = 0.33$–0.36, $p < 0.001$) and hyperactivity/impulsivity symptoms ($r = 0.22$–0.30, $p < 0.001$). However, Hanson and colleagues (2019) [136] found no correlation between binge eating and ADHD-Inattentive symptoms for men in their student sample ($r = 0.19$, $p > 0.05$).

Five studies conducted regression analyses. Woldeyoannes and colleagues (2015) [119] showed that correlates of BE reported by patients with mood disorder did not include symptomatology of current ADHD or retrospectively assessed childhood ADHD (adjusted Odds Ratio (aOR) = 1.33, 95% CI: 0.40–4.49, aOR = 1.05, 95% CI: 0.43–2.58 respectively). However, the other four studies (with students, patients with severe obesity or with ED) found a significant association between ADHD symptoms/diagnosis and addictive-like eating behavior such as binge eating [11,132,136], disordered eating, bulimic symptoms [134] and FA [11]. Ferre and colleagues (2017) [134] found that patients with ED and ADHD symptoms scored higher on the EAT-40 (assessing disordered eating), the BITE-symptomatology sub-scale (assessing binge eating symptomatology) and BITE-severity sub-scale (assessing binge eating severity). The predictive power of ADHD symptoms on these scales was 14%, 7% and 11% respectively.

Nielsen and colleagues (2017) and Brunault and colleagues (2019) reported that addictive-like eating was more strongly associated with adulthood than childhood ADHD ([117]: the correlation between ADHD symptoms and ED psychopathology scales was stronger for adulthood than childhood ADHD symptoms; [11]: ORs for the association between ADHD symptoms and FA or binge eating were higher for adulthood than childhood ADHD symptoms).

3.3. Indirect Association between ADHD and Disordered Eating through Negative Affectivity and Disrupted Emotion Self-Regulation

3.3.1. ADHD, Negative Affectivity, and Disrupted Emotion Self-Regulation

Children and Adolescents

Two studies conducted with children found that ADHD group had more adolescent with clinical internalizing (i.e., Strengths and Difficulties Questionnaire subscale investigating emotional symptoms and peer problems) (33.3% vs. 16.0%; OR: 2.6; 95% CI: 1.9–3.7) [106] and ADHD symptoms was significatively correlated with depressive symptoms (r = 0.49, $p < 0.0001$) [126].

Adults

Among the studies included in this review, twelve focused on the comorbidity of ADHD symptoms and negative affectivity. Many of them identified a high correlation between ADHD symptoms and anxiety (rated from 0.28, $p < 0.008$ to 0.42, $p < 0.001$) [9,120] and depressive symptoms (rated from 0.29, $p < 0.001$ to 0.38, $p < 0.001$) [9,120,132]. Both inattention and hyperactivity/impulsivity symptoms were correlated with anxiety (r = 0.68, $p < 0.0001$ and r = 0.57, $p < 0.0001$ respectively) and depressive symptoms (r = 0.56–0.63, $p < 0.001$ and r = 0.41–0.51, $p < 0.001$) [117,122]. As reported by several publications [109–111], Jacob and colleagues (2018) [108] showed that individuals who screened positive for adult ADHD (ASRS) had a greater risk for anxiety disorder (33.6% vs. 5.1%, $p < 0.001$), mood disorders such as major depressive disorder (17.1% vs. 2.1%, $p < 0.001$), as well as borderline personality disorder traits (24.0% vs. 2.7% $p < 0.001$). Gorlin and colleagues (2016) [110] did not find an association between ADHD diagnosis and higher anxiety and depressive disorders. However, that study was conducted with psychiatry outpatients who may have been under medication for mood and anxiety disorders.

ADHD symptomatology was also associated with a higher number of stressful life events (3 vs. 1.7 $p < 0.001$) and more frequent perceived stress (85.9% vs. 59.1%, $p < 0.001$) [108]. In addition, ED patients with ADHD symptoms had higher anxiety ($p = 0.02$) [124], higher perceived stress and lower life satisfaction and perceived social support than those with ADHD symptoms [134]. These results indicate high rates of negative affectivity for ADHD individuals. Both inattentive and hyperactivity-impulsivity symptoms were shown to be correlated negatively with effortful control-regulative temperament (inattention: r = −0.556, $p < 0.001$, hyperactivity: r = −0.348, $p < 0.001$ and impulsivity: r = −0.476, $p < 0.001$) [117], and positively with emotion regulation difficulties (r = 0.42, $p < 0.001$ for both inattentive and hyperactivity/impulsivity ADHD symptoms) [135].

3.3.2. Negative Affectivity and Disrupted Emotion Self-Regulation as Mediators in the Association between ADHD and Disordered Eating

Children and Adolescents

Tong and colleagues (2017) [126] clarified the association between ADHD symptoms and addictive-like eating behavior by introducing a potential mediating effect of depression in this relationship. Their data are in line with the hypothesis that ADHD is associated with bulimia and emotional overeating through depression. Koch and colleagues (2020) [137], who investigated the incidence of mental disorders in zero to three-year-old children, suggested that the associations between emotional and affective disorders and ED and ADHD respectively were stronger than the direct association between feeding and eating disorders and ADHD. Indeed, the comorbidity between feeding and eating disorders and ADHD was OR = 15.4 (95% CI 9.6–24.7), whereas the comorbidity between EAD (i.e., emotional and affective disorders) and feeding and eating disorders was OR = 66.8 (95% CI 42.6–104.7) and between EAD and ADHD was OR = 150.7 (95% CI 95.1–238.7). In a sample of overweight or obese children, Gowey and colleagues (2017) [112] found that negative affectivity mediated the relationship between ADHD symptoms and disordered eating. They found significant interactions between body dissatisfaction and both inattentive and hyperactivity/impulsivity ADHD symptoms with an effect on addictive-like eating behavior, especially food preoccupation and oral control.

Adults

A study with adults by Jacob and colleagues (2018) [108] reported a relationship between ADHD symptoms and possible ED, especially uncontrolled eating symptoms largely explained by anxiety disorder (40% for possible ED, 33% for uncontrolled eating) and stressful life events (28% for possible ED, 24% for uncontrolled eating). Another study found that the odds ratio of ADHD-ED association was considerably attenuated after adjusting for comorbid psychiatric disorders (such as mood and anxiety disorders), especially for BN (before adjusting for psychiatric disorders: OR: 28.24, 95% CI: 6.33–126.01; after adjusting for psychiatric disorders: OR: 5.04, 95% CI: 1.15–22.08) [109].

Similarly, Kaisari and colleagues (2018) [138] found that ADHD inattentive and hyperactivity/impulsivity symptoms were both directly and indirectly associated with binge eating through negative affectivity (anxiety, depression and perceived stress). Moreover, after controlling for depressive and anxiety symptoms, there was no longer a correlation between ADHD symptoms and BMI (inattention: $r = -0.031$; $p = 0.350$ and hyperactivity/impulsivity: $r = -0.05$; $p = 0.307$ respectively) [9].

Christian and colleagues (2020) [135] found that negative urgency and emotion self-regulation difficulties were associated with both bulimic and ADHD symptoms, highlighting a possible shared pathway to both ADHD and ED symptoms. Further investigations revealed an impact of negative urgency and emotion self-regulation difficulties in the association between ADHD and ED, especially bulimic symptoms. These results support the hypothesis that negative urgency and emotion dysregulation mediate the association between ADHD and disordered eating.

Williamson and colleagues (2017) [139] investigated the role of emotion self-regulation and ADHD symptoms in the weight loss of obesity patients after bariatric surgery. The interaction between ADHD symptomatology and emotion self-regulation accounted for 13% of the weight loss variance. The results also indicated an inverse association between ADHD symptoms and weight loss 12 months post-surgery among patients with low scores on emotion self-regulation (36.7% of the sample).

4. Discussion

The purpose of the present study was to investigate the association between ADHD symptomatology, disordered eating, especially addictive-like eating behavior, and emotion self-regulation. We noted a significant association with disordered eating (especially addictive-like eating behavior) in 38 publications, eight of them highlighting the mediator role of negative affectivity and emotion dysregulation. This trend was qualified in 19 publications; 16 publications reported differences depending on type of disordered eating behavior, gender or ADHD symptoms. The majority of results thus suggest that both childhood and adulthood ADHD symptomatology is associated with a higher risk of addictive-like eating behavior, especially binging and/or purging, loss of control overeating, emotional overeating and binge eating, bulimic symptoms, as well as a strong desire for food, food responsiveness and food preoccupation. Furthermore, some authors suggest that ADHD symptoms during early childhood lead to disordered eating during later childhood or adolescence.

Several authors found that severe obesity or ED comorbidities increased the strength of the association between ADHD and disordered eating, especially binge eating. Their results indicate that binge eating and purging behavior play a key role in this association, particularly the BN and AN binge/purging subtype. According to Granero and colleagues (2014) [19], this subtype has the highest rate of FA, supporting the hypothesis of a strong association between ADHD and FA. Other publications show that ADHD psychostimulant treatment can improve ED symptoms, suggesting that ADHD and disordered eating share pathways [131,140,141]. According to Zhang and colleagues (2020) [131], low grey matter volume in the orbitofrontal cortex is a mediator between ADHD symptoms and the development of purging, binging/purging behaviors and depression. Moreover, dopaminergic reward pathways are implicated in both ADHD and disordered eating. In ADHD, disruption of the dopaminergic system involves impulse control deficits, inattention and reward sensitivity. These features increase the risk of resorting to food, and even of FA, with palatable food seen as a natural reward [140].

Longitudinal studies demonstrate that a combination of high inattention and hyperactivity/impulsivity symptoms in childhood lead to increasing BMI in late childhood and to ED in adolescence through addictive-like eating behaviors. However, some publications reported that disordered eating is particularly linked to inattentive symptoms. It is not possible in this systematic review to draw clear conclusions about the involvement of inattentive and/or hyperactivity/impulsivity ADHD symptoms in the association between ADHD and addictive-like eating behavior, and further investigations are needed.

The second aim of this systematic review was to examine the mediator role of negative affectivity and emotion self-regulation in the association between ADHD and addictive-like eating behavior. We showed that high ADHD severity would be associated with a high risk of disrupted emotion regulation, negative affectivity (comorbid anxiety and mood disorders, and perceived stress), which mediate the link between ADHD symptomatology and disordered eating, especially addictive-like eating behavior. Some studies show that ADHD symptoms are associated with high emotion dysregulation [117,135], impacting the ability to cope with daily difficulties, and involving greater negative affectivity and a higher risk of mood disorder comorbidity. As expected, some studies indicated that negative affectivity and emotion dysregulation mediates the association between ADHD and addictive-like eating behavior [108,109,112,126,135,137–139], supported by publications which showed association between ADHD and emotional eating [126–128]. Negative affectivity and lack of emotion regulation, commonly observed in ADHD, would trigger food intake. Results also suggest that individuals with ADHD tend to act rashly when experiencing negative affectivity (negative urgency), which is associated with disordered eating, such as binging [135].

The studies included in this systematic review suggest a pattern of links between ADHD symptomatology, negative affectivity, emotion regulation, and addictive-like eating behaviors (Figure 3). ADHD symptomatology would lead to greater difficulty coping with daily life, due to emotion dysregulation. Due to their inability to regulate negative affectivity, people with ADHD tend to run away from them by seeking positive sensations such as eating. Impulsivity and negative urgency would further encourage disordered eating behaviors such as binge eating, leading to greater BMI. The urge to eat when in a negative affectivity indicates an addictive process involving similar dopaminergic pathways to ADHD.

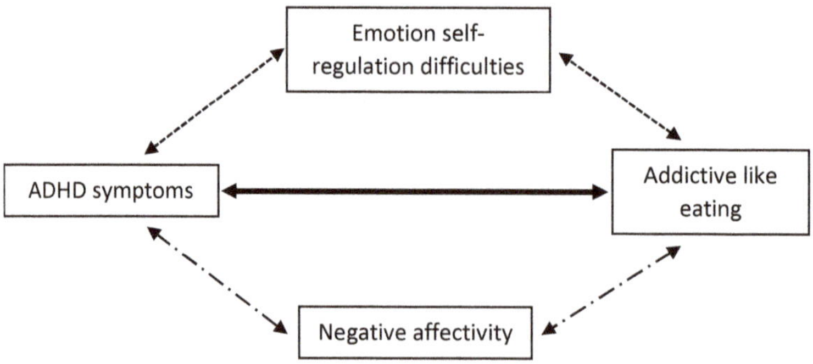

Figure 3. Model illustrating association between ADHD symptoms and disordered eating mediated by emotion self-regulation difficulties and negative affectivity. ⟷: [9,11,103,105,112,113,116,117,120–124,126,127,129–138]. ⇠⇢ (dashed): [135]. ⇠·⇢ (dash-dot): [108,109,112,126,137,138].

A better understanding of the mechanisms underlying the association between ADHD symptomatology and disordered eating suggests new approaches to psychological interventions. In view of the high incidence of disordered eating among people with ADHD, it seems important to identify any maladaptive eating behavior. Interventions aimed at assessing and targeting emotion dysregulation could be an appropriate way of preventing disordered eating behavior and FA, as well as comorbid anxiety and depression disorders. Integrative cognitive-affective therapy (ICAT) adapted to BN and BED targeting emotion regulation (identification of emotional states, especially negative ones, self-monitoring of eating patterns, behaviors and emotions) has been shown to be effective in reducing the frequency of binge eating [142]. Similarly, early detection of ADHD symptoms among people with disordered eating would enable suitable intervention programs to be set up, particularly to treat poor impulse control and emotion dysregulation. A number of personal characteristics that have a negative impact on ED therapy outcome should be identified, including the presence of ADHD symptoms. ADHD symptomology could be a predictor of the outcome of bariatric surgery in individuals with severe obesity [139]. It is thus essential to identify inattention and hyperactivity/impulsivity symptoms in order to provide appropriate joint interventions. For example, Cortese and colleagues (2007) advocated a dual intervention of medication (to reduce comorbid ADHD and ED symptomatology) and cognitive behavioral therapy (to control impulsive and maladaptive behavior, and emotion regulation) [123,143].

This review has a number of limitations. First, it does not provide any causal link. Indeed, as far as we know, no study investigated the effect of ADHD negative affectivity or emotion dysregulation therapeutic interventions on addictive like eating behavior. This link could be of interest for further studies. Moreover, this systematic review includes only qualitative and no quantitative analyses. The variety of populations studied (individuals diagnosed with ADHD, different types of disordered eating, severe obesity, students, etc.) and methods used to assess ADHD and disordered eating make it difficult to draw clear conclusions. In addition, some studies were based on ADHD diagnosis criteria of the DSM-IV-TR and others on DSM5 criteria, with a change of symptom onset from 7 to 12 years of age, making it difficult to compare results. Another limitation involves publications which did not provide necessary information about current medication. Indeed, medication can conceal symptoms of disrupted emotion and eating, so there is an impact on results of investigations. Furthermore, as only a few studies assessed food addiction directly, we included those involving addictive-like eating symptoms and various aspects of food addiction. It should be noted that the addictive nature of food is still under debate, notably whether features of substance addiction can be applied to food, the addictive power of palatable food, common features such as tolerance and withdrawal, and the distinction between food addiction and binge eating. However, people presenting with this type of pathological eating suffer in similar ways as those with substance use disorder, including "feelings of deprivation when the substance is withheld, a propensity to relapse during periods of abstinence, and consumption that persists despite awareness of negative health, social, financial, or other consequences" [144]. The publications reviewed have their own limitations. According to the DSM-5, childhood ADHD symptoms are used to diagnose adult ADHD. However, several studies involving adult ADHD did not investigate childhood symptoms. Some studies only used self-administered questionnaires to assess disordered eating and ADHD. This type of assessment is not as efficient as an interview with a clinician.

Future studies should investigate in greater depth emotion regulation difficulties in comorbid adult ADHD and addictive-like eating behavior, and the involvement of specific ADHD symptoms such as inattention, impulsivity and hyperactivity. This could clarify which emotion regulation strategies and ADHD symptoms should be targeted in clinical interventions. It would be interesting to investigate specific symptoms of ED in order to identify common sub-groups. The majority of studies of ADHD symptomatology in people with disordered eating were conducted with female samples, although some authors noted male-female differences in the relationship between ADHD and disordered eating. Future studies should thus investigate distinctive male characteristics in order to determine whether clinical interventions should be gender-specific. In addition, in order to identify causal links between ADHD symptomatology and addictive-like eating behavior, more longitudinal studies are needed. This would make it possible to set up early interventions with children with ADHD and investigate the impact on ADHD symptomatology, eating behaviors and risk of obesity in adolescence and adulthood. An important area of research would be to focus on the interplay between dysregulation of sleep, weight gain and emotional dysregulation, as it has been suggested by some authors [145] that alterations in sleep/arousal may be related to ADHD and weight gain/disordered eating and sleep deprivation may exacerbate emotional dysregulation [146].

5. Conclusions

Despite its limitations, this review provides information about the co-occurrence of ADHD symptoms and addictive-like eating behavior. It confirms the strong association between ADHD, emotion dysregulation and binge eating/addictive-like eating behavior in both clinical (i.e., people with ED or ADHD) and non-clinical populations. The data support the hypothesis of a mediating role of negative affectivity and emotion self-regulation difficulties in the association between addictive-like eating behavior and ADHD. This review paves the way for future therapeutic interventions that could improve clinical outcomes for people with ADHD and disordered eating.

Author Contributions: Writing—original draft preparation, S.E.A.; writing—review and editing, S.E.A., S.C., N.B., C.R., A.D.L., S.B., P.B.; study design and concept, S.B., P.B.; supervision, S.B., P.B. All authors have read and agreed to the published version of the manuscript.

Funding: This research received no external funding.

Conflicts of Interest: S.C. declares reimbursement for travel and accommodation expenses from the Association for Child and Adolescent Central Health (ACAMH) in relation to lectures delivered for ACAMH, and from Healthcare Convention for educational activity on ADHD. N.B. reports personal fees from Lundbeck, Astra-Zeneca and D&A Pharma, unrelated to the submitted work. P.B. reports personal fees and non-financial support from Lundbeck, personal fees from Astra-Zeneca and D&A Pharma, unrelated to the submitted work. S.E.A., C.R., A.D.L. and S.B. declares no conflict of interest.

References

1. Kooij, J.J.S.; Bijlenga, D.; Salerno, L.; Jaeschke, R.; Bitter, I.; Balázs, J.; Thome, J.; Dom, G.; Kasper, S.; Nunes Filipe, C.; et al. Updated European Consensus Statement on diagnosis and treatment of adult ADHD. *Eur. Psychiatry* **2019**, *56*, 14–34. [CrossRef] [PubMed]
2. Thomas, R.; Sanders, S.; Doust, J.; Beller, E.; Glasziou, P. Prevalence of attention-deficit/hyperactivity disorder: A systematic review and meta-analysis. *Pediatrics* **2015**, *135*, e994-e1001. [CrossRef] [PubMed]
3. Polanczyk, G.; De Lima, M.S.; Horta, B.L.; Biederman, J.; Rohde, L.A. The worldwide prevalence of ADHD: A systematic review and metaregression analysis. *Am. J. Psychiatry* **2007**, *164*, 942–948. [CrossRef] [PubMed]
4. Faraone, S.V.; Biederman, J.; Mick, E. The age-dependent decline of attention deficit hyperactivity disorder: A meta-analysis of follow-up studies. *Psychol. Med.* **2006**, *36*, 159–165. [CrossRef]
5. Cortese, S.; Adamo, N.; Del Giovane, C.; Mohr-Jensen, C.; Hayes, A.J.; Carucci, S.; Atkinson, L.Z.; Tessari, L.; Banaschewski, T.; Coghill, D.; et al. Comparative efficacy and tolerability of medications for attention-deficit hyperactivity disorder in children, adolescents, and adults: A systematic review and network meta-analysis. *The Lancet Psychiatry* **2018**, *5*, 727–738. [CrossRef]
6. Daley, D.; Van Der Oord, S.; Ferrin, M.; Cortese, S.; Danckaerts, M.; Doepfner, M.; Van Den Hoofdakker, B.J.; Coghill, D.; Thompson, M.; Asherson, P.; et al. Practitioner review: Current best practice in the use of parent training and other behavioural interventions in the treatment of children and adolescents with attention deficit hyperactivity disorder. *J. Child Psychol. Psychiatry Allied Discip.* **2017**, *59*, 932–947. [CrossRef]
7. Cortese, S.; Moreira-Maia, C.R.; St Fleur, D.; Morcillo-Peñalver, C.; Rohde, L.A.; Faraone, S.V. Association between ADHD and obesity: A systematic review and meta-analysis. *Am. J. Psychiatry* **2016**, *173*, 34–43. [CrossRef]

8. Nigg, J.T.; Johnstone, J.M.; Musser, E.D.; Long, H.G.; Willoughby, M.; Shannon, J. Attention-deficit/hyperactivity disorder (ADHD) and being overweight/obesity: New data and meta-analysis. *Clin. Psychol. Rev.* **2016**, *43*, 67–79. [CrossRef]
9. Nazar, B.P.; de Sousa Pinna, C.M.; Suwwan, R.; Duchesne, M.; Freitas, S.R.; Sergeant, J.; Mattos, P. ADHD Rate in Obese Women with Binge Eating and Bulimic Behaviors from a Weight-Loss Clinic. *J. Atten. Disord.* **2016**, *20*, 610–616. [CrossRef]
10. Hilbert, A.; Kurz, S.; Dremmel, D.; Weihrauch Blüher, S.; Munsch, S.; Schmidt, R. Cue reactivity, habituation, and eating in the absence of hunger in children with loss of control eating and attention-deficit/hyperactivity disorder. *Int. J. Eat. Disord.* **2018**, *51*, 223–232. [CrossRef]
11. Brunault, P.; Frammery, J.; Montaudon, P.; De Luca, A.; Hankard, R.; Ducluzeau, P.H.; Cortese, S.; Ballon, N. Adulthood and childhood ADHD in patients consulting for obesity is associated with food addiction and binge eating, but not sleep apnea syndrome. *Appetite* **2019**, *136*, 25–32. [CrossRef] [PubMed]
12. Gearhardt, A.N.; Corbin, W.R.; Brownell, K.D. Preliminary validation of the Yale Food Addiction Scale. *Appetite* **2009**, *52*, 430–436. [CrossRef] [PubMed]
13. Romo, L.; Ladner, J.; Kotbagi, G.; Morvan, Y.; Saleh, D.; Tavolacci, M.P.; Kern, L. Attention-deficit hyperactivity disorder and addictions (substance and behavioral): Prevalence and characteristics in a multicenter study in France. *J. Behav. Addict.* **2018**, *7*, 743–751. [CrossRef] [PubMed]
14. Gearhardt, A.N.; Corbin, W.R.; Brownell, K.D. Development of the Yale Food Addiction Scale Version 2.0. *Psychol. Addict. Behav.* **2016**, *30*, 113–121. [CrossRef]
15. Mills, J.G.; Thomas, S.J.; Larkin, T.A.; Deng, C. Overeating and food addiction in Major Depressive Disorder: Links to peripheral dopamine. *Appetite* **2020**, *148*, 104586. [CrossRef]
16. Mies, G.W.; Treur, J.L.; Larsen, J.K.; Halberstadt, J.; Pasman, J.A.; Vink, J.M. The prevalence of food addiction in a large sample of adolescents and its association with addictive substances. *Appetite* **2017**, *118*, 97–105. [CrossRef]
17. Mason, S.M.; Flint, A.J.; Roberts, A.L.; Agnew-Blais, J.; Koenen, K.C.; Rich-Edwards, J.W. Posttraumatic stress disorder symptoms and food addiction in women by timing and type of trauma exposure. *JAMA Psychiatry* **2014**, *71*, 1271–1278. [CrossRef]
18. American Psychiatric Association. *Diagnostic and Statistical Manual of Mental Disorders*, 5th ed.; American Psychiatric Association: Washington, DC, USA, 2013; ISBN 9782294743382.
19. Granero, R.; Hilker, I.; Agüera, Z.; Jiménez-Murcia, S.; Sauchelli, S.; Islam, M.A.; Fagundo, A.B.; Sánchez, I.; Riesco, N.; Dieguez, C.; et al. Food Addiction in a Spanish Sample of Eating Disorders: DSM-5 Diagnostic Subtype Differentiation and Validation Data. *Eur. Eat. Disorders Rev.* **2014**, *22*, 389–396. [CrossRef]
20. El Archi, S.; Brunault, P.; Ballon, N.; Réveillère, C.; Barrault, S. Differential association between food craving, food addiction and eating-related characteristics in persons at risk for eating disorders. *Rev. Eur. Psychol. Appl.* **2020**, *7*. [CrossRef]
21. Maxwell, A.L.; Gardiner, E.; Loxton, N.J. Investigating the relationship between reward sensitivity, impulsivity, and food addiction: A systematic review. *Eur. Eat. Disord. Rev.* **2020**, *22*, 368–384. [CrossRef]
22. Steadman, K.M.; Knouse, L.E. Is the Relationship between ADHD Symptoms and Binge Eating Mediated by Impulsivity? *J. Atten. Disord.* **2016**, *20*, 907–912. [CrossRef] [PubMed]
23. Ortal, S.; van de Geurt, G.; Johan, F.; Itai, B.; Nir, Y.; Iliyan, I.; van den Wim, B. The role of different aspects of impulsivity as independent risk factors for substance use disorders in patients with ADHD: A review. *Curr. Drug Abuse Rev.* **2015**, *8*, 119–133. [CrossRef]
24. Romo, L.; Rémond, J.J.; Coeffec, A.; Kotbagi, G.; Plantey, S.; Boz, F.; Kern, L.; Coeffec, Á.A.; Kotbagi, Á.G.; Plantey, Á.S.; et al. Gambling and Attention Deficit Hyperactivity Disorders (ADHD) in a Population of French Students. *J Gambl. Stud.* **2015**, *31*, 1261–1272. [CrossRef]
25. Whiteside, S.P.; Lynam, D.R. The five factor model and impulsivity: Using a structural model of personality to understand impulsivity. *Pers. Individ. Dif.* **2001**, *30*, 669–689. [CrossRef]
26. Egan, T.E.; Dawson, A.E.; Wymbs, B.T. Substance Use in Undergraduate Students with Histories of Attention-Deficit/Hyperactivity Disorder (ADHD): The Role of Impulsivity. *Subst. Use Misuse* **2017**, *52*, 1375–1386. [CrossRef]
27. Brunault, P.; Mathieu, S.; Faussat, C.; Barrault, S.; Varescon, I. Impulsivity facets and cognitive distortions associated with problem gambling: Differences between ADHD and non-ADHD gamblers. *Rev. Eur. Psychol. Appl.* **2020**, *70*, 100559. [CrossRef]

28. Van Emmerik-Van Oortmerssen, K.; Van De Glind, G.; Van Den Brink, W.; Smit, F.; Crunelle, C.L.; Swets, M.; Schoevers, R.A. Prevalence of attention-deficit hyperactivity disorder in substance use disorder patients: A meta-analysis and meta-regression analysis. *Drug Alcohol Depend.* **2012**, *122*, 11–19. [CrossRef] [PubMed]
29. Anker, E.; Bendiksen, B.; Heir, T. Comorbid psychiatric disorders in a clinical sample of adults with ADHD, and associations with education, work and social characteristics: A cross-sectional study. *BMJ Open* **2018**, *8*. [CrossRef]
30. Mathews, C.L.; Morrell, H.E.R.; Molle, J.E. Video game addiction, ADHD symptomatology, and video game reinforcement. *Am. J. Drug Alcohol Abuse* **2019**, *45*, 67–76. [CrossRef]
31. Evren, B.; Evren, C.; Dalbudak, E.; Topcu, M.; Kutlu, N. Relationship of internet addiction severity with probable ADHD and difficulties in emotion regulation among young adults. *Psychiatry Res.* **2018**, *269*, 494–500. [CrossRef]
32. Mestre-Bach, G.; Steward, T.; Potenza, M.N.; Granero, R.; Fernández-Aranda, F.; Mena-Moreno, T.; Magaña, P.; Vintró-Alcaraz, C.; del Pino-Gutiérrez, A.; Menchón, J.M.; et al. The Role of ADHD Symptomatology and Emotion Dysregulation in Gambling Disorder. *J. Atten. Disord.* **2019**. [CrossRef]
33. Grall-Bronnec, M.; Wainstein, L.; Augy, J.; Bouju, G.; Feuillet, F.; Vénisse, J.L.; Sébille-Rivain, V. Attention deficit hyperactivity disorder among pathological and at-risk gamblers seeking treatment: A hidden disorder. *Eur. Addict. Res.* **2011**, *17*, 231–240. [CrossRef] [PubMed]
34. Young, K.S. *Caught in the Net: How to Recognize the Sign of Internet Addiction and a Winning Strategy for Recovery*; John Wiley & Sons: New York, NY, USA, 1998; ISBN 0471191590.
35. Gross, J.J.; Thompson, R.A. Emotion Regulation: Conceptual Foundations. Available online: https://www.researchgate.net/publication/303248970_Emotion_Regulation_Conceptual_Foundations (accessed on 4 June 2020).
36. Sheppes, G.; Suri, G.; Gross, J.J. Emotion Regulation and Psychopathology. *Annu. Rev. Clin. Psychol.* **2015**, *11*, 379–405. [CrossRef] [PubMed]
37. Corbisiero, S.; Stieglitz, R.D.; Retz, W.; Rösler, M. Is emotional dysregulation part of the psychopathology of ADHD in adults? *ADHD Atten. Deficit Hyperact. Disord.* **2013**, *5*, 83–92. [CrossRef] [PubMed]
38. Fox, H.C.; Axelrod, S.R.; Paliwal, P.; Sleeper, J.; Sinha, R. Difficulties in emotion regulation and impulse control during cocaine abstinence. *Drug Alcohol Depend.* **2007**, *89*, 298–301. [CrossRef] [PubMed]
39. Prefit, A.-B.; Mirela Cândea, D.; Szentagotai-Tătar, A. Emotion regulation across eating pathology: A meta-analysis. *Appetite* **2019**, *143*, 104438. [CrossRef]
40. Masi, G.; Fantozzi, P.; Muratori, P.; Bertolucci, G.; Tacchi, A.; Villafranca, A.; Pfanner, C.; Cortese, S. Emotional dysregulation and callous unemotional traits as possible predictors of short-term response to methylphenidate monotherapy in drug-naïve youth with ADHD. *Compr. Psychiatry* **2020**, *100*, 152178. [CrossRef]
41. Lenzi, F.; Cortese, S.; Harris, J.; Masi, G. Pharmacotherapy of emotional dysregulation in adults with ADHD: A systematic review and meta-analysis. *Neurosci. Biobehav. Rev.* **2018**, *84*, 359–367. [CrossRef] [PubMed]
42. Beheshti, A.; Chavanon, M.L.; Christiansen, H. Emotion dysregulation in adults with attention deficit hyperactivity disorder: A meta-analysis. *BMC Psychiatry* **2020**, *20*, 120. [CrossRef] [PubMed]
43. Seymour, K.E.; Chronis-Tuscano, A.; Iwamoto, D.K.; Kurdziel, G.; MacPherson, L. Emotion regulation mediates the association between ADHD and depressive symptoms in a community sample of youth. *J. Abnorm. Child Psychol.* **2014**, *42*, 611–621. [CrossRef]
44. Bauer, B.W.; Gustafsson, H.C.; Nigg, J.; Karalunas, S.L. Working Memory Mediates Increased Negative Affect and Suicidal Ideation in Childhood Attention-Deficit/Hyperactivity Disorder. *J. Psychopathol. Behav. Assess.* **2018**, *40*, 180–193. [CrossRef] [PubMed]
45. Karalunas, S.L.; Fair, D.; Musser, E.D.; Aykes, K.; Iyer, S.P.; Nigg, J.T. Subtyping attention-deficit/hyperactivity disorder using temperament dimensions: Toward biologically based nosologic criteria. *JAMA Psychiatry* **2014**, *71*, 1015–1024. [CrossRef]
46. Emilsson, M.; Gustafsson, P.; Öhnström, G.; Marteinsdottir, I. Impact of personality on adherence to and beliefs about ADHD medication, and perceptions of ADHD in adolescents. *BMC Psychiatry* **2020**, *2*. [CrossRef] [PubMed]
47. Wong, C.F.; Silva, K.; Kecojevic, A.; Schrager, S.M.; Bloom, J.J.; Iverson, E.; Lankenau, S.E. Coping and emotion regulation profiles as predictors of nonmedical prescription drug and illicit drug use among high-risk young adults. *Drug Alcohol Depend.* **2013**, *132*, 165–171. [CrossRef]

48. Jacobs, D.F. An overarching theory of addictions: A new paradigm for understanding and treating addictive behaviours. In Proceedings of the National Academy of Sciences Meeting, Washington, DC, USA, 27 April 1998.
49. Ahn, J.S.; Min, S.; Kim, M.H. The role of uncontrolled eating and screen time in the link of attention deficit hyperactivity disorder with weight in late childhood. *Psychiatry Investig.* **2017**, *14*, 808–816. [CrossRef]
50. Donnchadha, S.; Bramham, J.; Greene, C. Rethinking the association between overweight/obesity and ADHD in children: A longitudinal and psychosocial perspective. *Ir. J. Psychol. Med.* **2020**, 1–14. [CrossRef]
51. Dreber, H.; Reynisdottir, S.; Angelin, B.; Hemmingsson, E. Who is the treatment-seeking young adult with severe obesity: A comprehensive characterization with emphasis on mental health. *PLoS ONE* **2015**, *1*. [CrossRef] [PubMed]
52. Eiffener, E.; Eli, K.; Ek, A.; Sandvik, P.; Somaraki, M.; Kremers, S.; Sleddens, E.; Nowicka, P. The influence of preschoolers' emotional and behavioural problems on obesity treatment outcomes: Secondary findings from a randomized controlled trial. *Pediatr. Obes.* **2019**, *14*, e12556. [CrossRef]
53. Fruchter, E.; Marom-Harel, H.; Fenchel, D.; Kapra, O.; Ginat, K.; Portuguese, S.; Weiser, M. Functioning of Young Adults with ADHD in the Military. *J. Atten. Disord.* **2019**, *23*, 1470–1474. [CrossRef]
54. Goulardins, J.B.; Rigoli, D.; Piek, J.P.; Kane, R.; Palácio, S.G.; Casella, E.B.; Nascimento, R.O.; Hasue, R.H.; Oliveira, J.A. The relationship between motor skills, ADHD symptoms, and childhood body weight. *Res. Dev. Disabil.* **2016**, *55*, 279–286. [CrossRef]
55. Granato, M.F.; Ferraro, A.A.; Lellis, D.M.; Casella, E.B. Associations between attention-deficit hyperactivity disorder (ADHD) treatment and patient nutritional status and height. *Behav. Neurol.* **2018**, *2*, 7341529. [CrossRef] [PubMed]
56. Guenzel, N.; Schober, D.J. Psychiatric Comorbidities and BMI: An Exploratory Analysis. *Issues Ment. Health Nurs.* **2017**, *38*, 698–704. [CrossRef] [PubMed]
57. Hagman, E.; Danielsson, P.; Brandt, L.; Svensson, V.; Ekbom, A.; Marcus, C. Childhood Obesity, Obesity Treatment Outcome, and Achieved Education: A Prospective Cohort Study. *J. Adolesc. Health* **2017**, *61*, 508–513. [CrossRef] [PubMed]
58. Hanć, T.; Słopień, A.; Wolańczyk, T.; Szwed, A.; Czapla, Z.; Durda, M.; Dmitrzak-Węglarz, M.; Ratajczak, J. Attention-deficit/hyperactivity disorder is related to decreased weight in the preschool period and to increased rate of overweight in school-age boys. *J. Child Adolesc. Psychopharmacol.* **2015**, *25*, 691–700. [CrossRef]
59. Inoue, Y.; Howard, A.G.; Stickley, A.; Yazawa, A.; Gordon-Larsen, P. Sex and racial/ethnic differences in the association between childhood attention-deficit/hyperactivity disorder symptom subtypes and body mass index in the transition from adolescence to adulthood in the United States. *Pediatr. Obes.* **2019**, *14*, e12498. [CrossRef]
60. Aguirre-Castaneda, R.L.; Kumar, S.; Voigt, R.G.; Leibson, C.L.; Barbaresi, W.J.; Weaver, A.L.; Killian, J.M.; Katusic, S.K. Childhood Attention-Deficit/Hyperactivity Disorder, Sex, and Obesity: A Longitudinal Population-Based Study. *Mayo Clin. Proc.* **2016**, *91*, 352–361. [CrossRef]
61. Jang, B.Y.; Bu, S.Y. Nutritional Status of Korean Children and Adolescents with Attention Deficit Hyperactivity Disorder (ADHD). *Clin. Nutr. Res.* **2017**, *6*, 112. [CrossRef]
62. Kummer, A.; Barbosa, I.G.; Rodrigues, D.H.; Rocha, N.P.; Da Silva Rafael, M.; Pfeilsticker, L.; Simõese Silva, A.C.; Teixeira, A.L. Frequency of overweight and obesity in children and adolescents with autism and attention deficit/hyperactivity disorder. *Rev. Paul. Pediatr.* **2016**, *34*, 71–77. [CrossRef]
63. Kwak, Y.S.; Jung, Y.E.; Kim, M.D. Prevalence and correlates of attention-deficit hyperactivity disorder symptoms in korean college students. *Neuropsychiatr. Dis. Treat.* **2015**, *11*, 797–802. [CrossRef]
64. Leib, S.; Gilon Mann, T.; Stein, D.; Vusiker, I.; Tokatly Latzer, I.; Ben-Ami, M.; Feigin, A.; Dubnov-Raz, G. High prevalence of attention-deficit/hyperactivity disorder in adolescents with severe obesity seeking bariatric surgery. *Acta Paediatr. Int. J. Paediatr.* **2019**, *109*, 581–586. [CrossRef]
65. Lentferink, Y.E.; Van De Garde, E.M.W.; Knibbe, C.A.J.; Van Der Vorst, M.M.J. Psychostimulants: Influence on body mass index and height in a pediatric population with attention-deficit/hyperactivity disorder? *J. Child Adolesc. Psychopharmacol.* **2018**, *28*, 530–536. [CrossRef]
66. Mellström, E.; Forsman, C.; Engh, L.; Hallerbäck, M.U.; Wikström, S. Methylphenidate and Reduced Overweight in Children with ADHD. *J. Atten. Disord.* **2020**, *24*, 246–254. [CrossRef]

67. Mohammadi, M.R.; Mostafavi, S.-A.; Hooshyari, Z.; Khaleghi, A.; Ahmadi, N. Body Mass Index Status across Different Psychiatric Disorders in a National Survey amongst Children and Adolescents: To Identify the Role of Gender. *Iran J Psychatry* **2019**, *14*, 253–264. [CrossRef]
68. Pauli-Pott, U.; Reinhardt, A.; Bagus, E.; Wollenberg, B.; Schroer, A.; Heinzel-Gutenbrunner, M.; Becker, K. Psychosocial risk factors underlie the link between attention deficit hyperactivity symptoms and overweight at school entry. *Eur. Child Adolesc. Psychiatry* **2017**, *26*, 67–73. [CrossRef] [PubMed]
69. Pérez-Bonaventura, I.; Granero, R.; Ezpeleta, L. The relationship between weight status and emotional and behavioral problems in Spanish preschool children. *J. Pediatr. Psychol.* **2015**, *40*, 455–463. [CrossRef]
70. Racicka, E.; Hanć, T.; Giertuga, K.; Bryńska, A.; Wolańczyk, T. Prevalence of Overweight and Obesity in Children and Adolescents with ADHD: The Significance of Comorbidities and Pharmacotherapy. *J. Atten. Disord.* **2018**, *22*, 1095–1108. [CrossRef] [PubMed]
71. Bisset, M.; Rinehart, N.; Sciberras, E. Body dissatisfaction and weight control behaviour in children with ADHD: A population-based study. *Eur. Child Adolesc. Psychiatry* **2019**, *28*, 1507–1516. [CrossRef] [PubMed]
72. Sciberras, E.; Bisset, M.; Hazell, P.; Nicholson, J.M.; Anderson, V.; Lycett, K.; Jongeling, B.; Efron, D. Health-related impairments in young children with ADHD: A community-based study. *Child. Care. Health Dev.* **2016**, *42*, 709–717. [CrossRef]
73. Skoglund, C.; Kopp Kallner, H.; Skalkidou, A.; Wikström, A.K.; Lundin, C.; Hesselman, S.; Wikman, A.; Sundström Poromaa, I. Association of Attention-Deficit/Hyperactivity Disorder with Teenage Birth among Women and Girls in Sweden. *JAMA Netw. Open* **2019**, *2*, e1912463. [CrossRef] [PubMed]
74. Taymur, I.; Budak, E.; Onen, S.; Bicer, B.; Dilektasll, E.; Cayci, M.; Demirci, H.; Gungor, B.B. The Relationship between Childhood and Adult Attention-Deficit-Hyperactivity Disorder and General Psychop athological Features in Individuals Who Apply for Bariatric Surgery. *Bariatr. Surg. Pract. Patient Care* **2016**, *11*, 116–122. [CrossRef]
75. Türkoğlu, S.; Çetin, F.H. The relationship between chronotype and obesity in children and adolescent with attention deficit hyperactivity disorder. *Chronobiol. Int.* **2019**, *36*, 1138–1147. [CrossRef]
76. Van Eck, K.; Morse, M.; Flory, K. The Role of Body Image in the Link between ADHD and Depression Symptoms among College Students. *J. Atten. Disord.* **2018**, *22*, 435–445. [CrossRef] [PubMed]
77. Wynchank, D.; Bijlenga, D.; Lamers, F.; Kooij, J.J.S.; Bron, T.I.; Beekman, A.T.F.; Penninx, B.W.J.H. The Association between Metabolic Syndrome, Obesity-Related Outcomes, and ADHD in Adults with Comorbid Affective Disorders. *J. Atten. Disord.* **2018**, *22*, 460–471. [CrossRef] [PubMed]
78. Zhang, A.; Li, S.; Zhang, Y.; Jiang, F.; Jin, X.; Ma, J. Nocturnal enuresis in obese children: A nation-wide epidemiological study from China. *Sci. Rep.* **2019**, *9*, 8414. [CrossRef]
79. Türkoğlu, S.; Bilgiç, A.; Akça, Ö.F. ADHD symptoms, breast-feeding and obesity in children and adolescents. *Pediatr. Int.* **2015**, *57*, 546–551. [CrossRef] [PubMed]
80. Bowling, A.B.; Tiemeier, H.W.; Jaddoe, V.W.V.; Barker, E.D.; Jansen, P.W. ADHD symptoms and body composition changes in childhood: A longitudinal study evaluating directionality of associations. *Pediatr. Obes.* **2018**, *13*, 567–575. [CrossRef] [PubMed]
81. Chen, Q.; Kuja-Halkola, R.; Sjölander, A.; Serlachius, E.; Cortese, S.; Faraone, S.V.; Almqvist, C.; Larsson, H. Shared familial risk factors between attention-deficit/hyperactivity disorder and overweight/obesity—A population-based familial coaggregation study in Sweden. *J. Child Psychol. Psychiatry Allied Discip.* **2017**, *58*, 711–718. [CrossRef]
82. Cho, Y.J.; Choi, R.; Park, S.; Kwon, J. Parental smoking and depression, and attention-deficit hyperactivity disorder in children and adolescents: Korean national health and nutrition examination survey 2005–2014. *Asia-Pac. Psychiatry* **2018**, *10*, e12327. [CrossRef]
83. Çöl, N.; Gökçen, C.; Demırcıoğlu Kiliç, B.; Karadağ, M. Prevalence of obesity/hypertension in children and adolescents with ADHD and evaluation of total body composition. *Anatol. J. Psychiatry* **2019**, *20*, 93–100. [CrossRef]
84. Çolpan, M.; Eray, Ş.; Eren, E.; Vural, A.P. Perceived Expressed Emotion, Emotional and Behavioral Problems and Self-Esteem in Obese Adolescents: A Case-Control Study. *J. Clin. Res. Pediatr. Endocrinol.* **2018**, *10*, 357–363. [CrossRef]
85. Cook, B.G.; Li, D.; Heinrich, K.M. Obesity, Physical Activity, and Sedentary Behavior of Youth with Learning Disabilities and ADHD. *J. Learn. Disabil.* **2015**, *48*, 563–576. [CrossRef] [PubMed]

86. Corvey, K.; Menear, K.S.; Preskitt, J.; Goldfarb, S.; Menachemi, N. Obesity, Physical Activity and Sedentary Behaviors in Children with an Autism Spectrum Disorder. *Matern. Child Health J.* **2016**, *20*, 466–476. [CrossRef] [PubMed]
87. Karjalainen, L.; Gillberg, C.; Råstam, M.; Wentz, E. Eating disorders and eating pathology in young adult and adult patients with ESSENCE. *Compr. Psychiatry* **2016**, *66*, 79–86. [CrossRef] [PubMed]
88. Linthavong, O.; O'Shea, T.M.; Allred, E.; Perrin, E.; Bauserman, M.; Joseph, R.M.; Leviton, A.; Heeren, T.C.; Kuban, K.C.K. Neurocognitive and Health Correlates of Overweight and Obesity among Ten-Year-Old Children Born Extremely Preterm. *J. Pediatr.* **2018**, *200*, 84–90e4. [CrossRef]
89. Vitelli, O.; Tabarrini, A.; Miano, S.; Rabasco, J.; Pietropaoli, N.; Forlani, M.; Parisi, P.; Villa, M.P. Impact of obesity on cognitive outcome in children with sleep-disordered breathing. *Sleep Med.* **2015**, *16*, 625–630. [CrossRef]
90. Bleck, J.R.; DeBate, R.D.; Olivardia, R. The Comorbidity of ADHD and Eating Disorders in a Nationally Representative Sample. *J. Behav. Health Serv. Res.* **2015**, *42*, 437–451. [CrossRef]
91. Do, E.K.; Haberstick, B.C.; Williams, R.B.; Lessem, J.M.; Smolen, A.; Siegler, I.C.; Fuemmeler, B.F. The role of genetic and environmental influences on the association between childhood ADHD symptoms and BMI. *Int. J. Obes.* **2019**, *43*, 33–42. [CrossRef]
92. Grant, J.E.; Redden, S.A.; Lust, K.; Chamberlain, S.R. Nonmedical use of stimulants is associated with riskier sexual practices and other forms of impulsivity. *J. Addict. Med.* **2018**, *12*, 474–480. [CrossRef]
93. Groß-Lesch, S.; Dempfle, A.; Reichert, S.; Jans, T.; Geissler, J.; Kittel-Schneider, S.; Nguyen, T.T.; Reif, A.; Lesch, K.P.; Jacob, C.P. Sex- and Subtype-Related Differences in the Comorbidity of Adult ADHDs. *J. Atten. Disord.* **2016**, *20*, 855–866. [CrossRef]
94. Guerdjikova, A.I.; Blom, T.J.; Mori, N.; Matthews, A.; Cummings, T.; Casuto, L.L.; McElroy, S.L. Lisdexamfetamine in Pediatric Binge Eating Disorder: A Retrospective Chart Review. *Clin. Neuropharmacol.* **2019**, *42*, 214–216. [CrossRef]
95. Güngör, S.; Celiloğlu, Ö.S.; Raif, S.G.; Özcan, Ö.Ö.; Selimoğlu, M.A. Malnutrition and Obesity in Children with ADHD. *J. Atten. Disord.* **2016**, *20*, 647–652. [CrossRef] [PubMed]
96. Miesch, M.; Deister, A. Attention-deficit/hyperactivity disorder (ADHD) in adult psychiatry: Data on 12-month prevalence, risk factors and comorbidity. *Fortschr. Neurol. Psychiatr.* **2019**, *87*, 32–38. [CrossRef]
97. Penninx, B.W.J.H.; Lange, S.M.M. Metabolic syndrome in psychiatric patients: Overview, mechanisms, and implications. *Dialogues Clin. Neurosci.* **2018**, *20*, 63–73.
98. Pugh, S.J.; Hutcheon, J.A.; Richardson, G.A.; Brooks, M.M.; Himes, K.P.; Day, N.L.; Bodnara, L.M. Gestational weight gain, prepregnancy body mass index and offspring attention-deficit hyperactivity disorder symptoms and behaviour at age 10. *BJOG An Int. J. Obstet. Gynaecol.* **2016**, *123*, 2094–2103 [CrossRef]
99. Munsch, S.; Dremmel, D.; Wilhelm, P.; Baierlé, S.; Fischer, S.; Hilbert, A. To eat or not to eat: Reward delay impulsivity in children with loss of control eating, attention deficit hyperactivity disorder, a double diagnosis, and healthy children. *PLoS ONE* **2019**, *14*, e0221814. [CrossRef] [PubMed]
100. Oswalt, S.B.; Lederer, A.M.; Chestnut-Steich, K.; Day, C.; Halbritter, A.; Ortiz, D. Trends in college students' mental health diagnoses and utilization of services, 2009–2015. *J. Am. Coll. Health* **2020**, *68*, 41–51. [CrossRef] [PubMed]
101. Welch, E.; Ghaderi, A.; Swenne, I. A comparison of clinical characteristics between adolescent males and females with eating disorders. *BMC Psychiatry* **2015**, *15*, 45. [CrossRef] [PubMed]
102. Lewis, E.; Gittelsohn, J.; Baker, C. More Frequent Family Meals Associated with Greater Risk of Obesity in Youth with and without ADHD (P11-115-19). *Curr. Dev. Nutr.* **2019**, *3*, 195190753. [CrossRef]
103. Reinblatt, S.P.; Leoutsakos, J.M.S.; Mahone, E.M.; Forrester, S.; Wilcox, H.C.; Riddle, M.A. Association between binge eating and attention-deficit/hyperactivity disorder in two pediatric community mental health clinics. *Int. J. Eat. Disord.* **2015**, *48*, 505–511. [CrossRef]
104. Wentz, E.; Björk, A.; Dahlgren, J. Is there an overlap between eating disorders and neurodevelopmental disorders in children with obesity? *Nutrients* **2019**, *11*, 2496. [CrossRef]
105. Reinblatt, S.P.; Mahone, E.M.; Tanofsky-Kraff, M.; Lee-Winn, A.E.; Yenokyan, G.; Leoutsakos, J.M.S.; Moran, T.H.; Guarda, A.S.; Riddle, M.A. Pediatric loss of control eating syndrome: Association with attention-deficit/hyperactivity disorder and impulsivity. *Int. J. Eat. Disord.* **2015**, *48*, 580–588. [CrossRef] [PubMed]

106. Bisset, M.; Rinehart, N.; Sciberras, E. DSM-5 eating disorder symptoms in adolescents with and without attention-deficit/hyperactivity disorder: A population-based study. *Int. J. Eat. Disord.* **2019**, *52*, 855–862. [CrossRef]
107. Capusan, A.J.; Yao, S.; Kuja-Halkola, R.; Bulik, C.M.; Thornton, L.M.; Bendtsen, P.; Marteinsdottir, I.; Thorsell, A.; Larsson, H. Genetic and environmental aspects in the association between attention-deficit hyperactivity disorder symptoms and binge-eating behavior in adults: A twin study. *Psychol. Med.* **2017**, *47*, 2866–2878. [CrossRef]
108. Jacob, L.; Haro, J.M.; Koyanagi, A. Attention deficit hyperactivity disorder symptoms and disordered eating in the English general population. *Int. J. Eat. Disord.* **2018**, *51*, 942–952. [CrossRef]
109. Ziobrowski, H.; Brewerton, T.D.; Duncan, A.E. Associations between ADHD and eating disorders in relation to comorbid psychiatric disorders in a nationally representative sample. *Psychiatry Res.* **2018**, *260*, 53–59. [CrossRef] [PubMed]
110. Gorlin, E.I.; Dalrymple, K.; Chelminski, I.; Zimmerman, M. Diagnostic profiles of adult psychiatric outpatients with and without attention deficit hyperactivity disorder. *Compr. Psychiatry* **2016**, *70*, 90–97. [CrossRef]
111. Porteret, R.; Bouchez, J.; Baylé, F.J.; Varescon, I. L'impulsivité dans le TDAH: Prévalence des troubles du contrôle des impulsions et autres comorbidités, chez 81 adultes présentant un trouble déficit de l'attention/hyperactivité (TDA/H). *Encephale* **2016**, *42*, 130–137. [CrossRef] [PubMed]
112. Gowey, M.A.; Stromberg, S.E.; Lim, C.S.; Janicke, D.M. The moderating role of body dissatisfaction in the relationship between ADHD symptoms and disordered eating in pediatric overweight and obesity. *Child. Health Care* **2017**, *46*, 15–33. [CrossRef]
113. Egbert, A.H.; Wilfley, D.E.; Eddy, K.T.; Boutelle, K.N.; Zucker, N.; Peterson, C.B.; Celio Doyle, A.; Le Grange, D.; Goldschmidt, A.B. Attention-Deficit/Hyperactivity Disorder Symptoms Are Associated with Overeating with and without Loss of Control in Youth with Overweight/Obesity. *Child. Obes.* **2018**, *14*, 50–57. [CrossRef]
114. Rojo-Moreno, L.; Arribas, P.; Plumed, J.; Gimeno, N.; García-Blanco, A.; Vaz-Leal, F.; Luisa Vila, M.; Livianos, L. Prevalence and comorbidity of eating disorders among a community sample of adolescents: 2-year follow-up. *Psychiatry Res.* **2015**, *227*, 52–57. [CrossRef]
115. Mohammadi, M.R.; Mostafavi, S.A.; Hooshyari, Z.; Khaleghi, A.; Ahmadi, N.; Molavi, P.; Armani Kian, A.; Safavi, P.; Delpisheh, A.; Talepasand, S.; et al. Prevalence, correlates and comorbidities of feeding and eating disorders in a nationally representative sample of Iranian children and adolescents. *Int. J. Eat. Disord.* **2019**, *53*, 349–361. [CrossRef] [PubMed]
116. Kim, K.M.; Lim, M.H.; Kwon, H.J.; Yoo, S.J.; Kim, E.J.; Kim, J.W.; Ha, M.; Paik, K.C. Associations between attention-deficit/hyperactivity disorder symptoms and dietary habits in elementary school children. *Appetite* **2018**, *127*, 274–279. [CrossRef] [PubMed]
117. Nielsen, F.; Georgiadou, E.; Bartsch, M.; Langenberg, S.; Müller, A.; De Zwaan, M. Attention Deficit Hyperactivity Disorder Prevalence and Correlates Pre- and Post-Bariatric Surgery: A Comparative Cross-Sectional Study. *Obes. Facts* **2017**, *10*, 1–11. [CrossRef]
118. Brewerton, T.D.; Duncan, A.E. Associations between Attention Deficit Hyperactivity Disorder and Eating Disorders by Gender: Results from the National Comorbidity Survey Replication. *Eur. Eat. Disord. Rev.* **2016**, *24*, 536–540. [CrossRef] [PubMed]
119. Woldeyohannes, H.O.; Soczynska, J.K.; Maruschak, N.A.; Syeda, K.; Wium-Andersen, I.K.; Lee, Y.; Cha, D.S.; Xiao, H.X.; Gallaugher, L.A.; Dale, R.M.; et al. Binge eating in adults with mood disorders: Results from the International Mood Disorders Collaborative Project. *Obes. Res. Clin. Pract.* **2015**, *10*, 531–543. [CrossRef]
120. Svedlund, N.E.; Norring, C.; Ginsberg, Y.; von Hausswolff-Juhlin, Y. Symptoms of Attention Deficit Hyperactivity Disorder (ADHD) among adult eating disorder patients. *BMC Psychiatry* **2017**, *17*, 19. [CrossRef]
121. Svedlund, N.E.; Norring, C.; Ginsberg, Y.; von Hausswolff-Juhlin, Y. Are treatment results for eating disorders affected by ADHD symptoms? A one-year follow-up of adult females. *Eur. Eat. Disord. Rev.* **2018**, *26*, 337–345. [CrossRef] [PubMed]
122. Halevy-Yosef, R.; Bachar, E.; Shalev, L.; Pollak, Y.; Enoch-Levy, A.; Gur, E.; Weizman, A.; Stein, D. The complexity of the interaction between binge-eating and attention. *PLoS ONE* **2019**, *14*, e0215506. [CrossRef] [PubMed]

123. Carlucci, S.; Ivanova, I.; Bissada, H.; Tasca, G.A. Validity and reliability of the attention deficit hyperactivity disorder self-report scale (ASRS-v1.1) in a clinical sample with eating disorders. *Eat. Behav.* **2017**, *26*, 148–154. [CrossRef] [PubMed]
124. Sala, L.; Martinotti, G.; Carenti, M.L.; Romo, L.; Oumaya, M.; Pham-Scottez, A.; Rouillon, F.; Gorwood, P.; Janiri, L. Attention-deficit/hyperactivity disorder symptoms and psychological comorbidity in eating disorder patients. *Eat. Weight Disord.* **2018**, *23*, 513–519. [CrossRef]
125. Kurz, S.; Schoebi, D.; Dremmel, D.; Kiess, W.; Munsch, S.; Hilbert, A. Satiety regulation in children with loss of control eating and attention-deficit/hyperactivity disorder: A test meal study. *Appetite* **2017**, *116*, 90–98. [CrossRef] [PubMed]
126. Tong, L.; Shi, H.; Li, X. Associations among ADHD, abnormal eating and overweight in a non-clinical sample of Asian children. *Sci. Rep.* **2017**, *7*, 2844. [CrossRef]
127. Leventakou, V.; Micali, N.; Georgiou, V.; Sarri, K.; Koutra, K.; Koinaki, S.; Vassilaki, M.; Kogevinas, M.; Chatzi, L. Is there an association between eating behaviour and attention-deficit/hyperactivity disorder symptoms in preschool children? *J. Child Psychol. Psychiatry Allied Discip.* **2016**, *57*, 676–684. [CrossRef] [PubMed]
128. Fuemmeler, B.F.; Sheng, Y.; Schechter, J.C.; Do, E.; Zucker, N.; Majors, A.; Maguire, R.; Murphy, S.K.; Hoyo, C.; Kollins, S.H. Associations between attention deficit hyperactivity disorder symptoms and eating behaviors in early childhood. *Pediatr. Obes.* **2020**, *15*, e12631. [CrossRef]
129. Yilmaz, Z.; Javaras, K.N.; Baker, J.H.; Thornton, L.M.; Lichtenstein, P.; Bulik, C.M.; Larsson, H. Association between Childhood to Adolescent Attention Deficit/Hyperactivity Disorder Symptom Trajectories and Late Adolescent Disordered Eating. *J. Adolesc. Health* **2017**, *61*, 140–146. [CrossRef]
130. Sonneville, K.R.; Calzo, J.P.; Horton, N.J.; Field, A.E.; Crosby, R.D.; Solmi, F.; Micali, N. Childhood hyperactivity/inattention and eating disturbances predict binge eating in adolescence. *Psychol. Med.* **2015**, *45*, 2511–2520. [CrossRef]
131. Zhang, Z.; Robinson, L.; Jia, T.; Quinlan, E.B.; Tay, N.; Chu, C.; Barker, E.D.; Banaschewski, T.; Barker, G.J.; Bokde, A.L.W.; et al. Development of Disordered Eating Behaviors and Comorbid Depressive Symptoms in Adolescence: Neural and Psychopathological Predictors. *Biol. Psychiatry* **2020**. In press. [CrossRef]
132. Nazar, B.P.; Trindade, A.P.; Leslie, M.; Malloy-Diniz, L.F.; Sergeant, J.; Treasure, J.; Mattos, P. Eating disorders impact on vigilance and decision making of a community sample of treatment naive attention-deficit/hyperactivity disorder young adults. *Front. Psychiatry* **2018**, *9*, 531. [CrossRef]
133. Van der Oord, S.; Braet, C.; Cortese, S.; Claes, L. Testing the dual pathway model of ADHD in obesity: A pilot study. *Eat. Weight Disord.* **2017**, *23*, 507–512. [CrossRef] [PubMed]
134. Ferre, F.; Cambra, J.; Ovejero, M.; Basurte-Villamor, I.; Navarrete, F.F. Influence of attention deficit hyperactivity disorder symptoms on quality of life and functionality in adults with eating disorders. *Actas Esp. Psiquiatr.* **2017**, *45*, 98–107.
135. Christian, C.; Martel, M.M.; Levinson, C.A. Emotion regulation difficulties, but not negative urgency, are associated with attention-deficit/hyperactivity disorder and eating disorder symptoms in undergraduate students. *Eat. Behav.* **2020**, *36*, 101344. [CrossRef] [PubMed]
136. Hanson, J.A.; Phillips, L.N.; Hughes, S.M.; Corson, K. Attention-deficit hyperactivity disorder symptomatology, binge eating disorder symptomatology, and body mass index among college students. *J. Am. Coll. Health* **2019**, *68*, 543–549. [CrossRef]
137. Koch, S.V.; Andersson, M.; Hvelplund, C.; Skovgaard, A.M. Mental disorders in referred 0–3-year-old children: A population-based study of incidence, comorbidity and perinatal risk factors. *Eur. Child Adolesc. Psychiatry* **2020**. [CrossRef]
138. Kaisari, P.; Dourish, C.T.; Rotshtein, P.; Higgs, S. Associations between core symptoms of attention deficit hyperactivity disorder and both binge and restrictive eating. *Front. Psychiatry* **2018**, *9*, 103. [CrossRef] [PubMed]
139. Williamson, T.M.; Campbell, T.S.; Telfer, J.A.; Rash, J.A. Emotion Self-Regulation Moderates the Association between Symptoms of ADHD and Weight Loss after Bariatric Surgery. *Obes. Surg.* **2017**, *28*, 1553–1561. [CrossRef] [PubMed]
140. Reinblatt, S.P. Are Eating Disorders Related to Attention Deficit/Hyperactivity Disorder? *Curr. Treat. Options Psychiatry* **2015**, *2*, 402–412. [CrossRef]

141. Keshen, A.; Ivanova, I. Reduction of Bulimia Nervosa Symptoms after Psychostimulant Initiation in Patients with Comorbid ADHD: Five Case Reports. *Eat. Disord.* **2013**, *21*, 360–369. [CrossRef]
142. Anderson, L.M.; Smith, K.M.; Schaefer, L.M.; Crosby, R.D.; Cao, L.; Engel, S.G.; Crow, S.J.; Wonderlich, S.A.; Peterson, C.B. Predictors and Moderators of Treatment Outcome in a Randomized Clinical Trial for Binge-Eating Disorder. *J. Consult. Clin. Psychol.* **2020**, *88*, 631–642. [CrossRef]
143. Cortese, S.; Bernardina, B.D.; Mouren, M.-C. Attention-Deficit/Hyperactivity Disorder (ADHD) and Binge Eating. *Nutr. Rev.* **2008**, *65*, 404–411. [CrossRef]
144. Fletcher, P.C.; Kenny, P.J. Food addiction: A valid concept? *Neuropsychopharmacology* **2018**, *43*, 2506–2513. [CrossRef]
145. Cortese, S.; Konofal, E.; Bernardina, B.D.; Mouren, M.C.; Lecendreux, M. Does excessive daytime sleepiness contribute to explaining the association between obesity and ADHD symptoms? *Med. Hypotheses* **2008**, *70*, 12–16. [CrossRef]
146. Quaranta, G.; Barbuti, M.; Alessandro, P.; Colombini, P.; Moriconi, M.; Gemmellaro, T.; Tripodi, B.; Palagini, L.; Schiavi, E.; Perugi, G. Relationships among Delayed Sleep Phase Disorder, Emotional Dysregulation, and Affective Temperaments in Adults with Attention Deficit Hyperactivity Disorder and Cyclothymia. *J. Nerv. Ment. Dis.* **2020**, 32769692. [CrossRef]

nutrients

MDPI

Review

Meeting of Minds around Food Addiction: Insights from Addiction Medicine, Nutrition, Psychology, and Neurosciences

Aymery Constant [1,2,†], Romain Moirand [1,3,†], Ronan Thibault [1,4,†] and David Val-Laillet [1,*,†]

1 INRAE, INSERM, University Rennes, NuMeCan, Nutrition Metabolisms Cancer, 35590 St Gilles, 35000 Rennes, France; aymery.constant@ehesp.fr (A.C.); romain.moirand@chu-rennes.fr (R.M.); ronan.thibault@chu-rennes.fr (R.T.)
2 EHESP, School of Public Health, 35043 Rennes, France
3 Unité d'Addictologie, CHU Rennes, 35000 Rennes, France
4 Unité de Nutrition, CHU Rennes, 35000 Rennes, France
* Correspondence: david.val-laillet@inrae.fr
† These authors contributed equally to this work.

Received: 9 October 2020; Accepted: 18 November 2020; Published: 20 November 2020

Abstract: This review, focused on food addiction (FA), considers opinions from specialists with different expertise in addiction medicine, nutrition, health psychology, and behavioral neurosciences. The concept of FA is a recurring issue in the clinical description of abnormal eating. Even though some tools have been developed to diagnose FA, such as the Yale Food Addiction Scale (YFAS) questionnaire, the FA concept is not recognized as an eating disorder (ED) so far and is even not mentioned in the Diagnostic and Statistical Manuel of Mental Disorders version 5 (DSM-5) or the International Classification of Disease (ICD-11). Its triggering mechanisms and relationships with other substance use disorders (SUD) need to be further explored. Food addiction (FA) is frequent in the overweight or obese population, but it remains unclear whether it could articulate with obesity-related comorbidities. As there is currently no validated therapy against FA in obese patients, FA is often underdiagnosed and untreated, so that FA may partly explain failure of obesity treatment, addiction transfer, and weight regain after obesity surgery. Future studies should assess whether a dedicated management of FA is associated with better outcomes, especially after obesity surgery. For prevention and treatment purposes, it is necessary to promote a comprehensive psychological approach to FA. Understanding the developmental process of FA and identifying precociously some high-risk profiles can be achieved via the exploration of the environmental, emotional, and cognitive components of eating, as well as their relationships with emotion management, some personality traits, and internalized weight stigma. Under the light of behavioral neurosciences and neuroimaging, FA reveals a specific brain phenotype that is characterized by anomalies in the reward and inhibitory control processes. These anomalies are likely to disrupt the emotional, cognitive, and attentional spheres, but further research is needed to disentangle their complex relationship and overlap with obesity and other forms of SUD. Prevention, diagnosis, and treatment must rely on a multidisciplinary coherence to adapt existing strategies to FA management and to provide social and emotional support to these patients suffering from highly stigmatized medical conditions, namely overweight and addiction. Multi-level interventions could combine motivational interviews, cognitive behavioral therapies, and self-help groups, while benefiting from modern exploratory and interventional tools to target specific neurocognitive processes.

Keywords: obesity; craving; reward circuit; motivation; cognition; behavior; therapy

1. Introduction

Even though the concept of food addiction (FA) was introduced more than sixty years ago [1], its definition and implications are still fiercely debated [2,3]. Highly palatable foods [4,5], such as processed foods with added sugars and fat, could be as addictive as drugs [6,7], acting via the same neurocognitive and hedonic processes [8,9]. Therefore, through the alteration of the neurocognitive systems involved in food intake control [10], FA could be involved in the obesity pathogenesis. However, at this time, the concept of FA is still debated [2,3,11] and is probably entangled with complex psychological factors and predispositions. The concept of sugar addiction is well defended in animal models [12], but in humans, some authors rather suggest the concept of "eating addiction", i.e., an addiction to the eating behavior instead of an addiction to palatable foods like sugar or saturated fat. Nevertheless, considering the alterations of the neurocognitive systems involved in food intake control [10], the diagnostic criteria of FA were based on the Diagnostic and Statistical Manuel of Mental Disorders version 5 (DSM-V) criteria for substance use disorders [13]. In clinical practice, the Yale Food Addiction Scale (YFAS) 2.0 is the only self-administered questionnaire validated to diagnose and estimate the number of symptoms of FA [14,15].

Our first aim is to better describe the place of FA in the current nosography and establish a parallel between FA and other ED or addictive disorders, while discussing possible transfer or continuity between disorders. Our second aim is to illustrate, on the basis of existing data, how FA is frequent in the general and obesity population and how it articulates with comorbidities in obese patients. We also discuss why and how FA should be handled in the preoperative management of obesity surgery patients. Our third aim is to highlight the relationship between FA and specific psychological features, within a continuum ranging from normal to disordered eating. Our fourth aim is to describe the neurocognitive and brain correlates with other substance use disorders (SUD), such as with drugs and alcohol, to support a neurobiological picture of FA. Finally, we discuss the concept of FA in the context of prevention, diagnostic, and treatment, with the aim to present existing or innovative strategies in the scope of interdisciplinary and personalized medicine. Perspectives in terms of cognitive and behavioral therapies, digital technologies, and neuromodulation interventions are also discussed.

1.1. From the Addiction Medicine Clinician Point of View: Towards a Definition of Food Addiction (FA)

Addiction remains a difficult-to-define concept. The American Society of Addiction Medicine defined addiction as "a treatable, chronic medical disease involving complex interactions among brain circuits, genetics, the environment, and an individual's life experiences. People with addiction use substances or engage in behaviors that become compulsive and often continue despite harmful consequences" [16]. DSM-5 (Table 1) and ICD-11 give similar sets of criteria defining addiction, while actually avoiding using the term, instead preferring "substance-related and addictive disorders" (including "substance use disorders" (SUD) and "gambling disorder") and "dependence", respectively. These criteria encompass the loss of control over consumption, increased motivation to consume, and persistent consumption despite negative consequences, as well as tolerance or adverse effects of acute withdrawal. In the last twenty years, there has been a growing interest in the possibility that in some patients, food, and especially highly palatable food, could produce behavioral symptoms that parallel those of addiction and could activate the same neural reward circuits as drugs of abuse [17]. However, this concept has been challenged, either on the addictive nature of some eating disorders (ED) [18,19] or by debating as to whether FA is akin to a behavioral addiction versus a SUD [20].

Diagnostic criteria for SUD represent a cluster of cognitive, behavioral, and physiological symptoms, and most of them can apply to some patients by replacing "substance" with "certain food" (for extensive reviews, see [4,7,21]). "Taking larger amounts of the substance for longer periods than intended" has been cited as one of the most commonly reported symptoms in overweight/obese or eating disorder patients. It can be in the form of binges, but also snacking [22], food compulsion, or excessive portion sizes. "Unsuccessful attempts to reduce food intake" is clinically obvious, as many patients are unable to maintain their diet and lose weight in the long term. "Craving" is a central concept in the field of addiction [23], and craving for food has been recognized for a long time [24–26]. Similarity between drug and food craving is supported by the findings of cue-reactivity research [24]. "Social/interpersonal problems related to use" is supported by the poor social functioning associated with overweight and obesity due to weight stigmatization. Negative consequences of overeating include obesity and its medical consequences, stigmatization, psychological distress induced by shame, hopelessness [27]; then, many patients exhibit "continuous use despite recurrent physical or psychological problem". These negative consequences lead to the "failure to fulfill major role obligation" and to "reduced activities". Tolerance can be suspected, given that some overweight patients increased consumption and portion size over time [21]. This was further supported by an innovative study in which 61 overweight carbohydrate-craving women were induced into a sad mood, then exposed, double-blind and in counterbalanced order, to taste-matched carbohydrate or protein beverages and asked to choose the drink that made them feel better. They overwhelmingly chose and liked the carbohydrate beverage, which was more efficient in reducing dysphoria, but the effect decreased with repetition, suggesting tolerance, while liking increased, suggesting sensitization [28]. At last, a specific food withdrawal syndrome, as can be observed with alcohol, opioids, or nicotine, has not clearly been demonstrated in humans. However, it should be noted that the physiological criteria of tolerance and withdrawal are not necessary for a diagnosis of SUD, as even with a potent substance such as alcohol, withdrawal syndrome is observed in no more than one third of patients.

Table 1. Diagnostic and Statistical Manuel of Mental Disorders version 5 (DSM-5) diagnostic criteria of substance use disorder (SUD). A person needs to meet at least 2 of these criteria and have significant impairment or distress from his pattern of substance use to be diagnosed with a SUD. The severity of addiction is determined by the number of criteria met: 2–3 mild; 4–5 moderate; ≥6 severe.

Broader Categories	SUD Criteria
Impaired control	Substance often taken in larger amounts or over a longer period than was intended Craving, or a strong desire or urge to use the substance Persistent desire or repeated unsuccessful attempts to quit and/or control substance use Great deal of time is spent in activities necessary to obtain or use the substance or recover from its effects
Social impairment	Continued use despite having persistent or recurrent social or interpersonal problems caused or exacerbated by the effects of the substance Recurrent substance use resulting in a failure to fulfill major role obligations at work, school, or home Important social, occupational, or recreational activities are given up or reduced because of substance use
Continued used despite risk	Recurrent substance use in situations in which it is physically hazardous Substance use is continued despite knowledge of having a persistent or recurrent physical or psychological problem that is likely to have been caused or exacerbated by the substance
Pharmacological criteria	Tolerance: Need for markedly increased amounts of the substance to achieve intoxication or desired effect or Markedly diminished effect with continued use of the same amount of the substance Withdrawal: Withdrawal syndrome (differs by substance) or Substance is taken to relieve or avoid withdrawal symptoms

Gearhardt and coworkers have boosted studies on FA by releasing the Yale Food Addiction Scale (YFAS), by transposing the DSM-IV [29], then DSM-5 (YFAS 2.0) [30] criteria for substance dependence, then SUD, simply by replacing substance with "certain food" in the criteria. In comparison to the first version of the YFAS, the YFAS 2.0 explores the following additional criteria, the first three issued from the previous DSM-IV diagnostic of abuse: continued consumption despite social or interpersonal problems, failure to fulfill major role or obligation, use in physically hazardous situations, and craving. YFAS 2.0 comprises 35 questions exploring the 11 criteria of DSM-5 SUD and the existence of significant impairment or distress and uses exactly the same thresholds to assign an FA diagnosis (Table 1). Of note, DSM-5 allows us to grade SUD as mild, moderate, and severe, according to the number of criteria, and only severe SUD fits well with the diagnosis of dependence in DSM-IV and ICD-11. In other words, sensitivity was increased, and it is questionable to say that a mild SUD responds to the criteria of addiction [31]. Numerous systematic reviews have explored validation, prevalence, and correlates of FA diagnosed by the YFAS [4,15,32–35]. Gordon et al. [4] evaluated empirical studies examining the construct of "FA" and their conclusions supported FA as a unique construct consistent with criteria for other SUD diagnoses. The highest scored symptom was generally "unsuccessful attempt to cut down"; tolerance and "use despite knowledge of adverse consequences" were also frequent, followed by "activities given up" and withdrawal symptoms [35]. In studies using YFAS 2.0, severe FA predominated upon mild and moderate forms [35]. YFAS number of symptoms was correlated with body mass index (BMI) in non-clinical samples; this was variable in obese or ED samples [35]. Significant positive correlations were found between FA and depression or anxiety [32,35]. FA, in obese patients or university students, was consistently associated with self-report and other measures of impulsivity; associations with reward sensitivity were inconsistent, depending on the questionnaires used [34]. At last, FA prevalence was generally increased in patients with other chemical or behavioral addictions as compared with non-clinical samples [36–40].

The triggering mechanism of FA has been extensively debated. Hebebrand et al. [20] have argued that FA may be a behavioral addiction: indeed, most substances of abuse, apart from alcohol, are agonists of specific brain receptors by mimicking endogenous ligands, and this is not applicable to food. Moreover, eating is necessary for survival and drug use is not; eating is intrinsically rewarding and reinforcing, and food consumption is well known to naturally activate the brain reward system. Most other authors, however, considered FA to fit better with SUD [4,7,13,37,41,42]. Two properties of food could participate in mediating liking, wanting, or craving: hedonic taste and metabolic shifts following ingestion. Food contains a variety of compounds that may serve as chemical or metabolic triggers, and all commonly suspected problem foods share nutritive properties [41]. It is highly unlikely that all foods may be addictive, and studies have aimed at identifying the specific foods or food attributes capable of triggering an addictive response. The evidence that sugar (sucrose) could be an addictive substance is mainly supported by animal studies, the interpretations of which are controversial [18,42,43], and sugar (sucrose, fructose) is not considered a direct cause of obesity [42]. In university students, symptoms of FA were in majority related to combined high-fat savory and high-fat sweet foods, and rarely for mainly sugar-containing food [2]. Concerning fat, which has its own metabolic, physiological, and nutritional profiles, human evidence is scarce and comes mostly from studies on FA: individuals with FA had higher dietary fat intake compared to those without FA; animal models suggested that fat addiction may have different mechanisms to sugar addiction [44]. Highly processed foods, with the addition of fat and/or refined carbohydrates (sugar, white flour), have drawn the most attention. Subjects' rating of food pictures varying in their chemical composition showed that highly processed, energy dense foods with high glycemic load and high fat content were most frequently associated with addiction-like eating behaviors, especially for individuals endorsing elevated symptoms of FA [5]. Another study showed that high processed food pictures were associated with greater loss of control, liking, pleasure, and craving, which assess the abuse liability of substances [45]. This was confirmed using a taste test task and ad libitum consumption period in obese women (39% had an FA diagnosis) [46]. Highly processed foods with high glycemic index cause rapid shifts in blood glucose, insulin, and other metabolic fuel and hormones [41], and this could be associated with their addictive potential.

The concept of FA has many implications for the addiction clinician. Given the prevalence of FA in people with addiction, screening for FA and other eating disorders has to be performed in patients with other addictive disorders. FA could explain, by a mechanism of addiction transfer, the increase in the consumption of chocolate and other sweets in recovering patients with alcohol use disorder [47] as well as the tragic increase in alcohol use disorders after obesity surgery [48]. Denial is a well-recognized, albeit badly explained, feature of patients with SUD [49], but has not been explored in the FA literature. Given the importance of social stigma attached both to overweight and addiction [50], it is very likely that denial does exist, particularly in obese/overweight samples seeking treatment. Such a behavioral feature could minimize FA prevalence. Denial is in part related to concerns about being stigmatized or rejected or to social interactions of an accusatory or judgmental nature [49]. There have been extensive discussions about the influence of the recognition of the FA construct on stigmatization, either from family/relatives and society (externalized stigma) or from the patient himself (internalized stigma) [50,51]. An FA explanation model of the lack of control in obesity could decrease stigmatization from others [52]. Studies on the effects on self-esteem and internalized stigmatization gave contradictory results [50,53]. Increasing self-esteem and confidence could then constitute a therapeutic goal in obesity treatment.

In conclusion, the analogy between severe SUD and some eating-related behaviors including FA is obvious from a clinical point of view, although more studies are needed to precisely determine the triggering mechanisms and the possibility of preventive or therapeutic interventions.

1.2. From the Clinical Nutritionist's Point of View: Food Addiction (FA) in the Context of Obesity Treatment

In this section, we aim to: (i) illustrate how FA is frequent in the general and obese population and how it articulates with comorbidities in obese patients; (ii) discuss why and how FA should be handled in the management of obese patients, especially those referred to obesity surgery.

Obesity is pandemic worldwide [54,55] and leads to well-known comorbidities [56], representing a significant socioeconomic burden [57–59] (Figure 1). Obesity medical treatments usually fail to achieve weight loss or maintain it in the long term [60], justifying the recourse to obesity surgery in some instances. Obesity surgery decreases mortality, cardiovascular events, and type-2 diabetes in comparison to conventional therapy [60–63]. Nevertheless, 20% to 30% of operated patients regain weight because of the reoccurrence of ED, i.e., binge eating disorders, hyperphagia, snacking, craving, food compulsion, or bulimia [64]. These symptoms could be related to FA.

As previously mentioned, highly palatable foods [4,5], such as processed foods with added sugars and fat, could be as addictive as drugs [6,7], acting via the same neurocognitive and hedonic processes [8,9]. Therefore, through the alteration of the neurocognitive systems involved in food intake control [10], FA could be involved in obesity pathogenesis (Figure 1). As there is currently no validated therapy against FA, FA is often underdiagnosed and untreated [13,19]. FA may partly explain the failure of obesity treatment.

1.2.1. Prevalence of FA in the General and Obese Population

FA is frequent in the general and obese populations. Through meta-analysis, the mean prevalence of FA diagnosis was found to be 16.2%, more frequent in obese/overweight patients (from 10% in normal weight to around 25% in people with obesity), with the greatest prevalence in patients with ED [15,32]. FA was more frequent in women than in men [35]. In the US general population, Schulte and Gearhardt [65] reported that 15% of people may have FA, regardless of BMI. Pursey et al. [15] found that 19.9% of overweight and obese patients had FA. Som et al. [66] found a large prevalence (almost 40%) of FA in patients eligible for obesity surgery. The analysis of 19 studies which assessed FA among pre- and/or post-obesity surgery patients revealed that the presence of pre-surgical FA was not associated with pre-surgical weight or post-surgical weight outcomes; yet pre-surgical FA was related to broad levels of psychopathology [33]. The prevalence of FA has been reported to be 16.5% [67], 17.2% [68], 25% [69], or 40% [66,70] in obese patients referred to obesity surgery. Prevalence could be even higher in the case of ED [25,71]: 57% in patients with bulimic hyperphagia [72], 41% [72,73] and up to 96% in patients with binge-eating disorders patients and bulimia, respectively [74]. Pursey et al.

found that FA prevalence was higher in overweight/obese patients (24.9%) than in subjects with normal weight (11.1%) [15], in accordance with other findings [12,13,65]. Kiyici et al. found that 32% of obese patients with a mean BMI of 41.6 and seeking treatment for weight loss had FA [75].

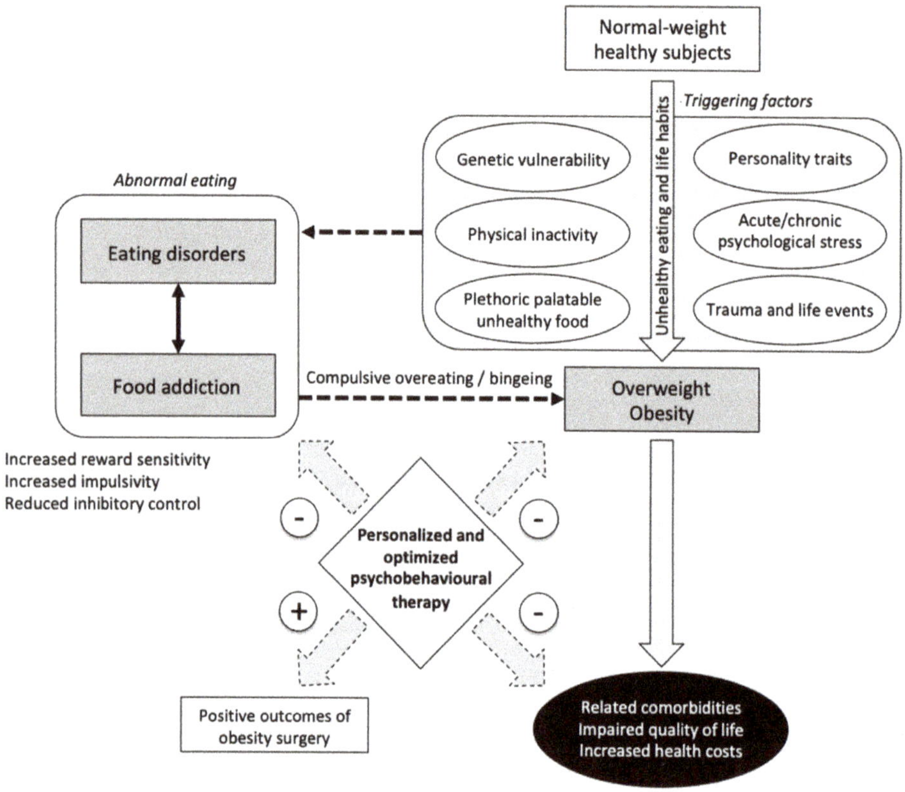

Figure 1. Food addiction as a causative or contributive factor for overweight and obesity. A personalized and optimized psychobehavioral therapy in patients with food addiction may help in preventing overweight and obesity, reducing their related comorbidities and related costs, and improving outcomes of obesity surgery. Dotted lines indicate connections for which published data are lacking or insufficient.

1.2.2. Association between Food Addiction and Obesity-Related Comorbidities

It remains unclear whether FA could be associated with or even favor obesity-related comorbidities. Kiyici et al. found that fasting plasma glucose level was lower in patients with FA, but serum insulin levels, homeostasis model assessment of insulin resistance, hemoglobin A1c, lipid parameters, and vascular adiposity index were comparable [75]. In obese patients with BMI ≥35 referred to obesity surgery, FA was not associated with obesity-related complications, such as cardiovascular diseases including arterial hypertension, obstructive sleep apnea syndrome (OSAS), type-2 diabetes, disabling osteoarticular disease, or non-alcoholic steatohepatitis [66]. Overall, FA could be considered as a potential contributing factor leading to obesity, but not to its complications, which are also driven by metabolic, environmental, or genetic factors.

1.2.3. Rationale for a Systematic Screening of FA in Obese Patients

Given the high prevalence of FA (almost 40% of patients referred to obesity surgery), evaluating FA should be part of the assessment of any obese patient, especially in patients referred to obesity surgery, as done for ED. However, there is no official recommendation about FA. Binge eating disorder and bulimia nervosa are a contraindication to obesity surgery because they increase the risk of postoperative complications, such as vomiting, esophagus dilation [76,77], addiction transfer [78], or weight regain. In children, FA in relation to psychological trauma was associated with a reduced likelihood of completing obesity surgery [79]. Only a few studies have looked at the relationship between FA and the success of behavioral or surgical obesity therapy. Pepino et al. [70] suggested that obesity surgery-induced weight loss induces remission of FA and improves several eating behaviors that are associated with FA. Lent et al. [80] reported that baseline FA status was not associated with weight loss 6 months after medical intervention in 178 adult obese patients. In a small sample size of 57 overweight or obese patients, but followed-up only for seven weeks, Burmeister et al. [81] found less weight loss in the case of FA. In morbidly obese patients, Som et al. did not find any relationship between baseline FA and weight loss in response to behavioral therapy [66]. Sevinçer et al. [60] found that the prevalence of FA of 58% in the preoperative period decreased significantly after obesity surgery to 7% and 14% at 6 months and 1 year, respectively.

Whereas obesity surgery could be beneficial for FA [70,82], there is an increased risk of "addiction transfer" from FA to another one. Indeed, obese patients may be at increased risk for SUD after obesity surgery [83]. The proportion of new substance users (alcohol, smoking, or drugs) after obesity surgery ranged from 34.3% to 89.5% [84]. Up to 20% of obese patients are diagnosed with alcohol use disorder after obesity surgery [85]. New-onset alcohol use disorder can represent more than 60% of alcohol use disorder in obese patients after obesity surgery [48]. This is why it is fundamental to diagnose FA and ED before obesity surgery. Future studies should demonstrate whether individualized cognitive behavioral therapy dedicated to the management of FA should prevent the occurrence of addiction transfer and optimize the postoperative outcomes after obesity surgery, especially in terms of the prevention of postoperative ED and weight regain [76,78]. For example, a recent review was aimed at assessing the outcomes of preoperative and post-operative psychosocial interventions for bariatric surgery patients, revealing mixed evidence but also the importance of acting early, before significant problematic eating behavior and weight regain occur [86].

We could also suggest studies assessing whether the YFAS 2.0 questionnaire could be integrated into obese patient phenotyping. If relevant, this could be included in the Edmonton Obesity Staging System (EOSS) [85,87] or French Obesity Staging System (FOSS) [88], which already integrate the psychological dimension. Limitations for the diagnosis of ED and FA are the evaluation by patient declaration, which could lead to potential information bias related to self-assessment objectivity or honesty. The proportion of patients with FA could be underestimated, as it is for all declarative information. This is why the development of complementary diagnostic strategies is required, in terms of biological, neurological, or behavioral markers, for example.

1.2.4. Proposed Therapy for Obese Patients with FA

So far, no psychological therapy has been validated in the management of FA in obese patients. In our opinion, this should be included in multidisciplinary programs combining empathic approaches and social support interventions, to help patients in coping with their daily life struggles and social stigma. In our center (Nutrition Department, CHU Rennes), all obese patients with BMI ≥35 are able to lose weight by following a one-year psychobehavioral program, including six multidisciplinary consultations (nutritionist physician, dietician, and nurse specialized in therapeutic education), in order to manage or prevent ED. This program includes at each meeting with the patient: motivational interview, advice for physical activity, and food education. In addition, all patients have two psychological consultations to help them consider the importance of emotions and stress in their eating behavior, but they are not psychotherapy sessions. All the patients expected to be eligible for obesity surgery participate in this program for an average of one year before obesity surgery.

In conclusion, as the prevalence of FA is high in obese patients, especially in patients referred for obesity surgery (40%), it is a relevant issue for the clinical practice of obesity care. Future studies should assess whether dedicated management of FA is associated with better outcomes, especially after obesity surgery. Given the prevalence of FA in the general population, public health policies should help in screening early and managing FA before it leads to obesity, which is a burden worldwide (Figure 1).

1.3. From the Health Psychologist's Point of View: Toward a More Comprehensive Psychological Approach to Food Addiction

Contrary to the categorical approach of the medical or psychiatric practices, psychology considers that eating behaviors can be mapped onto a continuum ranging from normal to disordered eating, prompted by multiple environmental, contextual, and individual factors [89,90]. For instance, environments constantly influence unhealthy food choices and overeating through food cues—sights, sounds, and smells—associated with palatable food [91], which may undermine the self-regulatory capacity in obesogenic environments [92]. It is recognized that problematic eating behaviors—such as binge eating episodes, overeating, and (failed) cognitive restriction—are not limited to psychological disorders and tend to increase over time in the general population [93].

Studies have found that 7.2% to 13% of the population currently engage in regular binge eating episodes [94]. Another study found that their prevalence increased six-fold from 1998 (2.7%) to 2015 (13.0%) in the adult general population [95]. In addition to these binge eating episodes, eating in response to specific emotional cues was investigated in relation to weight gain [96,97], ED [98,99], and psychiatric and addictive disorders [100,101]. However, this behavioral response is common in normal-weight women, as half of the female students participating in our study reported overeating in response to anxiety in the last 28 days, and 4 in 10 in response to loneliness, sadness, and happiness [102]. These intermittent overeating episodes were used as a time-limited response to emotional states and negatively correlated with alcohol use, which suggests two distinct and somewhat exclusive ways of coping for negative emotions. The Three-Factor Eating Questionnaire Revised, 18-item (TFEQ-R18), measures the cognitive and behavioral components of eating [103], which originate from obesity research but are present in other populations. It includes three subscales: (1) Cognitive Restraint (conscious restriction of food intake in order to control body weight or to promote weight loss) comprised of six items (e.g., "I consciously hold back at meals in order not to gain weight"), (2) Uncontrolled Eating (tendency to eat more than usual due to a loss of control over intake accompanied by subjective feelings of hunger), comprised of nine items (e.g., "When I see a real delicacy, I often get so hungry that I have to eat right away"), and (3) Emotional Eating (inability to resist emotional cues), comprised of three items (e.g., "When I feel blue, I often overeat"). In our study, inability to resist emotional cues outweighed other cognitive components of eating which, again, suggests that overeating is a common tendency to cope with negative emotions. Moreover, while problematic eating behaviors were initially approached independently, they may interact and/or co-exist in complex patterns.

In many cases, overeating may be a paradoxical consequence of attempts at caloric restriction [104,105], and overlaps exist with emotional overeating and binge eating episodes, as studies showed a direct relationship between binge eating disorder (BED), stress, anxiety, and anxiety proneness [106,107]. However, outside of bulimia nervosa studies, much of the theoretical and empirical binge eating research to date has not directly addressed the role of anxiety [108]; even less has addressed the role of other emotional states such as depression, boredom, or fatigue. Personality may also have a structural albeit overlooked role in problematic eating. For instance, a recent study provided a phenotypic characterization of the FA construct by conducting a clustering analysis of FA in patients with eating disorder and obesity [109]. They found the highest FA symptoms in the "dysfunctional clusters", characterized by more dysfunctional personality traits, greater impulsivity, and more general psychopathology. Conversely, the "adaptive" cluster presented with more functional personality traits and low levels of general psychopathology, as well as the lowest levels of FA. This suggests that FA in the adaptive cluster may be the result of different factors than in other clusters, which could have important implications for treatment. Another study showed that emotional eating was strongly positively associated with neuroticism, particularly impulsiveness and depression [110]. External eating was likewise mainly associated with the characteristics of impulsiveness (e.g., tendency to act impulsively under strong negative and positive affective experiences, to act on the spur of the moment without regard for the consequences, to enjoy activities that are exciting or novel, etc.) and lower self-discipline [111]. Restrained eating was, on the other hand, related to higher conscientiousness, extraversion and openness, and lower neuroticism. These results imply that poor self-control seen in impulsiveness and lower self-discipline was most important for eating due to negative emotions as well as in response to external food stimuli. Attempts to control food intake and body weight seen in restrained eating were associated with more character strengths and ambitions and also a more outgoing personality style with more stable emotions.

In this regard, the lack of mental stimulation could constitute a significant vulnerability factor for excessive eating [112] and drinking [113]. Som et al. found a higher proportion of food addiction in unemployed patients [66], and our previous work showed a greater predisposition to boredom in patients with excessive drinking [113]. One possible explanation is that some vulnerable people use these compulsive behaviors to cope with excessive lack of internal and/or external stimulation in their daily lives, which may increase the risk of addiction and jeopardize their social and professional functioning. This fits with the definition of one of the eleven diagnostic criteria of addiction in the DSM-5: giving up important social, occupational, or recreational activities because of substance use. Secondly, FA includes the negative feelings following compulsive eating, typically guilt and shame, which are also commonly reported amid overeating episodes in the general population [95]. One possible explanation is that negative feelings after overeating episodes come from the social stigma attached to weight issues [114] rather than from the overeating episode itself. This emotional response to internalized weight stigma could explain the high proportion of FA diagnosis in obese patients, although, in most cases, obesity is the result of poor dietary habits rather than compulsive eating [105,115]. Accordingly, a large part of the FA syndrome, as assessed by the YFAS 2.0, could be seen as a context-dependent pattern of problematic eating behaviors and negative feeling, existing in various forms and intensity, in the general population as well as in patients with chronic conditions, independently of any psychiatric disorders. Finally, tolerance and withdrawal are the only symptoms specific to addictive processes in FA, since they are unrelated to environmental and individual factors and therefore possibly those distinguishing FA from highly frequent problematic eating behaviors. This would be consistent with a review on FA [4], concluding that behavioral and substance-related aspects of FA appear to be intertwined, but the substance (highly palatable food) component may be more salient to the diagnostic classification of this phenomenon than the behavior (eating).

From this perspective, pharmacological criteria, namely craving towards palatable food and withdrawal symptoms, could constitute the main—if not the only—solid indicators of FA, possibly co-existing with varying patterns of problematic eating behaviors, negative feelings, and social disturbances. Investigating their relative contributions to FA, together with their interactions with social environments, unhealthy eating habits, and clinical outcomes, could contribute greatly to the understanding of FA developmental history. However, the YFAS conception study adopted confirmatory factor analysis (CFA), a theory-based approach that can only estimate the extent to which questionnaire data fit the theoretical single-factor structure derived from DSM-5 criteria. The main advantage of CFA lies in its ability to help researchers to bridge the frequent gap between theory and observation. One disadvantage of CFA is that secondary factor loadings are not part of the output [116]. This may lead to the assumptions that (1) all items belong to the same single-factor variable, by simply ignoring the other possible structural hypotheses, and (2) they are equally important for characterizing FA, even though their relative contributions (loadings) to the latent variable are quite heterogeneous in most studies. It must be noted that elevated loadings are expected in construct validation studies since summary scores (or, in the present case, symptoms scores) are computed for clinical or research purposes [117]. Most analyses were performed on the 11 binary diagnoses instead of the original 7-point Likert items, which limited the total variance to be analyzed. This approach is clearly suboptimal in the case of a newly devised instrument, being psychometrically investigated for the first time. Consequently, the extent to which each of these behavioral, psychological, and social disturbances contribute to FA is still unclear. Unraveling these complex relationships warrants data-driven approaches that establish the data's underlying structure by addressing a wide range of candidate hypotheses, i.e., exploratory factor analyses. This could allow for a more comprehensive description of FA as a biopsychosocial construct lying on a continuum from normal to disordered eating and therefore earlier identification of high-risk profiles.

1.4. *From the Behavioral Neuroscientist's Point of View: Is There a "Food Addict" Brain?*

Before answering the question "is there a 'food addict' brain?", it is necessary to remember how drug addiction is described in light of its brain phenotype. As stated earlier, the DSM-5 [118] does not recognize FA in itself, but it identifies different forms of substance-related and addictive disorders (including gambling) that can be used as a reference framework for our discussion. In this context, if we accept the existence of FA, then the neurobiological characteristics of substance-related and addictive disorders should reveal common patterns between food and drug abuse. Valuable recent review papers were aimed at describing common underlying neurobiological mechanisms contributing to drug and FA [119,120]. One of the main pitfalls of these overviews, which is honestly highlighted by the authors but often bypassed in general, is the fact that most human studies taken as support were performed in obese subjects and/or patients suffering from eating disorders (ED), especially bingeing ED-subtype patients. This bias is usually accepted because there are very few studies that aimed at characterizing the brain phenotype/responses of human patients who have been specifically diagnosed with FA. There is consequently a significant risk for circular reasoning: because we make the assumption that FA should resemble drug addiction and its associated brain phenotype, then the observation of this specific brain phenotype in obese and/or bingeing patients should be sufficient to defend the FA hypothesis. The point is that the FA construct must be supported by precise definitions, as well as dedicated neurobiological and neuroimaging studies. These definitions must be supported by concrete data and not only by shortcuts based on analogies with obesity or food abuse. Even though we must assume that substance addiction always starts with substance use, not all obese and/or bingeing ED-subtype patients have FA, and not all "food addicts" are obese.

The diagnosis of an SUD is based on a pathological pattern of behaviors related to use of the substance, and the DSM-5 assists this diagnosis with eleven criteria categorized under four groupings (Table 1). It is important to mention that only one criterion, related to craving (Criterion 4), refers to a specific brain pattern associated with this condition. Craving corresponds to an intense desire or urge for the substance and is described as being associated with the activation of specific reward structures in the brain.

The literature on the neurobiology of addiction provides a consensus on the fact that drugs of abuse, as well as particular excessive behavioral patterns (e.g., gambling), exert a direct activation of the brain reward system. On a chronic basis, they also induce profound neuronal plasticity changes in the corticostriatal and limbic systems. Initially, drugs of abuse trigger abnormal surges of dopamine in the nucleus accumbens, which promotes the direct striatal pathway and inhibits the indirect striato-cortical pathway [121]. Repeated drug consumption and/or administration induce mesolimbic sensitization [122], as well as neuroplasticity changes in the glutamatergic inputs to the striatum and midbrain dopamine neurons. These changes enhance the brain's reactivity to drugs and their associated cues that gain incentive salience, i.e., incentive sensitization [123], reduce the sensitivity to other types of reward, decrease cognitive control mechanisms, and increase the susceptibility to stress and emotional dysregulation [121]. Eventually, there is a transition between controlled to habitual and compulsive use or intake [124]. The precise neuropharmacological mechanisms involved in this transition may depend on the type of drugs used, but a recurrent feature of repeated exposure to substances of abuse is the downregulation of the dopaminergic system, especially the dopamine type-2 receptor (D2R) in the ventral and dorsal striatum. Similar observations have been made in humans [125] and animal models [126], but this is not the scope of our review. Here, we are rather interested in the very few studies that tried to describe the brain phenotype of patients who were specifically diagnosed with FA.

As stated by Fletcher and Kenny [18], information will be lost if we begin with the assumption that drug addiction processes explain food overconsumption and schedule our empirical endeavors exclusively toward a survey of similarities, some of which are superficial and imprecise. In light of the DSM-5 substance use nosology, many authors consider FA as a true addiction [13,127]. As previously stated, the most widely used and accepted tool to measure FA to date is the Yale Food Addiction Scale (YFAS), of which Version 2.0 has been validated in different languages in addition to English [30], including French, Spanish, and Japanese [101,128,129]. As a consequence, we decided to gather information on brain imaging studies that were aimed at describing anatomical and functional features that are characteristic of patients fitting the YFAS criteria for FA (Table 2).

Table 2. Studies investigating the brain anatomical or functional specificities associated with YFAS-diagnosed food addiction (FA) in the human.

Articles Titles	Subjects	Exploration Methods	Main Results	References
Neural correlates of inhibitory control in youth with symptoms of FA	76 young subjects (8.2–17.8 yo, 44 males)	Go/no-go task during BOLD fMRI	YFAS-positive subjects showed deactivation in three clusters: middle temporal gyrus/occipital gyrus, precuneus/calcarine sulcus, and inferior frontal gyrus	[130]
Neuroanatomical correlates of food addiction symptoms and body mass index in the general population	625 subjects (Leipzig Research Centre for Civilization Diseases LIFE-Adult study), 20–59 yo, 45% women	BMI, personality questionnaires including YFAS and TFEQ, and brain structure via high-resolution 3T MRI	Small, additional contribution of YFAS symptom score to lower right lateral orbitofrontal cortex thickness over the effect of BMI	[131]
Food cue reactivity in FA: A functional magnetic resonance imaging study	44 women with overweight or obesity, $n = 20$ with moderate-to-severe YFAS FA	YFAS, BOLD fMRI cue reactivity task	Subjects with FA exhibited modest, elevated responses in the sFG for highly processed food images and more robust, decreased activations for minimally processed food cues, whereas control subjects showed the opposite responses; Housefold items elicited greater activation than the food cues in regions associated with interoceptive awareness and visuospatial attention (e.g., INS, iFG, iPL)	[132]
FA distinguishes an overweight phenotype that can be reversed by low calorie diet	36 overweight women	YFAS, 18 FDG-PET	Greater activation in thalamus, hypothalamus, midbrain, putamen, and occipital cortex (reward), but not in prefrontal and orbitofrontal cortices (control/ reward receipt) in the high-YFAS versus low-YFAS group. In high-YFAS subjects, orbitofrontal responsiveness was inversely related to YFAS severity and hunger rating, and positive associations were observed between regional brain activation and lipid intake. A 3-month low-calorie diet abolished group differences in brain activation	[133]
Correlation of tryptophan metabolites with connectivity of extended central reward network in healthy subjects	63 healthy subjects with and without elevated BMI (29 men and 34 women)	Fecal sampling, HAD anxiety and YFAS questionnaires, functional and anatomical connectivity of the amygdala, nucleus accumbens, and anterior insula	Direct positive association of indole metabolites with BMI and indirect positive association with YFAS through functional connectivity of the nucleus accumbens	[134]

Table 2. *Cont.*

Articles Titles	Subjects	Exploration Methods	Main Results	References
FA is associated with impaired performance monitoring	34 YFAS-positive and 34 control subjects	YFAS, Eriksen flanker task, and EEG measurement	YAFS-positive subjects have reduced ERN and Pe waves and demonstrate a higher number of errors on the flanker task, suggesting impaired performance monitoring	[135]
Neural correlates of FA	49 healthy adolescent females ranging from lean to obese	YFAS, BOLD fMRI in response to receipt and anticipated receipt of palatable food (chocolate milkshake)	YFAS correlated with greater activation in the aCC, OFC, and amygdala in response to anticipated receipt of food. Participants with higher (n = 15) vs. lower (n = 11) YFAS showed greater activation in the DLPFC and CAU in response to anticipated receipt of food, but less activation in the lOFC in response to receipt of food	[136]

aCC, anterior cingulate cortex; BMI, body mass index; BOLD, blood-oxygen-level-dependent; CAU, caudate; DLPFC, dorsolateral prefrontal cortex; EEG, electroencephalography; ERN, error-related negativity; FA, food addiction; FDG, F-2-fluoro-2-deoxy-glucose; fMRI, functional magnetic resonance imaging; iFG, inferior frontal gyrus; INS, insula; iPL, inferior parietal lobe; lOFC, lateral orbitofrontal cortex; OFC, orbitofrontal cortex; Pe, error positivity; PET, positron emission tomography; TFEQ, three-factor eating questionnaire; YFAS, Yale Food Addiction Scale; yo, years old.

The first study of this kind was performed by Gearhardt herself in collaboration with American colleagues from different teams investigating the neural correlates of eating behavior [136]. These authors used the well-known "milkshake paradigm" of blood-oxygen-level-dependent (BOLD) functional magnetic resonance imaging (fMRI) in response to receipt and anticipated receipt of palatable food (i.e., chocolate milkshake). With this paradigm, it has been well described that obese compared to lean individuals show greater activation of the gustatory cortex and oral somatosensory regions in response to anticipated intake and consumption of palatable foods. Obese individuals also show increased activation in the orbitofrontal cortex and putamen in response to palatable food pictures (i.e., reward anticipation), as well as decreased activation in the caudate nucleus in response to consumption of milkshake vs. a tasteless solution (i.e., reward receipt) [137,138]. Gearhardt et al. [136] demonstrated that YFAS scores correlated with greater activation in the anterior cingulate cortex, orbitofrontal cortex, and amygdala in response to anticipated receipt of food. Subjects with higher vs. lower YFAS showed greater activation of the dorsolateral prefrontal cortex and caudate, in response to anticipated receipt of food, but less activation in the lateral orbitofrontal cortex in response to receipt of food. This enhanced anticipation of the rewarding properties of food resembles the reward surfeit theory of obesity, suggesting that individuals at risk for obesity initially show hyper-responsivity of reward circuitry to high-calorie food cues, which would further increase intake of such foods. The fact that some of the reward circuit responses are decreased after consumption of the palatable food illustrates a reward deficit that may drive further intake to fulfill the need for food pleasure. The loss of control over food intake is an important criterion for FA. This inhibitory control was specifically investigated in youth with symptoms of FA by Hardee et al. [130] using a dedicated go/no go task. They demonstrated that YFAS-positive subjects showed deactivation in three clusters of brain regions including the middle temporal gyrus/occipital gyrus, precuneus/calcarine sulcus, and inferior frontal gyrus. The inferior frontal gyrus, notably, has been regularly described as being involved in executive and motor control, and decreased activation of this structure is usually interpreted as a lack of inhibitory control during a go/no go task. The decreased activity in the other clusters was perhaps related to decreased sustained attention during the task; a lack of attention, notably towards interoceptive perceptions, might contribute to the difficulty of obese people to regulate calorie intake, as postulated by Volkow et al. [139]. These results are somewhat corroborated by another study demonstrating that people who meet the YFAS criteria for FA have impaired performance monitoring, both on the behavioral and neural levels, consequently sharing some neurocognitive characteristics with patients diagnosed with substance use disorder [135].

Surprisingly, Gearhardt et al. [136] found no correlation between YFAS scores and BMI, which suggests that FA can occur in subjects within different body weight categories (as confirmed by the prevalence data previously cited). Even though the authors showed limited differences in reward circuitry activation between high- and low-YFAS subjects during food intake, high-YFAS individuals exhibited patterns of neural activation associated with reduced inhibitory control, which might explain their difficulty to resist food craving. In our opinion, two main questions arise from this work, and they still require additional retrospective and prospective studies to provide answers. First, does a FA profile in normal-weight individuals increase the risk to further declare obesity or other nutritional disorders, as postulated by the reward surfeit theory? Second, since almost all imaging studies performed in obese subjects did not include the YFAS score as a factor in their analysis, what is the probability that the brain pattern associated with YFAS in some "undiagnosed" individuals had influenced the general patterns observed in obese people? Considering the high prevalence of FA in obese patients, there is a significant bias in most studies describing the brain patterns characteristic of obesity, simply because there are many forms and behavioral phenotypes of obesity. Most studies probably characterized brain responses in obese subjects with different clinical profiles, since YFAS was not part of their routine checking, and a fair proportion of their obese subjects might very well have been YFAS-positive. Consequently, this percentage of "undiagnosed" YFAS patients might have influenced and biased our knowledge of the "obesity brain phenotype".

Interestingly, Beyer et al. [131] showed that symptoms of FA were not associated with the major structural brain differences correlated with BMI in the general population, but they might rather explain additional variance towards a lower right lateral orbitofrontal cortex thickness. Whether this anatomical specificity in the orbitofrontal cortex is responsible for the functional differences observed in this particular structure after food reward [136] necessitates further validation. However, the criteria for manifest FA were met by only 6% of the general population in this study [131], which does not indicate what the effects of YFAS symptoms on brain anatomy would be in a large cohort exclusively composed of obese subjects with higher prevalence of FA.

Using (18) F-2-fluoro-2-deoxyglusose (18FDG) positron emission tomography (TEP) instead of BOLD fMRI, Guzzardi et al. [133] investigated in overweight women the brain responses to high-calorie sweet food pictures and found greater activation in the thalamus, hypothalamus, midbrain, putamen, and occipital cortex, but not in the prefrontal and orbitofrontal cortices, in high-YFAS compared to low-YFAS subjects. Interestingly, in high-YFAS women, metabolic responsiveness in the orbitofrontal cortex was progressively lower with increasing YFAS severity and hunger subjective ratings. The authors' conclusions were that inadequate activation in response to the rewarding food in brain regions involved in inhibitory control and reward processing, in spite of greater activation in brain areas involved in somatosensory stimuli processing, reward and memory of hedonic behavior, distinguishes overweight women with FA from women with similar overweight but not FA [133]. It is also very interesting to highlight that the same authors demonstrated that a 3-month low-calorie diet was sufficient to reverse these specific brain activation patterns, which suggests that weight loss (3.8 kg or 4.1% of initial body weight in high YFAS) can help in correcting the neurocognitive anomalies associated with FA [133], exactly as demonstrated for the brain anomalies associated with obesity in formerly obese women who have successfully lost weight [140]. However, restrictive diets are usually ineffective in the long term and should not be advocated alone as obesity treatment.

The nutritional environment consequently has a major role in sustaining or correcting brain anomalies related to FA. The regular consumption of palatable high-calorie foods profoundly modifies many cognitive processes related to food perception, valuation, and motivation. The incentive sensitization theory postulates an excessive amplification of the psychological "wanting" of food, but it also highlights the particular role of external triggering cues and specific attention to these cues in maintaining a vicious circle. Food cue reactivity was found to be modified in overweight or obese women with YFAS-diagnosed FA, towards modest, elevated responses in the superior frontal gyrus for highly processed food pictures and more robust, decreased activations for minimally processed food cues, these responses being opposite in control subjects with similar overweight or obesity [132]. Exteroceptive stimuli such as visual cues are therefore very important in triggering and maintaining the neurocognitive patterns of food addiction. As reminded by Gearhardt et al. [136], activation in the nucleus accumbens is associated with craving in SUD and the amygdala is commonly implicated in drug cue reactivity and craving. Interestingly, Osadchiy et al. [134] demonstrated in healthy subjects with or without elevated BMI that YFAS scores had positive associations with functional connectivity between the amygdala and nucleus accumbens. In the same study, gut microbiota-derived indole metabolites were found to have a direct positive association with BMI and an indirect positive association with YFAS through functional connectivity of the nucleus accumbens [134], which might suggest a role of the gut microbiota in hedonic food intake in the context of FA. Both exteroceptive (e.g., related to food and environment) and interoceptive cues (e.g., related to the internal state and gut microbiota metabolites) are consequently important to understand how the neurocognitive patterns of FA emerge and establish in the long term, with the possibility to increase the risk for further psychological and metabolic disorders.

All these data support the existence of a specific FA brain phenotype that can be detected in normal-weight, overweight, or obese individuals and that is characterized by anomalies in the reward and inhibitory control processes, with likely corollary consequences in the limbic/emotional and cognitive/attentional spheres (Figure 2). Even though a recent meta-analysis of fMRI studies defends an addiction model of obesity, characterized by reduced cognitive control and interoceptive brain responses [141], this vision is probably restricted to part of the obesity spectrum and cannot be generalized to all forms of obesity. Further research is needed to better phenotype the neurobehavioral patterns of YFAS-positive subjects and disentangle their complex relationships and overlap with other diseases including obesity and other forms of addiction. Such work is mandatory to improve medical care because a better understanding of the patients' specificities leads to better treatment. As reminded by Ho et al. [142], post-obesity surgery patients are at increasing risk for developing alcohol and SUD, which likely represents an "addiction transfer" from food to other means of fulfilling the individuals' drives for pleasure or comfort. This risk could be especially increased if the presence of an FA profile has not been diagnosed and treated beforehand. The YFAS 2.0 questionnaire is a useful tool to predict continued emotional and binge eating behavior following obesity surgery [143] and might be used to identify subpopulations of patients with higher risk for unsuccessful obesity surgery. However, as a questionnaire, this method remains limited by the usual constraints and uncertainties of declarative diagnostic methods, which necessitates the development of additional diagnostic tools and markers, derived from brain imaging or biological measurements at the gut–microbiota–brain level, for example.

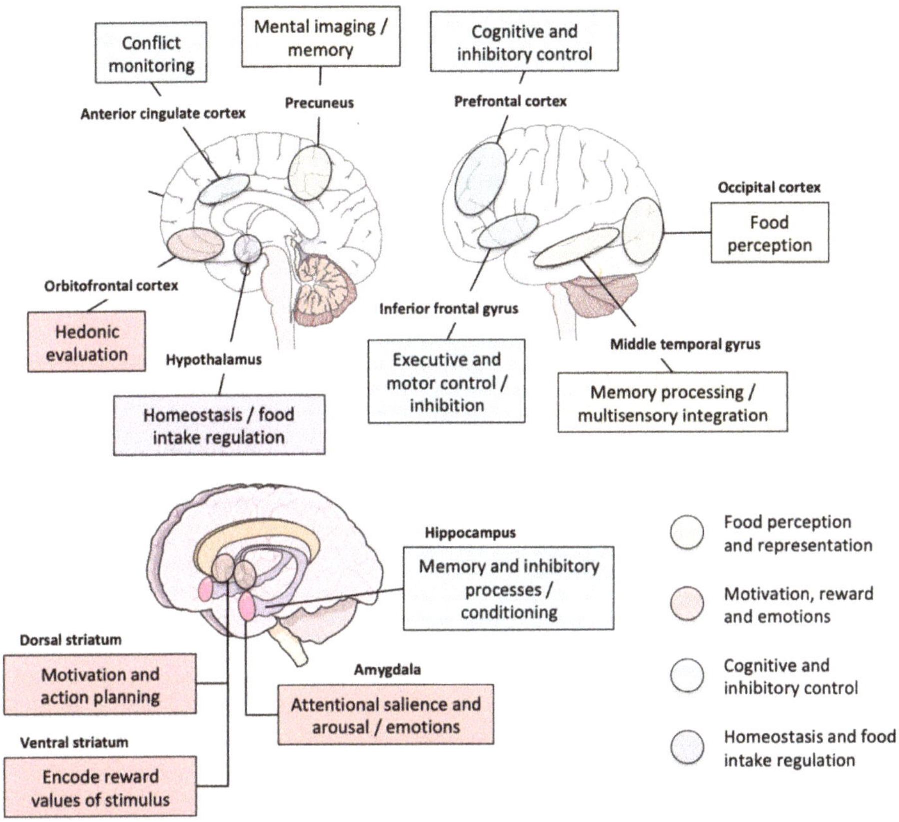

Figure 2. Neurocognitive functions and brain areas that are impacted by food addiction and for which people who meet the YFAS criteria for food addiction have different brain activity, metabolism, or functional connectivity compared to normal subjects. Please refer to Table 2 for details on results and imaging modalities used. Brain schematic representations were collected from Servier Medical Art (Suresnes, France; http://www.servier.fr).

2. General Discussion and Conclusions

The aim of this review was to enlighten the current concept of FA through four different angles and the spectrum of complementary disciplines: addiction medicine, nutrition in the context of obesity management, health psychology, and behavioral neurosciences. In our opinion, only a multidisciplinary perspective can render the complexity of FA and how it relates to environmental, social, and individual factors, while being inscribed in a continuum ranging from normal to disordered eating. This multifactorial comprehension of FA is needed to better organize prevention, diagnostic, and treatment, through the implementation of existing but also future strategies in the scope of personalized medicine. Even though the concept of personalized medicine sometimes appears hackneyed nowadays, it is particularly important and relevant in the context of FA and obesity treatment, considering the variety of individual profiles or situations, as well as the complex combination of environmental and individual factors at their origin.

Evaluation of FA is very unusual during first-line medical management and is not even systematic during obesity consultation. Practices are highly variable and dependent on the medical services and clinicians in charge of these consultations. As mentioned earlier, there is still no recognized psychological therapy validated for the management of FA in obese and/or ED patients, even though these patients represent high-risk profiles for FA. The role of FA in favoring body weight management problems and related comorbidities is still unclear, but early detection and treatment of FA might prevent the onset of further medical problems. The use of the YFAS questionnaire, when disordered eating is suspected, should become widespread in general medicine and specialized consultations, before referring the patient to a person with expertise in the management of ED and FA. This could be facilitated by the existence of a short version which is easier to fill. Because questionnaire studies are always subject to the biases and limitations of declarative methods, and denial could be present in some patients, there is a need for objective markers for which modern neuroimaging might represent an asset. Other biological markers might be explored—for example, at the metabolome and gut microbiota levels [134]—since the relationship between the gut microbiota and some neurocognitive processes has been extensively demonstrated.

Moreover, the FA construct has important treatment implications [21,51]. The standard approach to weight loss involves maintaining a healthy diet and physical exercise and is often associated with poor adherence and success rates. In the range of existing strategies are cognitive interventions [21], psychobehavioral interviews, and counseling via medical staff specialized in therapeutic education and nutrition, but also consultations with psychotherapists or psychiatrists. Addressing the psychological impact of internalized social stigma on patients remains pivotal, as several authors raised concerns that a diagnosis of food addiction could result in a double or additive stigma [50]. This emphasizes the need for empathic approaches and social support interventions in patients' management programs. For instance, the implementation of a self-help support group through a structured program could promote mutual support between persons with FA, break isolation, and create a space for sharing experiences [144]. Some authors also reviewed the beneficial input of online support options for food addiction, as well as other forms of self-help groups and sessions [145,146]. The restriction, or even the relative reduction, of some specific foods seen as addictive for a specific patient could be an option, contrary to the current view which aims at reducing dysfunctional dieting in favor of regular eating with flexible and moderate food consumption with no forbidden foods [51]. Such an approach is advocated by anonymous group meetings such as Overeaters Anonymous (OA), based directly on the 12-step program developed by Alcoholic Anonymous, which might help patients to break social isolation and the vicious circle at the origin of some forms of overeating [147], but we still lack perspective on the long-term success of such a strategy. Other initiatives can be applied in the context of FA, such as acceptance and commitment therapy (ACT) [148]. A wide range of motivational interviewing and cognitive behavioral therapies (CBT), which requires patients to critically evaluate the thoughts, feelings, and behaviors resulting in maladaptive responses and helps them to find their own solutions, adapted to their daily lives, can also be implemented in the context of FA and have already demonstrated their usefulness [149]. Such results must be replicated on larger cohorts and in the long term, and different types of CBT should be compared to provide recommendations about matching strategies to individual profiles, depending on their personality traits and susceptibilities, for example.

Several authors praised the use of innovative neuromodulation strategies to treat obesity, ED, but also FA [21,150], with the aim to modulate if not normalize some brain activities and neurocognitive processes involved in food intake control. If we put aside invasive strategies such as deep brain stimulation (DBS), there are still several candidates in the scope of minimally invasive strategies, such as transcranial direct current stimulation (tDCS), transcranial magnetic stimulation (TMS), and real-time fMRI neurofeedback. All these techniques can be used to stimulate or inhibit specific brain regions. The tDCS and TMS, via external electric or magnetic stimulation, are rather restricted to superficial (i.e., cortical) brain areas, such as the prefrontal cortex, which plays an important role in the cognitive control of eating. The rtfMRI can be applied to any brain area (including the deep striatal component of the reward circuit), since this method relies on the ability of the subject to voluntarily modify his/her brain activity on the basis of real-time feedback on this activity (e.g., via a visual gauge) combined with explicit or implicit tasks or mindfulness techniques. This approach has already been validated, with promising outcomes in healthy, overweight, and obese women, with the aim to reduce hunger and cravings [151,152].

Prevention, diagnosis, and treatment should also benefit from new developments in the scope of information and communication digital technologies. Innovative smart devices, smartphone applications, and online counseling platforms might provide potent tools for phenotyping individual profiles, adjusting eating habits on a daily basis, and providing information to both patients and actors of medical care. Preliminary data suggested the effectiveness of a mobile health app based on FA to treat young obese people [153]. The faster disordered eating and FA are detected, the easier corrective measures can be applied in order to prevent the onset of a vicious circle and complete loss of control over food intake, further leading to obesity and numerous comorbidities.

Author Contributions: All authors contributed to the introduction, general discussion and conclusions. R.M. was responsible for the "addiction medicine" section, R.T. for the "nutrition" section, A.C. for the "psychology" section, and D.V.-L. for the "neurosciences" section. All authors have read and agreed to the published version of the manuscript.

Funding: This research received no external funding.

Conflicts of Interest: The authors declare no conflict of interest.

References

1. Randolph, T.G. The descriptive features of food addiction; addictive eating and drinking. *Q. J. Stud. Alcohol* **1956**, *17*, 198–224. [CrossRef]
2. Markus, C.R.; Rogers, P.J.; Brouns, F.; Schepers, R. Eating dependence and weight gain; no human evidence for a 'sugar-addiction' model of overweight. *Appetite* **2017**, *114*, 64–72. [CrossRef]
3. Rogers, P.J.; Smit, H.J. Food craving and food "addiction": A critical review of the evidence from a biopsychosocial perspective. *Pharmacol. Biochem. Behav.* **2000**, *66*, 3–14. [CrossRef]
4. Gordon, E.L.; Ariel-Donges, A.H.; Bauman, V.; Merlo, L.J. What Is the Evidence for "Food Addiction?" A Systematic Review. *Nutrients* **2018**, *10*, 477. [CrossRef]
5. Schulte, E.M.; Avena, N.M.; Gearhardt, A.N. Which foods may be addictive? The roles of processing, fat content, and glycemic load. *PLoS ONE* **2015**, *10*, e0117959. [CrossRef]
6. Davis, C.; Carter, J.C. Compulsive overeating as an addiction disorder. A review of theory and evidence. *Appetite* **2009**, *53*, 1–8. [CrossRef]
7. Rogers, P.J. Food and drug addictions: Similarities and differences. *Pharmacol. Biochem. Behav.* **2017**, *153*, 182–190. [CrossRef]
8. Levine, A.S.; Kotz, C.M.; Gosnell, B.A. Sugars: Hedonic aspects, neuroregulation, and energy balance. *Am. J. Clin. Nutr.* **2003**, *78*, 834S–842S. [CrossRef]
9. Volkow, N.D.; Wang, G.J.; Fowler, J.S.; Tomasi, D.; Baler, R. Food and drug reward: Overlapping circuits in human obesity and addiction. *Curr. Top. Behav. Neurosci.* **2012**, *11*, 1–24. [CrossRef]
10. Schulte, E.M.; Yokum, S.; Potenza, M.N.; Gearhardt, A.N. Neural systems implicated in obesity as an addictive disorder: From biological to behavioral mechanisms. *Prog. Brain Res.* **2016**, *223*, 329–346. [CrossRef]
11. Carter, A.; Hendrikse, J.; Lee, N.; Yucel, M.; Verdejo-Garcia, A.; Andrews, Z.B.; Hall, W. The Neurobiology of "Food Addiction" and Its Implications for Obesity Treatment and Policy. *Annu. Rev. Nutr.* **2016**, *36*, 105–128. [CrossRef] [PubMed]
12. Ahmed, S.H.; Guillem, K.; Vandaele, Y. Sugar addiction: Pushing the drug-sugar analogy to the limit. *Curr. Opin. Clin. Nutr. Metab. Care* **2013**, *16*, 434–439. [CrossRef] [PubMed]
13. Meule, A.; Gearhardt, A.N. Food addiction in the light of DSM-5. *Nutrients* **2014**, *6*, 3653–3671. [CrossRef] [PubMed]
14. Gearhardt, A.N.; Roberto, C.A.; Seamans, M.J.; Corbin, W.R.; Brownell, K.D. Preliminary validation of the Yale Food Addiction Scale for children. *Eat. Behav.* **2013**, *14*, 508–512. [CrossRef] [PubMed]
15. Pursey, K.M.; Stanwell, P.; Gearhardt, A.N.; Collins, C.E.; Burrows, T.L. The prevalence of food addiction as assessed by the Yale Food Addiction Scale: A systematic review. *Nutrients* **2014**, *6*, 4552–4590. [CrossRef] [PubMed]

16. American Society of Addition Medicine. Definition of Addiction. Available online: https://www.asam.org/quality-practice/definition-of-addiction (accessed on 15 September 2019).
17. Hauck, C.; Cook, B.; Ellrott, T. Food addiction, eating addiction and eating disorders. *Proc. Nutr. Soc.* **2020**, *79*, 103–112. [CrossRef]
18. Fletcher, P.C.; Kenny, P.J. Food addiction: A valid concept? *Neuropsychopharmacol. Off. Publ. Am. Coll. Neuropsychopharmacol.* **2018**, *43*, 2506–2513. [CrossRef]
19. Long, C.G.; Blundell, J.E.; Finlayson, G. A Systematic Review of the Application and Correlates of YFAS-Diagnosed 'Food Addiction' in Humans: Are Eating-Related 'Addictions' a Cause for Concern or Empty Concepts? *Obes. Facts* **2015**, *8*, 386–401. [CrossRef]
20. Hebebrand, J.; Albayrak, Ö.; Adan, R.; Antel, J.; Dieguez, C.; de Jong, J.; Leng, G.; Menzies, J.; Mercer, J.G.; Murphy, M.; et al. "Eating addiction", rather than "food addiction", better captures addictive-like eating behavior. *Neurosci. Biobehav. Rev.* **2014**, *47*, 295–306. [CrossRef]
21. Adams, R.C.; Sedgmond, J.; Maizey, L.; Chambers, C.D.; Lawrence, N.S. Food Addiction: Implications for the Diagnosis and Treatment of Overeating. *Nutrients* **2019**, *11*, 2086. [CrossRef]
22. Bonder, R.; Davis, C.; Kuk, J.L.; Loxton, N.J. Compulsive "grazing" and addictive tendencies towards food. *Eur. Eat. Disord. Rev. J. Eat. Disord. Assoc.* **2018**, *26*, 569–573. [CrossRef] [PubMed]
23. Tiffany, S.T.; Wray, J.M. The clinical significance of drug craving. *Ann. N. Y. Acad. Sci.* **2012**, *1248*, 1–17. [CrossRef] [PubMed]
24. Boswell, R.G.; Kober, H. Food cue reactivity and craving predict eating and weight gain: A meta-analytic review. *Obes. Rev. Off. J. Int. Assoc. Study Obes.* **2016**, *17*, 159–177. [CrossRef] [PubMed]
25. Gearhardt, A.N.; Rizk, M.T.; Treat, T.A. The association of food characteristics and individual differences with ratings of craving and liking. *Appetite* **2014**, *79*, 166–173. [CrossRef] [PubMed]
26. Joyner, M.A.; Gearhardt, A.N.; White, M.A. Food craving as a mediator between addictive-like eating and problematic eating outcomes. *Eat. Behav.* **2015**, *19*, 98–101. [CrossRef] [PubMed]
27. Hurt, R.T.; Edakkanambeth Varayil, J.; Mundi, M.S.; Martindale, R.G.; Ebbert, J.O. Designation of obesity as a disease: Lessons learned from alcohol and tobacco. *Curr. Gastroenterol. Rep.* **2014**, *16*, 415. [CrossRef]
28. Spring, B.; Schneider, K.; Smith, M.; Kendzor, D.; Appelhans, B.; Hedeker, D.; Pagoto, S. Abuse potential of carbohydrates for overweight carbohydrate cravers. *Psychopharmacology* **2008**, *197*, 637–647. [CrossRef]
29. Gearhardt, A.N.; Corbin, W.R.; Brownell, K.D. Preliminary validation of the Yale Food Addiction Scale. *Appetite* **2009**, *52*, 430–436. [CrossRef]
30. Gearhardt, A.N.; Corbin, W.R.; Brownell, K.D. Development of the Yale Food Addiction Scale Version 2.0. *Psychol. Addict. Behav. J. Soc. Psychol. Addict. Behav.* **2016**, *30*, 113–121. [CrossRef]
31. Saunders, J.B.; Degenhardt, L.; Reed, G.M.; Poznyak, V. Alcohol Use Disorders in ICD-11: Past, Present, and Future. *Alcohol. Clin. Exp. Res.* **2019**, *43*, 1617–1631. [CrossRef]
32. Burrows, T.; Kay-Lambkin, F.; Pursey, K.; Skinner, J.; Dayas, C. Food addiction and associations with mental health symptoms: A systematic review with meta-analysis. *J. Hum. Nutr. Diet. Off. J. Br. Diet. Assoc.* **2018**, *31*, 544–572. [CrossRef]
33. Ivezaj, V.; Wiedemann, A.A.; Grilo, C.M. Food addiction and bariatric surgery: A systematic review of the literature. *Obes. Rev. Off. J. Int. Assoc. Study Obes.* **2017**, *18*, 1386–1397. [CrossRef] [PubMed]
34. Maxwell, A.L.; Gardiner, E.; Loxton, N.J. Investigating the relationship between reward sensitivity, impulsivity, and food addiction: A systematic review. *Eur. Eat. Disord. Rev. J. Eat. Disord. Assoc.* **2020**, *28*, 368–384. [CrossRef] [PubMed]
35. Penzenstadler, L.; Soares, C.; Karila, L.; Khazaal, Y. Systematic Review of Food Addiction as Measured with the Yale Food Addiction Scale: Implications for the Food Addiction Construct. *Curr. Neuropharmacol.* **2019**, *17*, 526–538. [CrossRef] [PubMed]
36. Canan, F.; Karaca, S.; Sogucak, S.; Gecici, O.; Kuloglu, M. Eating disorders and food addiction in men with heroin use disorder: A controlled study. *Eat. Weight Disord. EWD* **2017**, *22*, 249–257. [CrossRef]
37. Hauck, C.; Schipfer, M.; Ellrott, T.; Cook, B. The relationship between food addiction and patterns of disordered eating with exercise dependence: In amateur endurance athletes. *Eat. Weight Disord. EWD* **2019**, *25*, 1573–1582. [CrossRef]
38. Li, W.Q.; McGeary, J.; Cho, E.; Flint, A.; Wu, S.; Ascherio, A.; Rimm, E.; Field, A.; Qureshi, A. Indoor tanning bed use and risk of food addiction based on the modified Yale Food Addiction Scale. *J. Biomed. Res.* **2016**, *31*, 31–39. [CrossRef]

39. Mies, G.W.; Treur, J.L.; Larsen, J.K.; Halberstadt, J.; Pasman, J.A.; Vink, J.M. The prevalence of food addiction in a large sample of adolescents and its association with addictive substances. *Appetite* **2017**, *118*, 97–105. [CrossRef]
40. Tinghino, B.; Lugoboni, F.; Amatulli, A.; Biasin, C.; Bramani Araldi, M.; Cantiero, D.; Cremaschini, M.; Galimberti, G.L.; Giusti, S.; Grosina, C.; et al. The FODRAT study (FOod addiction, DRugs, Alcohol and Tobacco): First data on food addiction prevalence among patients with addiction to drugs, tobacco and alcohol. *Eat. Weight Disord. EWD* **2020**. [CrossRef]
41. Lennerz, B.; Lennerz, J.K. Food Addiction, High-Glycemic-Index Carbohydrates, and Obesity. *Clin. Chem.* **2018**, *64*, 64–71. [CrossRef]
42. Wiss, D.A.; Avena, N.; Rada, P. Sugar Addiction: From Evolution to Revolution. *Front. Psychiatry* **2018**, *9*, 545. [CrossRef] [PubMed]
43. Westwater, M.L.; Fletcher, P.C.; Ziauddeen, H. Sugar addiction: The state of the science. *Eur. J. Nutr.* **2016**, *55*, 55–69. [CrossRef] [PubMed]
44. Sarkar, S.; Kochhar, K.P.; Khan, N.A. Fat Addiction: Psychological and Physiological Trajectory. *Nutrients* **2019**, *11*, 2785. [CrossRef] [PubMed]
45. Schulte, E.M.; Smeal, J.K.; Gearhardt, A.N. Foods are differentially associated with subjective effect report questions of abuse liability. *PLoS ONE* **2017**, *12*, e0184220. [CrossRef]
46. Schulte, E.M.; Sonneville, K.R.; Gearhardt, A.N. Subjective experiences of highly processed food consumption in individuals with food addiction. *Psychol. Addict. Behav. J. Soc. Psychol. Addict. Behav.* **2019**, *33*, 144–153. [CrossRef]
47. Junghanns, K.; Veltrup, C.; Wetterling, T. Craving shift in chronic alcoholics. *Eur. Addict. Res.* **2000**, *6*, 64–70. [CrossRef]
48. King, W.C.; Chen, J.Y.; Mitchell, J.E.; Kalarchian, M.A.; Steffen, K.J.; Engel, S.G.; Courcoulas, A.P.; Pories, W.J.; Yanovski, S.Z. Prevalence of alcohol use disorders before and after bariatric surgery. *JAMA* **2012**, *307*, 2516–2525. [CrossRef]
49. Howard, M.; McMillen, C.; Nower, L.; Elze, D.; Edmond, T.; Bricout, J. Denial in addiction: Toward an integrated stage and process model—Qualitative findings. *J. Psychoact. Drugs* **2002**, *34*, 371–382. [CrossRef]
50. Cassin, S.E.; Buchman, D.Z.; Leung, S.E.; Kantarovich, K.; Hawa, A.; Carter, A.; Sockalingam, S. Ethical, Stigma, and Policy Implications of Food Addiction: A Scoping Review. *Nutrients* **2019**, *11*, 710. [CrossRef]
51. Meule, A. A Critical Examination of the Practical Implications Derived from the Food Addiction Concept. *Curr. Obes. Rep.* **2019**, *8*, 11–17. [CrossRef]
52. O'Brien, K.S.; Puhl, R.M.; Latner, J.D.; Lynott, D.; Reid, J.D.; Vakhitova, Z.; Hunter, J.A.; Scarf, D.; Jeanes, R.; Bouguettaya, A.; et al. The Effect of a Food Addiction Explanation Model for Weight Control and Obesity on Weight Stigma. *Nutrients* **2020**, *12*, 294. [CrossRef] [PubMed]
53. Ruddock, H.K.; Christiansen, P.; Jones, A.; Robinson, E.; Field, M.; Hardman, C.A. Believing in food addiction: Helpful or counterproductive for eating behavior? *Obesity (Silver Spring Md.)* **2016**, *24*, 1238–1243. [CrossRef]
54. Matta, J.; Zins, M.; Feral-Pierssens, A.L.; Carette, C.; Ozguler, A.; Goldberg, M.; Czernichow, S. Prévalence du surpoids, de l'obésité et des facteurs de risque cardio-métaboliques dans la cohorte Constances. *Bull. Epidémiol. Hebd.* **2016**, *35–36*, 640–646.
55. The State of Obesity. Available online: http://healthyamericans.org/reports/stateofobesity2016/ (accessed on 20 September 2020).
56. Malnick, S.D.; Knobler, H. The medical complications of obesity. *QJM* **2006**, *99*, 565–579. [CrossRef] [PubMed]
57. Effertz, T.; Engel, S.; Verheyen, F.; Linder, R. The costs and consequences of obesity in Germany: A new approach from a prevalence and life-cycle perspective. *Eur. J. Health Econ.* **2016**, *17*, 1141–1158. [CrossRef] [PubMed]
58. Schwander, B.; Hiligsmann, M.; Nuijten, M.; Evers, S. Systematic review and overview of health economic evaluation models in obesity prevention and therapy. *Expert Rev. Pharm. Outcomes Res.* **2016**, *16*, 561–570. [CrossRef] [PubMed]
59. Blouin, C.; Hamel, D.; Vandal, N.; Barry, A.D.; Lo, E.; Lacroix, G.; Laguë, J.; Langlois, M.-F.; Martel, S.; Michaud, P.-C.; et al. The economic consequences of obesity and overweight among adults in Quebec. *Can. J. Public Health* **2017**, *107*, 507–513. [CrossRef] [PubMed]

60. Sjostrom, L.; Narbro, K.; Sjostrom, C.D.; Karason, K.; Larsson, B.; Wedel, H.; Lystig, T.; Sullivan, M.; Bouchard, C.; Carlsson, B.; et al. Effects of bariatric surgery on mortality in Swedish obese subjects. *N. Engl. J. Med.* **2007**, *357*, 741–752. [CrossRef]
61. Eliasson, B.; Liakopoulos, V.; Franzen, S.; Naslund, I.; Svensson, A.M.; Ottosson, J.; Gudbjornsdottir, S. Cardiovascular disease and mortality in patients with type 2 diabetes after bariatric surgery in Sweden: A nationwide, matched, observational cohort study. *Lancet Diabetes Endocrinol.* **2015**, *3*, 847–854. [CrossRef]
62. Sjostrom, L.; Peltonen, M.; Jacobson, P.; Ahlin, S.; Andersson-Assarsson, J.; Anveden, A.; Bouchard, C.; Carlsson, B.; Karason, K.; Lonroth, H.; et al. Association of bariatric surgery with long-term remission of type 2 diabetes and with microvascular and macrovascular complications. *JAMA* **2014**, *311*, 2297–2304. [CrossRef]
63. Sjostrom, L.; Peltonen, M.; Jacobson, P.; Sjostrom, C.D.; Karason, K.; Wedel, H.; Ahlin, S.; Anveden, A.; Bengtsson, C.; Bergmark, G.; et al. Bariatric surgery and long-term cardiovascular events. *JAMA* **2012**, *307*, 56–65. [CrossRef] [PubMed]
64. Lenoir, M.; Serre, F.; Cantin, L.; Ahmed, S.H. Intense sweetness surpasses cocaine reward. *PLoS ONE* **2007**, *2*, e698. [CrossRef] [PubMed]
65. Schulte, E.M.; Gearhardt, A.N. Associations of Food Addiction in a Sample Recruited to Be Nationally Representative of the United States. *Eur. Eat. Disord. Rev. J. Eat. Disord. Assoc.* **2018**, *26*, 112–119. [CrossRef] [PubMed]
66. Som, M.; Val-Laillet, D.; Constant, A.; Moirand, R.; Thibault, R. Prevalence of food addiction (FA) diagnosed by the Yale Food Addiction Scale version 2.0 (YFAS 2.0): Prospective study in a referral centre for obesity surgery. *Clin. Nutr.* **2018**, *37*, S264. [CrossRef]
67. Brunault, P.; Ducluzeau, P.H.; Bourbao-Tournois, C.; Delbachian, I.; Couet, C.; Reveillere, C.; Ballon, N. Food Addiction in Bariatric Surgery Candidates: Prevalence and Risk Factors. *Obes. Surg.* **2016**, *26*, 1650–1653. [CrossRef]
68. Hauck, C.; Weiss, A.; Schulte, E.M.; Meule, A.; Ellrott, T. Prevalence of 'Food Addiction' as Measured with the Yale Food Addiction Scale 2.0 in a Representative German Sample and Its Association with Sex, Age and Weight Categories. *Obes. Facts* **2017**, *10*, 12–24. [CrossRef]
69. Davis, C.; Curtis, C.; Levitan, R.D.; Carter, J.C.; Kaplan, A.S.; Kennedy, J.L. Evidence that 'food addiction' is a valid phenotype of obesity. *Appetite* **2011**, *57*, 711–717. [CrossRef]
70. Pepino, M.Y.; Stein, R.I.; Eagon, J.C.; Klein, S. Bariatric surgery-induced weight loss causes remission of food addiction in extreme obesity. *Obesity (Silver Spring Md.)* **2014**, *22*, 1792–1798. [CrossRef]
71. Davis, C. A commentary on the associations among 'food addiction', binge eating disorder, and obesity: Overlapping conditions with idiosyncratic clinical features. *Appetite* **2017**, *115*, 3–8. [CrossRef]
72. Gearhardt, A.N.; White, M.A.; Masheb, R.M.; Morgan, P.T.; Crosby, R.D.; Grilo, C.M. An examination of the food addiction construct in obese patients with binge eating disorder. *Int. J. Eat. Disord.* **2012**, *45*, 657–663. [CrossRef]
73. Gearhardt, A.N.; White, M.A.; Masheb, R.M.; Grilo, C.M. An examination of food addiction in a racially diverse sample of obese patients with binge eating disorder in primary care settings. *Compr. Psychiatry* **2013**, *54*, 500–505. [CrossRef] [PubMed]
74. de Vries, S.K.; Meule, A. Food Addiction and Bulimia Nervosa: New Data Based on the Yale Food Addiction Scale 2.0. *Eur. Eat. Disord. Rev. J. Eat. Disord. Assoc.* **2016**, *24*, 518–522. [CrossRef] [PubMed]
75. Kiyici, S.; Koca, N.; Sigirli, D.; Aslan, B.B.; Guclu, M.; Kisakol, G. Food Addiction Correlates with Psychosocial Functioning More Than Metabolic Parameters in Patients with Obesity. *Metab. Syndr. Relat. Disord.* **2020**, *18*, 161–167. [CrossRef] [PubMed]
76. Busetto, L.; Segato, G.; De Luca, M.; De Marchi, F.; Foletto, M.; Vianello, M.; Valeri, M.; Favretti, F.; Enzi, G. Weight loss and postoperative complications in morbidly obese patients with binge eating disorder treated by laparoscopic adjustable gastric banding. *Obes. Surg.* **2005**, *15*, 195–201. [CrossRef] [PubMed]
77. Wolnerhanssen, B.K.; Peters, T.; Kern, B.; Schotzau, A.; Ackermann, C.; von Flue, M.; Peterli, R. Predictors of outcome in treatment of morbid obesity by laparoscopic adjustable gastric banding: Results of a prospective study of 380 patients. *Surg. Obes. Relat. Dis.* **2008**, *4*, 500–506. [CrossRef]
78. Steffen, K.J.; Engel, S.G.; Wonderlich, J.A.; Pollert, G.A.; Sondag, C. Alcohol and Other Addictive Disorders Following Bariatric Surgery: Prevalence, Risk Factors and Possible Etiologies. *Eur. Eat. Disord. Rev. J. Eat. Disord. Assoc.* **2015**, *23*, 442–450. [CrossRef]

79. Holgerson, A.A.; Clark, M.M.; Ames, G.E.; Collazo-Clavell, M.L.; Kellogg, T.A.; Graszer, K.M.; Kalsy, S.A.; Grothe, K. Association of Adverse Childhood Experiences and Food Addiction to Bariatric Surgery Completion and Weight Loss Outcome. *Obes. Surg.* **2018**, *28*, 3386–3392. [CrossRef]
80. Lent, M.R.; Eichen, D.M.; Goldbacher, E.; Wadden, T.A.; Foster, G.D. Relationship of food addiction to weight loss and attrition during obesity treatment. *Obesity (Silver Spring Md.)* **2014**, *22*, 52–55. [CrossRef]
81. Burmeister, J.M.; Hinman, N.; Koball, A.; Hoffmann, D.A.; Carels, R.A. Food addiction in adults seeking weight loss treatment. Implications for psychosocial health and weight loss. *Appetite* **2013**, *60*, 103–110. [CrossRef]
82. Sevincer, G.M.; Konuk, N.; Bozkurt, S.; Coskun, H. Food addiction and the outcome of bariatric surgery at 1-year: Prospective observational study. *Psychiatry Res.* **2016**, *244*, 159–164. [CrossRef]
83. Conason, A.; Teixeira, J.; Hsu, C.H.; Puma, L.; Knafo, D.; Geliebter, A. Substance use following bariatric weight loss surgery. *JAMA Surg.* **2013**, *148*, 145–150. [CrossRef] [PubMed]
84. Li, L.; Wu, L.T. Substance use after bariatric surgery: A review. *J. Psychiatr. Res.* **2016**, *76*, 16–29. [CrossRef]
85. Reslan, S.; Saules, K.K.; Greenwald, M.K.; Schuh, L.M. Substance misuse following Roux-en-Y gastric bypass surgery. *Subst. Use Misuse* **2014**, *49*, 405–417. [CrossRef]
86. David, L.A.; Sijercic, I.; Cassin, S.E. Preoperative and post-operative psychosocial interventions for bariatric surgery patients: A systematic review. *Obes. Rev. Off. J. Int. Assoc. Study Obes.* **2020**, *21*, e12926. [CrossRef]
87. Sharma, A.M.; Kushner, R.F. A proposed clinical staging system for obesity. *Int. J. Obes.* **2009**, *33*, 289–295. [CrossRef]
88. Padwal, R.S.; Pajewski, N.M.; Allison, D.B.; Sharma, A.M. Using the Edmonton obesity staging system to predict mortality in a population-representative cohort of people with overweight and obesity. *CMAJ* **2011**, *183*, E1059–E1066. [CrossRef] [PubMed]
89. Lowe, M.R.; Gleaves, D.H.; DiSimone-Weiss, R.T.; Furgueson, C.; Gayda, C.A.; Kolsky, P.A.; Neal-Walden, T.; Nelsen, L.A.; McKinney, S. Restraint, dieting, and the continuum model of bulimia nervosa. *J. Abnorm. Psychol.* **1996**, *105*, 508–517. [CrossRef] [PubMed]
90. Meule, A.; Vogele, C. The psychology of eating. *Front. Psychol.* **2013**, *4*, 215. [CrossRef] [PubMed]
91. Burton, P.; Smit, H.J.; Lightowler, H.J. The influence of restrained and external eating patterns on overeating. *Appetite* **2007**, *49*, 191–197. [CrossRef] [PubMed]
92. Berthoud, H.R. The neurobiology of food intake in an obesogenic environment. *Proc. Nutr. Soc.* **2012**, *71*, 478–487. [CrossRef] [PubMed]
93. Reba-Harrelson, L.; Von Holle, A.; Hamer, R.M.; Swann, R.; Reyes, M.L.; Bulik, C.M. Patterns and prevalence of disordered eating and weight control behaviors in women ages 25-45. *Eat. Weight Disord. EWD* **2009**, *14*, e190–e198. [CrossRef] [PubMed]
94. Burton, A.L.; Abbott, M.J. Processes and pathways to binge eating: Development of an integrated cognitive and behavioural model of binge eating. *J. Eat. Disord.* **2019**, *7*, 18. [CrossRef] [PubMed]
95. Mitchison, D.; Touyz, S.; Gonzalez-Chica, D.A.; Stocks, N.; Hay, P. How abnormal is binge eating? 18-Year time trends in population prevalence and burden. *Acta Psychiatr. Scand.* **2017**, *136*, 147–155. [CrossRef] [PubMed]
96. Koenders, P.G.; van Strien, T. Emotional eating, rather than lifestyle behavior, drives weight gain in a prospective study in 1562 employees. *J. Occup. Environ. Med.* **2011**, *53*, 1287–1293. [CrossRef] [PubMed]
97. Masheb, R.M.; Grilo, C.M. Emotional overeating and its associations with eating disorder psychopathology among overweight patients with binge eating disorder. *Int. J. Eat. Disord.* **2006**, *39*, 141–146. [CrossRef]
98. De Young, K.P.; Zander, M.; Anderson, D.A. Beliefs about the emotional consequences of eating and binge eating frequency. *Eat. Behav.* **2014**, *15*, 31–36. [CrossRef]
99. Pinaquy, S.; Chabrol, H.; Simon, C.; Louvet, J.P.; Barbe, P. Emotional eating, alexithymia, and binge-eating disorder in obese women. *Obes. Res.* **2003**, *11*, 195–201. [CrossRef]
100. Van Strien, T.; Schippers, G.M.; Cox, W.M. On the relationship between emotional and external eating behavior. *Addict. Behav.* **1995**, *20*, 585–594. [CrossRef]
101. Brunault, P.; Courtois, R.; Gearhardt, A.N.; Gaillard, P.; Journiac, K.; Cathelain, S.; Reveillere, C.; Ballon, N. Validation of the French Version of the DSM-5 Yale Food Addiction Scale in a Nonclinical Sample. *Can. J. Psychiatry Rev. Can. Psychiatr.* **2017**, *62*, 199–210. [CrossRef]

102. Constant, A.; Gautier, Y.; Coquery, N.; Thibault, R.; Moirand, R.; Val-Laillet, D. Emotional overeating is common and negatively associated with alcohol use in normal-weight female university students. *Appetite* **2018**, *129*, 186–191. [CrossRef]
103. de Lauzon, B.; Romon, M.; Deschamps, V.; Lafay, L.; Borys, J.M.; Karlsson, J.; Ducimetiere, P.; Charles, M.A.; Fleurbaix Laventie Ville Sante Study, G. The Three-Factor Eating Questionnaire-R18 is able to distinguish among different eating patterns in a general population. *J. Nutr.* **2004**, *134*, 2372–2380. [CrossRef] [PubMed]
104. Pankevich, D.E.; Teegarden, S.L.; Hedin, A.D.; Jensen, C.L.; Bale, T.L. Caloric restriction experience reprograms stress and orexigenic pathways and promotes binge eating. *J. Neurosci.* **2010**, *30*, 16399–16407. [CrossRef] [PubMed]
105. Brogan, A.; Hevey, D. Eating styles in the morbidly obese: Restraint eating, but not emotional and external eating, predicts dietary behaviour. *Psychol. Health* **2013**, *28*, 714–725. [CrossRef] [PubMed]
106. Rosenbaum, D.L.; White, K.S. The relation of anxiety, depression, and stress to binge eating behavior. *J. Health Psychol.* **2015**, *20*, 887–898. [CrossRef]
107. Jung, J.Y.; Kim, K.H.; Woo, H.Y.; Shin, D.W.; Shin, Y.C.; Oh, K.S.; Shin, E.H.; Lim, S.W. Binge eating is associated with trait anxiety in Korean adolescent girls: A cross sectional study. *BMC Womens Health* **2017**, *17*, 8. [CrossRef]
108. Rosenbaum, D.L.; White, K.S. The Role of Anxiety in Binge Eating Behavior: A Critical Examination of Theory and Empirical Literature. *Health Psychol. Res.* **2013**, *1*, e19. [CrossRef] [PubMed]
109. Jimenez-Murcia, S.; Aguera, Z.; Paslakis, G.; Munguia, L.; Granero, R.; Sanchez-Gonzalez, J.; Sanchez, I.; Riesco, N.; Gearhardt, A.N.; Dieguez, C.; et al. Food Addiction in Eating Disorders and Obesity: Analysis of Clusters and Implications for Treatment. *Nutrients* **2019**, *11*, 2633. [CrossRef]
110. Elfhag, K.; Morey, L.C. Personality traits and eating behavior in the obese: Poor self-control in emotional and external eating but personality assets in restrained eating. *Eat. Behav.* **2008**, *9*, 285–293. [CrossRef]
111. Lynam, D.R.; Smith, G.T.; Whiteside, S.P.; Cyders, M.A. *Assessing Five Personality Pathways to Impulsive Behavior (Technical Report)*; Purdue University: West Lafayette, Indiana, 2006.
112. Moynihan, A.B.; van Tilburg, W.A.; Igou, E.R.; Wisman, A.; Donnelly, A.E.; Mulcaire, J.B. Eaten up by boredom: Consuming food to escape awareness of the bored self. *Front. Psychol.* **2015**, *6*, 369. [CrossRef]
113. Constant, A.; Val-Laillet, D.; Joubert, A.; Foret, K.; Thibault, R.; Moirand, R. Depressive symptoms are related to boredom proneness in patients receiving hospital care, regardless of alcohol status, lifestyle, or social support. *J. Health Psychol.* **2019**. [CrossRef]
114. Tomiyama, A.J.; Carr, D.; Granberg, E.M.; Major, B.; Robinson, E.; Sutin, A.R.; Brewis, A. How and why weight stigma drives the obesity 'epidemic' and harms health. *BMC Med.* **2018**, *16*, 123. [CrossRef] [PubMed]
115. Rogers, P.J. Obesity—Is food addiction to blame? *Addiction* **2011**, *106*, 1213–1214; discussion 1219–1220. [CrossRef] [PubMed]
116. Barrett, P. Structural equation modeling: Adjudging model fit. *Personal. Individ. Differ.* **2007**, *42*, 815–824. [CrossRef]
117. Brown, T.A. *Confirmatory Factor Analysis for Applied Research*, 2nd ed.; The Guilford Press: New York, NY, USA, 2015; p. xvii, 462.
118. American Psychiatric Association. *Diagnostic and Statistical Manual of Mental Disorders*, 5th ed.; American Psychiatric Publishing: Washington, DC, USA, 2013.
119. Volkow, N.D.; Michaelides, M.; Baler, R. The Neuroscience of Drug Reward and Addiction. *Physiol. Rev.* **2019**, *99*, 2115–2140. [CrossRef]
120. Lindgren, E.; Gray, K.; Miller, G.; Tyler, R.; Wiers, C.E.; Volkow, N.D.; Wang, G.J. Food addiction: A common neurobiological mechanism with drug abuse. *Front. Biosci.* **2018**, *23*, 811–836. [CrossRef]
121. Volkow, N.D.; Morales, M. The Brain on Drugs: From Reward to Addiction. *Cell* **2015**, *162*, 712–725. [CrossRef]
122. Berridge, K.C.; Robinson, T.E. Liking, wanting, and the incentive-sensitization theory of addiction. *Am. Psychol.* **2016**, *71*, 670–679. [CrossRef]
123. Robinson, T.E.; Berridge, K.C. Review. The incentive sensitization theory of addiction: Some current issues. *Philos. Trans. R. Soc. Lond. Ser. B Biol. Sci.* **2008**, *363*, 3137–3146. [CrossRef]
124. Everitt, B.J.; Robbins, T.W. Neural systems of reinforcement for drug addiction: From actions to habits to compulsion. *Nat. Neurosci.* **2005**, *8*, 1481–1489. [CrossRef]

125. Volkow, N.D.; Baler, R.D. Addiction science: Uncovering neurobiological complexity. *Neuropharmacology* **2014**, *76 Pt B*, 235–249. [CrossRef]
126. Novelle, M.G.; Dieguez, C. Food Addiction and Binge Eating: Lessons Learned from Animal Models. *Nutrients* **2018**, *10*, 71. [CrossRef] [PubMed]
127. Pai, N.; Vella, S.L.; Richardson, K. Is food addiction a valid phenomenon through the lens of the DSM-5? *Aust. N. Z. J. Psychiatry* **2014**, *48*, 216–218. [CrossRef] [PubMed]
128. Granero, R.; Hilker, I.; Aguera, Z.; Jimenez-Murcia, S.; Sauchelli, S.; Islam, M.A.; Fagundo, A.B.; Sanchez, I.; Riesco, N.; Dieguez, C.; et al. Food addiction in a Spanish sample of eating disorders: DSM-5 diagnostic subtype differentiation and validation data. *Eur. Eat. Disord. Rev. J. Eat. Disord. Assoc.* **2014**, *22*, 389–396. [CrossRef] [PubMed]
129. Khine, M.T.; Ota, A.; Gearhardt, A.N.; Fujisawa, A.; Morita, M.; Minagawa, A.; Li, Y.; Naito, H.; Yatsuya, H. Validation of the Japanese Version of the Yale Food Addiction Scale 2.0 (J-YFAS 2.0). *Nutrients* **2019**, *11*, 687. [CrossRef] [PubMed]
130. Hardee, J.E.; Phaneuf, C.; Cope, L.; Zucker, R.; Gearhardt, A.; Heitzeg, M. Neural correlates of inhibitory control in youth with symptoms of food addiction. *Appetite* **2020**, *148*, 104578. [CrossRef] [PubMed]
131. Beyer, F.; Garcia-Garcia, I.; Heinrich, M.; Schroeter, M.L.; Sacher, J.; Luck, T.; Riedel-Heller, S.G.; Stumvoll, M.; Villringer, A.; Witte, A.V. Neuroanatomical correlates of food addiction symptoms and body mass index in the general population. *Hum. Brain Mapp.* **2019**, *40*, 2747–2758. [CrossRef]
132. Schulte, E.M.; Yokum, S.; Jahn, A.; Gearhardt, A.N. Food cue reactivity in food addiction: A functional magnetic resonance imaging study. *Physiol. Behav.* **2019**, *208*, 112574. [CrossRef]
133. Guzzardi, M.A.; Garelli, S.; Agostini, A.; Filidei, E.; Fanelli, F.; Giorgetti, A.; Mezzullo, M.; Fucci, S.; Mazza, R.; Vicennati, V.; et al. Food addiction distinguishes an overweight phenotype that can be reversed by low calorie diet. *Eur. Eat. Disord. Rev. J. Eat. Disord. Assoc.* **2018**, *26*, 657–670. [CrossRef]
134. Osadchiy, V.; Labus, J.S.; Gupta, A.; Jacobs, J.; Ashe-McNalley, C.; Hsiao, E.Y.; Mayer, E.A. Correlation of tryptophan metabolites with connectivity of extended central reward network in healthy subjects. *PLoS ONE* **2018**, *13*, e0201772. [CrossRef]
135. Franken, I.H.A.; Nijs, I.M.T.; Toes, A.; van der Veen, F.M. Food addiction is associated with impaired performance monitoring. *Biol. Psychol.* **2018**, *131*, 49–53. [CrossRef]
136. Gearhardt, A.N.; Yokum, S.; Orr, P.T.; Stice, E.; Corbin, W.R.; Brownell, K.D. Neural correlates of food addiction. *Arch. Gen. Psychiatry* **2011**, *68*, 808–816. [CrossRef] [PubMed]
137. Stice, E.; Spoor, S.; Bohon, C.; Veldhuizen, M.G.; Small, D.M. Relation of reward from food intake and anticipated food intake to obesity: A functional magnetic resonance imaging study. *J. Abnorm. Psychol.* **2008**, *117*, 924–935. [CrossRef] [PubMed]
138. Stice, E.; Spoor, S.; Ng, J.; Zald, D.H. Relation of obesity to consummatory and anticipatory food reward. *Physiol. Behav.* **2009**, *97*, 551–560. [CrossRef] [PubMed]
139. Volkow, N.D.; Wang, G.J.; Tomasi, D.; Baler, R.D. Obesity and addiction: Neurobiological overlaps. *Obes. Rev. Off. J. Int. Assoc. Study Obes.* **2013**, *14*, 2–18. [CrossRef] [PubMed]
140. Le, D.S.; Pannacciulli, N.; Chen, K.; Salbe, A.D.; Del Parigi, A.; Hill, J.O.; Wing, R.R.; Reiman, E.M.; Krakoff, J. Less activation in the left dorsolateral prefrontal cortex in the reanalysis of the response to a meal in obese than in lean women and its association with successful weight loss. *Am. J. Clin. Nutr.* **2007**, *86*, 573–579. [CrossRef] [PubMed]
141. Brooks, S.J.; Cedernaes, J.; Schioth, H.B. Increased prefrontal and parahippocampal activation with reduced dorsolateral prefrontal and insular cortex activation to food images in obesity: A meta-analysis of fMRI studies. *PLoS ONE* **2013**, *8*, e60393. [CrossRef]
142. Ho, A.L.; Sussman, E.S.; Pendharkar, A.V.; Azagury, D.E.; Bohon, C.; Halpern, C.H. Deep brain stimulation for obesity: Rationale and approach to trial design. *Neurosurg. Focus* **2015**, *38*, E8. [CrossRef]
143. Clark, S.M.; Martens, K.; Smith-Mason, C.E.; Hamann, A.; Miller-Matero, L.R. Validation of the Yale Food Addiction Scale 2.0 among a Bariatric Surgery Population. *Obes. Surg.* **2019**, *29*, 2923–2928. [CrossRef]
144. Rodriguez-Martin, B.C.; Gallego-Arjiz, B. Overeaters Anonymous: A Mutual-Help Fellowship for Food Addiction Recovery. *Front. Psychol.* **2018**, *9*, 1491. [CrossRef]
145. McKenna, R.A.; Rollo, M.E.; Skinner, J.A.; Burrows, T.L. Food Addiction Support: Website Content Analysis. *JMIR Cardio* **2018**, *2*, e10. [CrossRef]

146. Hilker, I.; Sanchez, I.; Steward, T.; Jimenez-Murcia, S.; Granero, R.; Gearhardt, A.N.; Rodriguez-Munoz, R.C.; Dieguez, C.; Crujeiras, A.B.; Tolosa-Sola, I.; et al. Food Addiction in Bulimia Nervosa: Clinical Correlates and Association with Response to a Brief Psychoeducational Intervention. *Eur. Eat. Disord. Rev. J. Eat. Disord. Assoc.* **2016**, *24*, 482–488. [CrossRef] [PubMed]
147. Schroder, R.; Sellman, D.; Elmslie, J. Addictive overeating: Lessons learned from medical students' perceptions of Overeaters Anonymous. *N. Z. Med. J.* **2010**, *123*, 15–21. [PubMed]
148. Cattivelli, R.; Pietrabissa, G.; Ceccarini, M.; Spatola, C.A.; Villa, V.; Caretti, A.; Gatti, A.; Manzoni, G.M.; Castelnuovo, G. ACTonFOOD: Opportunities of ACT to address food addiction. *Front. Psychol.* **2015**, *6*, 396. [CrossRef] [PubMed]
149. Cassin, S.; Leung, S.; Hawa, R.; Wnuk, S.; Jackson, T.; Sockalingam, S. Food Addiction Is Associated with Binge Eating and Psychiatric Distress among Post-Operative Bariatric Surgery Patients and May Improve in Response to Cognitive Behavioural Therapy. *Nutrients* **2020**, *12*, 2905. [CrossRef] [PubMed]
150. Val-Laillet, D.; Aarts, E.; Weber, B.; Ferrari, M.; Quaresima, V.; Stoeckel, L.E.; Alonso-Alonso, M.; Audette, M.; Malbert, C.H.; Stice, E. Neuroimaging and neuromodulation approaches to study eating behavior and prevent and treat eating disorders and obesity. *NeuroImage Clin.* **2015**, *8*, 1–31. [CrossRef] [PubMed]
151. Ihssen, N.; Sokunbi, M.O.; Lawrence, A.D.; Lawrence, N.S.; Linden, D.E.J. Neurofeedback of visual food cue reactivity: A potential avenue to alter incentive sensitization and craving. *Brain Imaging Behav.* **2017**, *11*, 915–924. [CrossRef]
152. Kohl, S.H.; Veit, R.; Spetter, M.S.; Gunther, A.; Rina, A.; Luhrs, M.; Birbaumer, N.; Preissl, H.; Hallschmid, M. Real-time fMRI neurofeedback training to improve eating behavior by self-regulation of the dorsolateral prefrontal cortex: A randomized controlled trial in overweight and obese subjects. *Neuroimage* **2019**, *191*, 596–609. [CrossRef]
153. Pretlow, R.A.; Stock, C.M.; Roeger, L.; Allison, S. Treatment of the sensory and motor components of urges to eat (eating addiction?): A mobile-health pilot study for obesity in young people. *Eat. Weight Disord. EWD* **2020**. [CrossRef]

MDPI
St. Alban-Anlage 66
4052 Basel
Switzerland
Tel. +41 61 683 77 34
Fax +41 61 302 89 18
www.mdpi.com

Nutrients Editorial Office
E-mail: nutrients@mdpi.com
www.mdpi.com/journal/nutrients

www.ingramcontent.com/pod-product-compliance
Lightning Source LLC
Chambersburg PA
CBHW040825250926
51850CB00007BA/1264
9783036574660